Farmer Innovation in Africa

A Source of Inspiration for Agricultural Development

Farmer Innovation in Africa

A Source of Inspiration for Agricultural Development

Edited by
Chris Reij and Ann Waters-Bayer

earthscan
from Routledge

First published by Earthscan in the UK and USA in 2001

For a full list of publications please contact:
Earthscan
2 Park Square, Milton Park, Abingdon, Oxon OX14 4RN
Simultaneously published in the USA and Canada by Earthscan
711 Third Avenue, New York, NY 10017

Earthscan is an imprint of the Taylor & Francis Group, an informa business

A catalogue record for this book is available from the British Library

ISBN: 978-1-85383-816-3 (pbk)

Typesetting by PCS Mapping & DTP, Newcastle upon Tyne

Cover design by Danny Gillespie
Cover photos (clockwise from top left): Emmanuel Kamgono, Cameroon © Paul
Tchawa; Tsige GebreAbezgi, Central Tigray © Fetien Abay; Farmers' trial plot, West
Cameroon © Chris Reij; Barthelémy Djambou, West Cameroon © Chris Reij

A catalogue record for this book is available from the British Library

Library of Congress Cataloging-in-Publication Data

Farmer innovation in Africa : a source of inspiration for agricultural development /
edited by Chris Reij and Ann Waters-Bayer.
      p. cm.
   Includes bibliographical references (p.).

   1. Agricultural innovations. 2. Agricultural productivity. 3. Farmers. I. Reij, Chris,
1948- II. Waters-Bayer, Ann.

S494.5.I5 F35 2001
338.1'6'096—dc21

                                                                    2001004778

Contents

List of Acronyms and Abbreviations   *xii*
List of Tables, Figures, Boxes, Plates and Maps   *xv*
Foreword   *xix*
Acknowledgements   *xxi*

PART 1 INTRODUCING THE FARMER INNOVATION APPROACH AND SOME REMARKABLE INNOVATORS

1   Entering research and development in land husbandry through
    farmer innovation *Chris Reij and Ann Waters-Bayer*   3
    Background   3
    Introducing the farmer innovation programmes   6
    Components of the farmer innovation methodology   8
    Major strengths   20

2   The career and influence of Barthelémy Kameni Djambou in
    Cameroon *Paul Tchawa, Noubissié Tchiagam Jean-Baptiste
    and Yossa Bonneau*   23
    Seeking new ideas   24
    Sharing with other farmers   25
    Teaching in his fields   26

3   Ayelech Fikre: an outstanding woman farmer in Amhara Region,
    Ethiopia *Million Alemayehu*   28
    Construction of stone bunds   29
    Management of soil fertility   30
    Harvesting of rainwater   30
    Maintaining biodiversity for multiple purposes   31
    The problem was her teacher   32
    Recognition of her accomplishments   32

4   Pits for trees: how farmers in semi-arid Burkina Faso increase and
    diversify plant biomass *Hamado Sawadogo, Fidèle Hien,
    Adama Sohoro and Frédéric Kambou*   35
    Location and methods of the study   36
    From farming to regenerating natural vegetation   38

Impact of *zaï* and related innovations   40
Differences in motives and strategies of the innovators   42
Wider context   45

PART 2 BUILDING PARTNERSHIPS FOR INNOVATION IN LAND HUSBANDRY

5   Forging partnership between farmers, extension and research in
    Tanzania *O T Kibwana*   49
    Building institutional partnership   50
    Preparing innovative farmers for partnership   53
    Negotiating joint experimentation   54
    Partnership in action   55
    Lessons learnt   56

6   Joining forces to discover and celebrate local innovation in land
    husbandry in Tigray, Ethiopia *Mitiku Haile, Fetien Abay and*
    Ann Waters-Bayer   58
    Wide involvement in search   59
    Methods of seeking innovators   61
    Compiling and analysing the findings   65
    Identifying innovations applicable by the poor   67
    Celebrating local knowledge and creativity   67
    Enhancing farmer experimentation   70
    Favourable conditions for promoting local innovation   71
    Inclusive approach   72

PART 3 FARMER INNOVATION: PROCESS, EVIDENCE AND ANALYSIS

7   An initial analysis of farmer innovators and their innovations
    Chris Reij and Ann Waters-Bayer   77
    What do the identified innovators have in common?   77
    Motivation for innovation   83
    Sources of inspiration for innovation   85
    One of many possible typologies of local innovations in SWC   87
    The suitability of innovations for easy dissemination   89

8   Why do farmers innovate and why don't they innovate more?
    Insights from a study in East Africa *Flemming Nielsen*   92
    Survey of farmer innovation   94
    Three study sites   95
    Variety of innovations and variations in innovativeness   96

Explaining innovation: the emic view 98
Explaining innovation: the etic view 100
Conclusions 102

9 **Chain of innovations by farmers in Cameroon** *Paul Tchawa* 104
Night paddock manuring 104
New harvesting tool 105
Irrigation in response to market demand 106
Social innovation and mutual inspiration 107
Spreading innovation 108

10 **Women and innovation: experiences from Promoting Farmer Innovation in East Africa** *Milcah Ong'ayo, Janet Njoroge and Will Critchley* 110
Early days: gender imbalance 111
Getting to the root of the problem: gender analysis 112
Moving forward: gender sensitization workshops 113
Signs of change 115
Conclusion 120

11 **Innovators in land husbandry in arid areas of Tunisia** *Noureddine Nasr, Bellachheb Chahbani and Chris Reij* 122
Harvesting water for agriculture 122
A participatory approach to the development of rainfed farming 124
Role of scientists 128
Institutionalizing the approach 130

12 **Women's innovations in rural livelihood systems in arid areas of Tunisia** *Noureddine Nasr, Bellachheb Chahbani and Radhia Kamel* 132
Identifying women innovators 132
Characteristics of women innovators 133
Spheres of innovation 134
Potential for the spread of women's innovations 135

13 **Namwaya Sawadogo: the ecologist of Touroum, Burkina Faso** *Jean-Baptiste Taonda, Fidèle Hien and Constant Zango* 137
Introducing Namwaya Sawadogo 137
Motivation to innovate 138
The process of innovation 139
Wide range of innovations 140
Benefits derived from innovation 142

14 **Outwitters of water: outstanding Irob innovation in northern Ethiopia** *Asfaha Zigta and Ann Waters-Bayer* 144
Irob: a land of extremes 144
Zigta Gebremedhin robs the water of its silt 146

Yohannes Tesfaye exploits the river's strength to hold walls   150
Site-specific innovations   153

15 A challenge and an opportunity: innovation by women farmers in
    Tigray *Fetien Abay, Mamusha Lemma, Pauline O'Flynn and*
    Ann Waters-Bayer   155
    Approach to recognizing women innovators   155
    Examples of women's innovation in land husbandry   158
    Initial analysis of the nature of women's innovations   163
    Impact of ISWC-Ethiopia on innovation by women   165
    Conclusions and further challenges   166

PART 4 FARMERS' EVALUATION AND EXTENSION OF LOCAL INNOVATIONS

16 Community assessment of local innovators in northern Ethiopia
    Yohannes GebreMichael   171
    Challenges in identifying innovators   172
    Characteristics of innovative farmers   173
    Innovators and community values   175

17 Stimulating creativity among East African farmers: from isolated
    individuals to interactive groups *Will Critchley, Patrick Lameck,*
    Alex Lwakuba, Charles Mburu and Dan Miiro   178
    Identifying farmer innovators   179
    Networking between innovators   181
    Lessons learned thus far   182
    Concluding remarks   183

18 Facilitating farmer-to-farmer communication about innovation in
    Tigray *Fetien Abay, Belay Teshome, Mengistu Hailu and*
    Mamusha Lemma   185
    Innovators' travelling seminar   185
    Farmers' Fora on water management   192
    Comparison and conclusion   196

19 Innovation and impact: a preliminary assessment in Kabale,
    Uganda *Dan Miiro, Will Critchley, Alie van der Wal and*
    Alex Lwakuba   198
    Background and rationale   198
    Objectives, focus and methodology of impact assessment   201
    Results and discussion   203
    Conclusions and recommendations   209

20   Three models of extension by farmer innovators in Burkina Faso
     Aly Ouedraogo and Hamado Sawadogo   213
     The 'market day' model   214
     The '*zaï* school' model   215
     The 'teacher–student' model   216
     Towards food security and wealth   217

PART 5 STIMULATING AND SUPPORTING JOINT EXPERIMENTATION

21   Participatory Technology Development on soil fertility improvement
     in Cameroon *Paul Tchawa, Pierre Kamga, Christopher Ndi,*
     Christopher Vitsuh, Samuel Toh and Antoine Mvondo Zé   221
     Background   221
     Preparing for joint experimentation   222
     Setting up and carrying out the experiment   225
     Initial results   226
     Supportive on-station research   228
     Add-on experiment by Christopher Vitsuh   230
     General conclusions   233

22   Farmer-led experimentation in the drylands of Central Tigray
     GebreEgziabher Miruts and Fetien Abay   234
     Participatory situation analysis   235
     Participatory planning of experiments   239
     Monitoring and evaluation of the trial   242
     Impact of the process   245

23   Farmer innovation and plant breeding: the case of maize K525
     developed by Emmanuel Kamgouo of Bandjoun, West Cameroon
     Paul Tchawa, Noubissié Tchiagam Jean-Baptiste,
     Antoine Mvondo Zé and Eric Mujih   248
     The innovator and the history of maize K525   249
     Joint experimentation on maize K525   251
     Next steps   254

24   Joint analysis of the sustainability of a local SWC technique in
     Burkina Faso *Fidèle Hien and Aly Ouedraogo*   256
     The research framework   256
     Methodology   257
     Assessment of agrobiological resources at village level   258
     Assessment of organic matter flows at farm levels   260
     On-farm experimentation   262
     Socioeconomic evaluation   264
     Discussion and methodological issues   265

25 **Sowing maize in pits: farmer innovation in southern Tanzania**
    Zacharia Malley, Anderson Temu, Norbert Kinabo,
    Salome Mwigune and Anunciata Mwageni   267
    Scientists encounter farmer innovators   268
    The maize pit innovation   270
    Farmer-led experimentation   271
    Scientists' analysis of farmers' data   272
    Farmers' assessment of the pit technique   274
    Joint experimentation in the 1999–2000 season   276
    Farmers' reflections and conclusions   277

PART 6 RAISING AWARENESS, POLICY LOBBYING AND MAINSTREAMING

26 **Understanding and influencing policy processes for SWC**
    James Keeley   281
    Which policies matter?   282
    How does policy change?   284
    Strategies to effect policy change   287
    Conclusion   291

27 **A bridge between local innovation, development and research:**
    the regional radio of Gafsa, Tunisia *Noureddine Nasr,*
    El Ayech Hdaidi and Ali Ben Ayed   293
    Bringing the stakeholders together   294
    Letters to the radio   296
    Impact of extension by radio   296
    Mass media and innovation   298

28 **Mainstreaming participatory approaches to SWC in Zimbabwe**
    Kudakwashe Murwira, Jürgen Hagmann and Edward Chuma   300
    Background   300
    Creating alternatives at local level   303
    Mainstreaming the approach   307
    Constraints and opportunities   307
    Conclusion   309

29 **Liberating local creativity: building on the 'best farming practices'**
    extension approach from Tigray's struggle for liberation
    Berhane Hailu and Mitiku Haile   310
    Agricultural extension during the struggle for liberation   311
    A vision of participatory agricultural development   318
    Concluding remarks   324

30 Impact of the farmer innovation approach on the attitudes of
stakeholders in agricultural development in Tunisia
Noureddine Nasr 325
From scepticism to conviction 325
Learning by jointly solving farmers' problems 327
Growing enthusiasm 328
Final remark 330

31 Learning for sustainability: incorporating participatory approaches
into education for rural development in Ethiopia and Tanzania
O T Kibwana, Mitiku Haile and Firew Tegegne 331
Development-oriented education at Mekelle University 331
Development-oriented education at Cooperative College Moshi 340
The need to broaden the incorporation 344

32 **An encouraging beginning** *Ann Waters-Bayer, Laurens van
Veldhuizen and Chris Reij* 347
Promoting innovation, not just innovations 348
Working with individuals, groups or communities 349
Dissemination of both technology and methodology 350
Assessment of innovations and processes 351
Institutionalizing the farmer innovation approach 352
An innovative approach at programme level 353

Index 355

List of Acronyms and Abbreviations

ADDA	Adigrat Diocese Development Action
ADRK	Association pour le Développement de la Région de Kaya (Association for the Development of Kaya Region)
AGRITEX	Department of Agricultural and Technical Extension Services
AHI	African Highlands Initiative
ARI	Agricultural Research Institute
BoA	Bureau of Agriculture
BoANR	Bureau of Agriculture and Natural Resources
CBA	cost-benefit analysis
CCD	Convention to Combat Desertification
CCM	Cooperative College Moshi
CDCS	International Cooperation Centre (Vrije Universiteit, Amsterdam, The Netherlands)
CIAL	Community-operated local agricultural research committee
CIPCRE	Cercle International pour la Promotion de la Création (International Circle for the Promotion of Creation)
Contil	Conservation Tillage Project
CORDEP	Community-Oriented Development Project
CPTP	Comprehensive Participatory Training Process
CRDA	Centre Régional de Développement Agricole (Regional Centre for Agricultural Development)
CWSSE	Conserve Water to Save Soil and Environment
DA	development agent
DALDO	District Agriculture and Livestock Development Office(r)
DAP	diammonium phosphate
DEO	Divisional Extension Officer
DFE	Directorate of Field Education
DG	Director General
DGIS	Directorate General for International Cooperation
DNR	Department of Natural Resources
DoA	Department of Agriculture
DRCS	Directorate of Research and Consultancy Services
DRSS	Department of Research and Specialist Services
DSP	Directorate of Studies and Programmes
FAO	Food and Agriculture Organization of the United Nations
FFS	Farmer Field School
FI	farmer innovator

FMPR	Farmer Managed Participatory Research
FYM	farmyard manure
GTZ	German Agency for Technical Cooperation (Deutsche Gessellschaft für technische Zusammenarbeit)
HIMA	*Hifadhi Mazingira*
HYV	high-yielding variety
IA	impact assessment
ICARDA	International Centre for Agricultural Research in Dry Areas
ICRAF	International Centre for Research on Agroforestry
IDS	Institute for Development Studies (University of Sussex, Brighton, UK)
IES	Institute of Environmental Studies
IFA	International Fertilizer Industry Association
IFAD	International Fund for Agricultural Development
IFDC	International Fertilizer Development Center
IFPRI	International Food Policy Research Institute
IIED	International Institute for Environment and Development
IITA	International Institute for Tropical Agriculture
ILEIA	Information Centre for Low-External-Input and Sustainable Agriculture
INADES	Institut Africain pour le Développement Economique et Social
INERA	Institut National d'Etudes et de Recherches Agricoles (National Agricultural Research Institute)
IRA	Institut des Régions Arides (Institute for Arid Regions)
IRAD	Institut de Recherche Agricole pour le Développement (Institute of Agricultural Research for Development)
ISCO	International Soil Conservation Organizations
ISWC	Indigenous Soil and Water Conservation
ITDG	Intermediate Technology Development Group
ITK	Indigenous Technical Knowledge
IUCN	International Union for the Conservation of Nature
KEKUFAG	Kedjom Ketingoh Union Farmers Group
LEISA	Low-External-Input and Sustainable Agriculture
M&E	monitoring and evaluation
MAAIF	Ministry of Agriculture, Animal Industry and Fisheries
MARTI	Ministry of Agriculture Research and Training Institute
MCDANR	Mekelle College of Dryland Agriculture and Natural Resources
MoA	Ministry of Agriculture
MoE	Ministry of Education
MRC	Mekelle Research Centre
MU	Mekelle University
MUC	Mekelle University College
MVIWATA	*Mtandao wa Vikundi vya Wakulima Tanzania* (Network of Farmers Groups in Tanzania)
NORAD	Norwegian Agency for Development
NGO	non-governmental organization

NRM	natural resource management
NSC	National Steering Committee
NV	network visit
OM	organic matter
ORFA	Organisation Formation Appui au Développement des Communautés de Base (Organization for Development Training Support for Grassroots Communities)
PA	Peasant Association
PAP	Practical Attachment Programme
PEA	Participatory Extension Approach
PFI	Promoting Farmer Innovation in Rainfed Agriculture
PM&E	participatory monitoring and evaluation
PNVRA	Programme National de la Vulgarisation et de la Recherche Agricole (National Agricultural Extension and Research Programme)
PPE	Pilot Projects and Experimentation Department
PPP	People's Participation Programme
PRA	Participatory Rural Appraisal
PTD	Participatory Technology Development
RELMA	Regional Land Management Unit
REPIH	Réseau des Paysans Innovateurs du Haut-Nkam (Network of Farmer Innovators in Haut-Nkam)
REST	Relief Society of Tigray
SC	Steering Committee
SCRP	Soil Conservation Research Project
SNV	Netherlands Volunteer Programme
SOM	soil organic matter
SPSS	Statistical Package for the Social Sciences
SSI	semi-structured interview(s)
ST	study tour
SUA	Sokoine University of Agriculture, Tanzania
SWC	soil and water conservation
SWOT	strengths, weaknesses, opportunities, threats
T&V	Training and Visit
TIP	Traditional Irrigation Improvement Project
TLU	tropical livestock unit
ToT	transfer of technology
TPLF	Tigray People's Liberation Front
UNDP	United Nations Development Programme
UNSO	Office to Combat Desertification and Drought
VEO	Village Extension Officer

List of Tables, Figures, Boxes, Plates and Maps

TABLES

4.1 Farm data for four farmer innovators in Burkina Faso   37

4.2 Yields of sorghum and millet obtained by four innovators using the *zaï* technique   40

4.3 Some tree species in the fields of Yacouba Sawadogo   41

8.1 Types of innovation during the previous 12 months on 505 farms in Kenya and Tanzania   96

8.2 Farmers' reasons for their innovations over the previous 12 months   98

8.3 The obstacles that farmers see to innovations   99

8.4 Number of innovations and innovative households according to ethnic group   101

10.1 Gender studies and sensitization workshops under PFI   112

10.2 Findings and recommendations of gender studies and external review of PFI   114

10.3 Gender ratio of farmer innovators in PFI before and after gender sensitization workshops   115

19.1 Characterization of farmer innovators and adopters   204

19.2 Characterization of the innovations   205

19.3 Results of SWOT analysis of network visits and study tours   207

19.4 New initiatives of 32 farmer innovators since joining ISWC-Uganda activities   208

21.1 Bags of nightshade leaves harvested by Samuel Toh   227

21.2 Bags of nightshade leaves harvested by Christopher Vitsuh   227

21.3 Additional data collected in the parallel experiments on-station   228

21.4 Estimated quantities of dung and urine deposited on the farmers' experimental plots   230

21.5 Bags of nightshade harvested on Vitsuh's add-on experimental plot   232

22.1 Number and gender of farmers attending *tabia*-level meetings for situation analysis in Ahferom district   236

22.2 Suggested solutions for the problems discussed by subgroups of farmer researchers in the selected *tabias*   238

22.3 Division of tasks between research partners   242

23.1 Roles of the different partners in the joint experimentation   252

23.2 Extrapolated average maize yields (t/ha) on each treatment   253
24.1 Four stages in the joint research with farmers in Yatenga province 259
24.2 Theoretical livestock carrying capacity of Kao and Somyaga village
  areas at the end of the 1999 rainy season   260
24.3 Quantity of compost and number of pits per hectare   263
25.1 Results from seven farmers growing maize in pits   273
25.2 Changes in soil conditions as affected by pitting   274
25.3 Farmers' SWOT analysis of sowing maize in pits   275
25.4 Yields of maize grown in two systems of sowing in Njombe and
  Iringa in the 1999–2000 season   276
26.1 Types of policy significant for land husbandry   283

FIGURES

8.1 Number of innovations by number of households over the
  previous 12 months   97
17.1 Ten steps in PFI's field activities   180
18.1 Sites visited during innovators' travelling seminar in Tigray   187
21.1 Layout of the trial plots in the farmer-led research   225
24.1 Flow of organic matter and other inputs entering the *zaï*:
  schematic presentation based on the flows identified by 12 farmers 261
24.2 Number of animals, quantity of compost produced and potential
  area of *zaï* per farm household   262
24.3 Average yields of millet according to the amount of compost
  applied to fields recently rehabilitated with *zaï*   263
24.4 Overview of labour inputs (family and hired) for transporting
  harvest from Ousséni Zoromé's fields   264

BOXES

1.1 Participatory Technology Development in land husbandry   5
6.1 Guidelines for seeking local innovation in land husbandry   62
7.1 The integrated farm system of Cleopas Banda, farmer innovator
  in Zimbabwe   84
7.2 Grace Bura: a local expert in gully rehabilitation   89
9.1 Chain of innovations by Babanki farmers   108
10.1 Kakundi Kiteng'u and her sugarcane pitting technique in Kenya   116
10.2 Florence Akol harvests water for bananas in Uganda   117
10.3 Susanna Sylvester and her novel composting technique in Tanzania 119

PLATES

1.1 The first activity of ISWC Tunisia was to organize a national
training workshop in PRA/PTD (October 1997)   10
2.1 Barthelémy Djambou, a key innovator in West Cameroon   25
2.2 Outsiders listen to and learn from Barthelémy Djambou
(Cameroon)   27
3.1 Ayelech Fikre explaining her innovations to researchers and
extension staff (North Shewa, Ethiopia)   33
4.1 Yacouba Sawadogo with the forest he created in the background   39
4.2 Ousséni Zoromé has systematically protected natural
regeneration in his fields (Burkina Faso)   42
4.3 Ousséni Kindo in his rehabilitated fields (Burkina Faso)   44
5.1 Team members of ISWC-Tanzania discuss compost trials with a
farmer innovator   56
6.1 Researchers, development agents and farmers in Tigray (Ethiopia)
jointly discussing innovations   64
7.1 Cleopas Banda with mango tree on a raised bed with buried clay
pots   84
7.2 Sustainable increases in yields are an important motivating factor   86
7.3 Improved soil fertility management leads to better harvests
(Burkina Faso)   88
9.1 Philip Ndong invented a simple and low-cost harvesting tool
(Cameroon)   106
10.1 Women innovators in Uganda look at bananas grown in mulched
plots   117
10.2 Susanna Sylvester is a specialist in composting (Tanzania)   119
11.1 Abbès Sandi grows a wide range of fruit trees in a pre-desert region,
and irrigates them from a small dam he constructed (Tunisia)   126
12.1 Mbirika Chokri incubates chicken eggs in dry cattle dung (Tunisia)   135
13.1 Namwaya Sawadogo in his eucalyptus plantation (Burkina Faso)   140
14.1 Zigta Gebremedhin explaining the functioning of his checkdams
(Tigray, Ethiopia)   147
15.1 By ploughing herself, a woman in Tigray, Ethiopia challenges
local cultural norms   160
17.1 A meeting of a cluster of farmer innovators (Uganda)   181
18.1 The travelling seminar was filmed by the Bureau of Agriculture
and Natural Resources (Ethiopia)   186
21.1 One of the farmers' trial plots (West Cameroon)   226
21.2 Christopher Vitsuh records trial data himself   232
22.1 Discussing group experimentation in Central Tigray (Ethiopia)   244
23.1 Emmanuel Kamgouo (middle) with researchers on his
experimental plot   252
25.1 Wilbert Mville digging and manuring a maize pit (Tanzania)   269
27.1 Mbirika Chokri presenting her innovation in the studio of the
regional radio of Gafsa (Tunisia)   295

30.1 Meeting of the ISWC team at the Institut des Régions Arides
     (Tunisia)   326
31.1 Farmer innovator in Tigray, Ethiopia, demonstrating an improved
     plough to students and staff of Mekelle University   337

MAPS

*The maps in this book have been prepared solely for the convenience of the reader
and in no way imply an opinion about the status of the boundaries*

1.1   African partner countries in the ISWC 2 and PFI programmes   7
2.1   Cameroon   24
3.1   Ethiopia (Northern Shewa action area)   29
4.1   Burkina Faso (Yatenga and Zondoma action areas)   36
5.1   Tanzania   50
6.1   Ethiopia (Tigray action area)   59
8.1   Research sites in Kenya and Tanzania   93
9.1   Cameroon   105
10.1 PFI action areas in Kenya, Tanzania and Uganda   111
11.1 ISWC action areas in Tunisia   123
12.1 ISWC action areas in Tunisia   133
13.1 Burkina Faso   138
14.1 Ethiopia (showing Irobland)   145
15.1 Ethiopia (Tigray action area)   156
16.1 Ethiopia (Southern Wello and Northen Shewa action areas)   172
17.1 Kenya, Tanzania and Uganda   179
19.1 Uganda (showing Kabale district)   199
20.1 Burkina Faso (Yatenga and Zondoma action areas)   214
21.1 Cameroon   222
22.1 Ethiopia (Axum action area)   235
23.1 Western province of Cameroon   249
24.1 Burkina Faso (Yatenga and Zondoma action areas)   257
25.1 Tanzania (Njombe district action area)   268
27.1 Tunisia   294
28.1 Action areas, ISWC-Zimbabwe   301

Foreword

This book is about positive developments in Africa that have never drawn headlines in newspapers. Yet they are remarkable and newsworthy. They oblige those who are unfamiliar with the African countryside and farming communities to reconsider some common assumptions about Africa and about African smallholder farmers. The book describes a wide range of innovations in African agriculture. These innovations were not introduced by 'experts' or by field agents, but rather by the farmers themselves. Teams of African scientists and field agents in eight countries managed to identify farmers (both men and women) who, on their own initiative, but based on shared community knowledge and technologies, have tried to improve their practices of managing land and water resources in order to overcome the immediate difficulties they were facing. Almost 1000 innovators were identified within only two years, and the contributors to this book argue that these innovators constitute just the tip of the proverbial iceberg.

Two Dutch-funded programmes that focused on farmer innovation in land husbandry – Indigenous Soil and Water Conservation in Africa (ISWC) and Promoting Farmer Innovation in Rainfed Agriculture (PFI) – did not content themselves with merely identifying innovative farmers. They took the process several steps onward by involving scientists, field agents and other farmers in joint experimentation to improve the innovations still further, following a research agenda set by the farmers. The farmer innovators were given opportunities to visit other innovators, and this gave them new ideas that they could try out on their own fields. The two programmes show how processes of innovation can be stimulated, with one new idea spawning the next. And they show how scientists, field agents, local authorities and development planners can give the support and create the conditions that encourage farmers to accelerate innovation.

This book argues that the conventional 'transfer-of-technology' paradigm, in which scientists develop technologies on research stations and extension workers pass these technologies on to farmers, is producing disappointing results. Much that is proposed to smallholder farmers in Africa is not acceptable to them because it is too costly or does not suit their farming conditions or, in some other way, is unacceptable to them. The working hypothesis of the authors is that one should look first at what farmers themselves are experimenting with and then use this as a starting point for joint research and development by farmers and scientists.

In both programmes, scientists soon began appreciating and admiring the knowledge and skills of farmer innovators who apply their fund of accumulated knowledge and technologies to solve a specific problem or to grasp a new opportunity. Scientists, field agents and farmers learned to communicate with one another as equals. This led to a complete reversal of roles: scientists and field agents began to listen to and learn from farmers.

Within the framework of these two programmes, farmer innovators also had opportunities to present their innovations to others through various mass media, including newspapers, rural radio and national television. The reaction of the journalists who witnessed farmer innovation was often: 'This is fascinating! Why did we not know about this before?'

What is most striking in this book is the picture it gives of the creativity and the perseverance of African farmers. So often we hear 'doom and gloom' stories from Africa – about war, drought and hunger, about growing poverty and increasing desertification. But here we have a book full of positive accounts about farmers who, when confronted with huge problems that threatened their very survival, had the courage and capacity to experiment and to innovate. The case studies in this book depict men and women farmers who managed to expand and improve their resource base. It took them years to do so, but the end result is that they have improved the food security of their families, as well as the environment. This is immediately relevant to all those working on the United Nations Convention to Combat Desertification (CCD), as one of its key aspects is that it recognizes local knowledge and seeks to build upon it.

It is also impressive that many of the farmer innovators who managed to improve their living conditions did not keep their knowledge to themselves. Rather, they made considerable efforts to share it with others. They invested not only their time and energy but, in some cases, even their own money in training other farmers. This was done on their own initiative and without any external support, by farmers who live and work as members of closely knit communities that nurture the innovations and help to develop them.

One conclusion that can be drawn at this stage is that Africa still has a major resource waiting to be tapped, and that is the creativity of its farmers. This is hardly surprising for anyone who knows Africa: it is the local communities that sustain Africa – in spite of and not because of the inept governments.

This book is a confirmation of my knowledge of Africa and my confidence in its future. It will be vital reading and, indeed, should be required reading for all policy-makers, scientists and development workers who are concerned with sustainable development in Africa and elsewhere, but who have been bombarded by the current superficial negative views about Africa.

Tewolde Berhan Gebre Egziabher
Addis Ababa, Ethiopia

Acknowledgements

The enthusiasm, creativity and collaboration of the many African researchers, development agents and farmer innovators involved in the ISWC 2 programme made this book possible. A small number of European colleagues worked together with the African partners and, in the process, the working relationships became more than just that. We came to know each other well and, in many cases, also came to know each other's families. The annual review meetings, each in a different country in Africa, took on the nature of a warm reunion of friends.

The programme, which aimed to link scientists, development agents and farmer innovators in joint research and extension activities, has benefited all involved as it triggered a process of continuous learning. We have all developed a great admiration for the farmer innovators with whom we are working: men and women who do not accept land degradation as inevitable, but struggle to do something about it. To us, they are a real source of inspiration and we feel indebted to them.

The success of the farmer innovation approach to Participatory Technology Development (PTD) and farmer-to-farmer extension owes a great deal to the dynamism of the ISWC 2 national coordinators and the PTD specialists: Fidèle Hien and Matthieu Ouedraogo in Burkina Faso, Paul Tchawa in Cameroon, Bellachheb Chahbani and Noureddine Nasr in Tunisia, Mitiku Haile and Fetien Abay, and later Mamusha Lemma, in Ethiopia, O T Kibwana in Tanzania, Dan Miiro in Uganda and Edward Chuma and Kudakwashe Murwira in Zimbabwe. Sometimes researchers fear that working directly with farmers in the field could hinder their professional careers, as this type of work is not included in the conventional qualifications for promotion. Some examples from ISWC 2 may help to dispel such fears. In March 2000, Dr Mitiku Haile, coordinator of ISWC-Ethiopia, became President of Mekelle University. In November 2000, Dr Fidèle Hien, national coordinator in Burkina Faso, was appointed Minister of Environment and Water. These are excellent positions from which to promote the institutionalization of the farmer innovation approach.

We would like to thank our colleagues Laurens van Veldhuizen, Jean-Marie Diop and Lida Zuidberg from ETC Ecoculture, Camilla Toulmin, Bara Gueye and Thea Hilhorst from the Drylands Programme of the International Institute for Environment and Development (IIED), Ian Scoones and James Keeley from the Institute for Development Studies (IDS) at the University of

Sussex, Brighton and Will Critchley from CDCS (International Cooperation Centre), Vrije Universiteit, Amsterdam, for the enthusiastic support they have given within the European Consortium, as thematic advisers and as backstoppers to the seven African countries involved.

The ISWC 2 programme was generously funded by the Environment Programme of the Directorate General for International Cooperation (DGIS) of The Netherlands Ministry of Foreign Affairs. We are particularly indebted to Fea Boegborn for her critical and constructive support.

Secretarial support was ably provided by Lia de Groot (CDCS) and project management and other support by Alie van de Wal (CDCS). Especially since their recent departure, we realize the tremendous efforts that they put into making the ISWC 2 programme run smoothly and keeping the information flowing freely. We also thank Wolfgang Bayer and Sibylle Pich in Göttingen, Germany, for their valuable comments on the manuscript and their assistance in editing this book. Piet Kostense of Wageningen University drew the maps.

Finally, we extend special thanks to Tewolde Berhan Gebre Egziabher who was recently awarded the Alternative Nobel Prize by the Right Livelihood Foundation in Sweden for his defence of biodiversity on behalf of developing countries. We greatly appreciate the time he gave in participating in the Anglophone African ISWC workshop on farmer innovation held in Mekelle, Ethiopia, and in writing the Foreword to this book.

Chris Reij and Ann Waters-Bayer
February 2001

Part 1

Introducing the farmer innovation approach and some remarkable innovators

Entering research and development in land husbandry through farmer innovation

*Chris Reij and Ann Waters-Bayer**

Since 1997, the second phase of the action-research programme Indigenous Soil and Water Conservation (ISWC 2) has been operating in seven countries in anglophone and francophone Africa, while its sister programme, Promoting Farmer Innovation in Rainfed Agriculture (PFI), has been operating in East Africa. Both have taken a similar approach to agricultural research and development through building on local innovation and encouraging farmer experimentation and Participatory Technology Development (PTD). This chapter outlines the main components of this farmer innovation approach.

BACKGROUND

Despite much rhetoric about the need for more demand-driven and participatory approaches to agricultural research and development, the transfer-of-technology (ToT) model continues to dominate in most countries in Africa (Bauer et al, 1998). This model implies that scientists generate new or improved technologies which are then transferred by extension agents to farmers. However, many of the technologies generated and promoted in this way are too expensive for the hundreds of millions of small-scale farmers who cannot afford to invest in the packages of required inputs, such as introduced seed, fertilizers and pesticides. Moreover, these packages are often standard-

* Chris Reij is a geographer at CDCS, Vrije Universiteit, Amsterdam, and international coordinator of ISWC 2; Ann Waters-Bayer is an agricultural sociologist with ETC Ecoculture, The Netherlands, and advises ISWC 2 on participatory research and extension methods

ized and promoted countrywide, without regard to agroecological differences, and poorly suited to the diverse and variable conditions of smallholders in semi-arid and other marginal areas (eg Howard et al, 1998). Many of these farmers have therefore been reluctant to adopt the technologies offered by conventional research and extension, despite sometimes massive 'encouragement' for them to do so.

For years, the World Bank strongly pushed a form of ToT called Training and Visit (T&V). This was meant to have an inbuilt feedback loop: through the extension agents, farmers could convey their assessment of the technologies back to the scientists. In actuality, however, this feedback rarely took place (Bauer et al, 1998). Reflections on the T&V system by the World Bank and the various countries involved led to suggestions to strengthen the voice of the farmer (Technical Centre for Agricultural Rural Cooperation (CTA), 1996). The dissatisfaction with conventional extension triggered the development of new approaches, such as Farmer Field Schools (Röling and van de Fliert, 1994). Some international and national research institutes have started to explore more participatory ways of working, such as through greater use of on-farm trials and more involvement of farmers in problem analysis and evaluation of trial results. The focus has been primarily on farmers' testing of technologies generated by scientists. Basically it concerns applying participatory approaches to improving ToT, but gives little or no attention to technologies generated by farmers or to strengthening farmers' capacities to develop and adapt technologies.

With growing population pressure and growing awareness of environmental degradation, farmers are seeking more productive ways to use the available resources without depleting them. They have to adjust rapidly to changing conditions. If agriculture is to be sustainable, farmers must be capable of actively and continuously creating new local knowledge (Röling and Brouwers, 1999).

For some time, publications by discerning development professionals have reported on farmers' own experimentation and innovation (eg Johnson, 1972). In the 1980s, books by Robert Chambers and his colleagues (1983, 1989) and Paul Richards (1985, 1986) drew wider attention to this. Roland Bunch (1985) argued, on the basis of the experience of the non-governmental organization (NGO) World Neighbors that worked with peasant farmers in Central America, that strengthening their capacities to experiment and solve their own problems was the key to sustainable agriculture. This call was taken up by various other NGOs, especially the Information Centre for Low-External-Input and Sustainable Agriculture (ILEIA) (eg Reijntjes and Hiemstra, 1989, Haverkort et al, 1991). Rhoades and Bebbington (1995) refer to a large number of experiences in supporting farmer experimentation and building on local initiatives to solve problems. However, these experiences remained marginal to mainstream research and development programmes which continued with the ToT approach. Both ISWC 2 and PFI decided to make a concerted effort to build on the potentials of farmer innovators and their innovations in Africa, with the ultimate aim of institutionalizing the promotion of farmer innovation and PTD (see Box 1.1) within national agricultural research and extension systems.

Box 1.1 Participatory Technology Development
in Land Husbandry

Participatory Technology Development (PTD) refers to collaboration between farmers, development agents and scientists in a manner that combines their knowledge and skills. The heart of PTD is farmer-led experimentation to find better ways of using available resources to improve the well-being of families and communities. The purpose of supporting farmer experimentation is to strengthen farmers' capacities to seek and try out new ideas so that they are better able to experiment and to adjust to changing conditions. The purpose is not to convince farmers to adopt a new technology, but rather to encourage them to test new possibilities and choose what is right for their circumstances or adapt the new ideas to their conditions.

A close look at the many good examples of this kind of interaction between farmers and 'outsiders' reveals a common pattern, which consists of six main clusters of activity (van Veldhuizen et al, 1997):

*   *Getting started:* establishing contact between farmers and 'outsiders' and agreeing to take this approach to improved land husbandry.
*   *Analysing the situation:* farmers and 'outsiders' seek a joint understanding of local problems, resources and opportunities, often using tools of Participatory Rural Appraisal (PRA).[1]
*   *Looking for things to try:* joint identification of possible solutions or new opportunities for improving land husbandry and agreeing on what to try out first.
*   *Trying things out:* experimenting with and adapting the new idea(s) in trials planned and implemented by farmers, with the support of development agents and scientists, and monitored and evaluated jointly according to agreed criteria.
*   *Sharing the results:* letting other farmers, development agents and scientists know what came out of the experiments and how they were done.
*   *Sustaining the process:* helping farmers to organize themselves to continue this kind of interaction and to obtain new ideas and inputs with which to experiment, and creating a political and institutional environment that fosters PTD.

Farmers' innovations show how local resources can be used to address problems and opportunities that farmers have recognized by doing their own informal situation analysis. The options that farmers involved in PTD may consider for testing include:

*   local farmers' innovations in order to validate them or test them more widely;
*   other ideas developed by farmers in other areas to address the same problem or opportunity;
*   ideas from formal research and extension to address the same problem or opportunity.

Thus, identifying farmer innovations can be a door to PTD at the point of 'Looking for things to try'.

INTRODUCING THE FARMER INNOVATION PROGRAMMES

Indigenous Soil and Water Conservation in Africa

The first phase of the ISWC research programme focused on analysing the dynamics of indigenous soil and water conservation (SWC) practices in Africa. Twenty-seven case studies from 15 countries showed that many indigenous practices are being maintained and expanded by farmers, in contrast with many modern SWC techniques promoted by development projects (Reij et al, 1996). The second phase (ISWC 2), which operates in Burkina Faso, Cameroon, Ethiopia, Tanzania, Tunisia, Uganda and Zimbabwe (see map on page 7), aims to improve the effectiveness of both indigenous and modern SWC practices through a process of joint experimentation involving farmers, scientists and development agents. It promotes research on local practices and innovations in SWC, assists in disseminating the results, and supports lobbying platforms to show policy-makers that building on ISWC practices is a promising path to development.

An unknown number of farmers are experimenting with SWC techniques on their own initiative. ISWC 2 pays special attention to these innovators who are often overlooked as a source of inspiration for development. Identification of farmer innovators who are already in the midst of informal experiments is an entry point into a process of PTD. The ISWC 2 approach involves training scientists and extensionists in PRA and PTD, identifying farmer innovators and their innovations, networking between farmer innovators, participatory research to develop and validate improved techniques and systems of land husbandry, and disseminating ideas and methods through farmer-to-farmer exchange.

In each collaborating country, a 'lead agency' concerned with agricultural research or development maintains links with other local research, development and teaching institutions interested and/or experienced in participatory approaches to improving land husbandry. A coordinator attached to the lead agency manages the programme activities. A steering committee, composed mainly of representatives from the collaborating organizations, plans and evaluates activities funded by the programme. All seven ISWC teams learn from each other during annual review meetings hosted by different partner countries in Africa.[2]

The ISWC 2 programme defined farmer innovators as those farmers who spontaneously try out new things, without the direct support of formal research and extension. They are not the 'model' or 'progressive' farmers who have often been selected by projects to test new crop varieties or packages of external inputs. The programme regards the farmers and their innovations, not the scientists and their technologies, as the starting point for development. This does not deny the important role that scientists can play. ISWC 2 seeks to link scientists, extensionists and farmer innovators in joint efforts to improve local and introduced practices of managing land and water.

Map 1.1 *African partner countries in the ISWC 2 and PFI programmes*

Promoting Farmer Innovation in Rainfed Agriculture

The PFI programme was developed by the United Nations Development Programme (UNDP) Office to Combat Desertification and Drought (UNSO) in the context of implementing the Convention to Combat Desertification (CCD). It operates in Kenya, Uganda and Tanzania and is, like ISWC 2, funded by the Directorate General for International Cooperation (DGIS). The goal is to harness the energies, ideas and rich experiences of innovative farmers for improving rainfed agriculture. Innovations in SWC are identified and disseminated.

Methodology development

A methodological approach to promoting farmer innovation was developed and refined during the course of PFI and ISWC 2. Here, we explain the major components of this approach and illustrate them with examples from experiences in the various countries, making cross-references to the relevant chapters in this book.

This presentation of ten components does not mean that these were or should be neatly implemented one after the other. In practice, several components run parallel to each other or are repeated several times during the process. Nor does it mean that each country involved in ISWC 2 and PFI has given the same attention to each component. The vastly differing socioeconomic and agroecological conditions in each country have led to different emphases being placed.

The three country programmes under PFI followed a series of ten steps (see Figure 17.1) fairly closely, whereas each of the seven countries under ISWC 2 designed its programme in the light of its own perceptions, experience and history, as well as the capacity and orientation of the partner organizations.

Within each country, the ISWC 2 and PFI programmes concentrated their initial work in selected regions – for example, the Southern Highlands in Tanzania, parts of the Central Plateau of Burkina Faso, the Tigray region in northern Ethiopia, the Kabale district in south-west Uganda. An exception was the Zimbabwe programme: it had already considerable experience in participatory research and extension in Masvingo province before ISWC 2 began and sought to mainstream the approach throughout the country (see Chapter 28). Most of the components described below refer only to activities in the action areas, but the lobbying for policy change and institutionalizing the farmer innovation approach (see Components 9 and 10) reach out beyond the action areas, with a view to scaling up the activities.

COMPONENTS OF THE FARMER INNOVATION METHODOLOGY

Component 1: Training in PRA and PTD

Although there are outstanding exceptions, many scientists and extension agents have difficulties in communicating eye-to-eye with farmers. Their conventional training and vision of development through the transfer of modern 'improved' technologies have imbued them with a sense of superiority. They tend to regard themselves as the teachers of farmers, as the agents of change drawing farmers out of the stagnancy of tradition. Most of them cannot imagine that they can also learn from (illiterate) farmers.

With a view to changing these attitudes, the ISWC 2 programme organized training-of-trainer workshops in PRA and PTD. Two courses were held on a regional level for the anglophone and francophone countries, in Zimbabwe and Burkina Faso respectively. The trainees from each country programme came from research, extension and development organizations, both governmental and non-governmental. They then collaborated in planning and facilitating PRA/PTD training and awareness-raising workshops at national and/or provincial (regional) level.

Some countries have invested more time in PRA/PTD training than others. For example, ISWC-Ethiopia strengthened local capacities through a series of workshops, building on each other and including new people as co-facilitators. It could thus train 121 development agents (DAs) from government services throughout Tigray, in addition to numerous people from NGOs and bilateral projects. The philosophy behind this was to create a broad basis for identifying local innovation and for supporting farmer-led experimentation and farmer-to-farmer exchange, rather than concentrating on isolated PTD experiments involving only a few farmers and DAs (see Chapter 6).

In Tanzania, PRA/PTD training capacities have been firmly established: already ten professionals from research, extension and training institutions are well practised in facilitating participatory training. ISWC-Tanzania developed a sequence of training events focused mainly on the scientists, extensionists and farmers directly involved in the programme. As in Ethiopia, introductory training in PRA/PTD was followed by refresher workshops to review experiences and increase insights. After this came training in farmer experimentation and then review and planning workshops based on the PTD experiences. Thus, training and learning have been closely integrated with exchange, joint review and planning, with the latter gradually assuming more importance than training per se.

In the case of ISWC-Tunisia, the scientists were less experienced in participatory research. For the first national PRA/PTD training workshop held in southern Tunisia in October 1997, trainers were invited from Senegal and Burkina Faso – an innovation in itself in that experts from south of the Sahara were training scientists in North Africa. The workshop was an eye-opener for most of the participants who were astounded at the level of local knowledge they discovered during the fieldwork. After this workshop, many participants found it difficult to apply their new knowledge and skills in an unchanged institutional setting, but some did continue to identify interesting innovations and innovators and this had a decisive, positive impact on the future of ISWC-Tunisia (see Chapter 30).

Not only in Tunisia but in all seven countries in the ISWC 2 programme, the success of the training in changing the participants' attitudes depended on the personalities of the individuals and the extent to which the institutional contexts in which they operate allow them space to use participatory approaches. The coordinators and backstoppers often observed that, when scientists and extensionists were interacting directly with farmers in the field after the training, they treated the farmers with greater respect. In some cases, they gave the innovators nicknames, such as 'professor', to express admiration for the farmers' wealth of knowledge and creativity and for their capacity to express what they know and do. Indeed, it became evident during field days and travelling seminars that some farmer innovators have oratory and presentational skills far superior to those of many scientists, college lecturers and professors.

Component 2: Identifying and verifying farmer innovators

The process of identifying farmer innovators is not easy and straightforward because farmers are not necessarily aware that they are experimenting and innovating. For most farmers, the process of generating knowledge through experimentation is part of their everyday agricultural activities, not separated from them as it is in the scientific knowledge system (den Biggelaar, 1996).

Credit: Noureddine Nasr

Plate 1.1 *The first activity of ISWC Tunisia was to organize a national training workshop in PRA/PTD (October 1997)*

In each country, all partners first had to agree on the concept of innovation and innovators. For example, as a working definition for an innovator, ISWC-Ethiopia used 'someone who develops or tries out new ideas without the support of formal extension services'. 'New' was defined as 'something that has been started within the lifetime of the farmers – not something that s/he has inherited from parents or grandparents' (see Chapter 6). This definition helped to focus the search on the dynamics of local knowledge, not just looking at what is traditional but seeking endogenous change. In contrast, ISWC-Tunisia decided also to include technologies inherited from parents in the inventory of local innovations in land husbandry.

An innovation was generally defined as something new to the particular locality, but not necessarily new to the world. For example, a refugee who saw a *shadouf* (an ancient Egyptian device for lifting water) in Sudan and who, upon his return to Tigray, developed a *shadouf*-based irrigation system was regarded as an innovator in his home area. In the PFI programme, farmers who simply copied what another farmer in the same village had developed or introduced were regarded as 'second-generation' innovators (adopters), even though what they did was new for their particular farms. An initial round of identifying farmer innovators by the PFI programme was therefore followed by verification of whether their fellow villagers indeed regarded them as 'first-generation' innovators.

In the ISWC 2 programme, the verification process focused more on verifying whether a local innovation was useful rather than whether the particular individual was a genuine innovator, ie the focus was on identifying new ideas that had been developed or tried by a few farmers but could be of interest to many. To aid in assessing whether an innovation was likely to be more widely accepted, some ISWC 2 partners studied the social setting and resource endowments of the innovators.

Many types of actors were involved in identifying farmer innovators and verifying their innovations: extension agents, NGO field staff, village leaders, students, teaching staff, scientists and farmers. Rarely were 'ordinary' farmers involved in the identification process, unlike the cases reported by Nielsen (see Chapter 8) and Yohannes (see Chapter 16), both based on doctoral research conducted outside the ISWC and PFI programmes. Sometimes, the field agents in NGOs and government agencies could make important contributions by naming innovators whom they had already observed in their day-to-day work. However, this was not always the case. Some field agents initially had eyes only for the technologies that they were promoting and for the 'progressive' farmers who adopted them. For example, NGO staff members in Cameroon were surprised to hear of the existence of local innovators; one of them said: 'We are always in the field and we have never seen them.' However, once they became familiar with the concept of farmer innovation and had actually met some local innovators and seen their accomplishments, the same staff members began to recognize innovators regularly. Some commented that they now noticed small differences in farming practices almost every time they went to the field.

Not only was a wide range of actors involved in the identification process, but a wide range of methods was also used: direct observation, interviews with farmer groups and with key informants, contests, peer identification and invitations for self-identification through radio stations. In Tunisia, for example, men and women farmers responded to the weekly radio programme on agricultural innovation by writing to the station that they, too, had innovated and wanted to tell others about it (see Chapter 27).

A total of over 800 farmer innovators have been identified in the seven countries involved in ISWC 2 and the innovations of most of them have been documented in reports, database entries, photographs and video films. Not only scientists but also field agents have been preparing this documentation. These few hundred innovators represent only the tip of the iceberg because the programme covers relatively small parts of each country. Most innovators and their innovations continue to be hidden, overlooked or ignored.

The main aim of the programme is not, however, to identify only outstanding innovators. It is to stimulate the innovative capacities of all farmers, by encouraging as many scientists and extensionists as possible to recognize local innovation and to work with farmers in ways that promote innovation, as outlined in the following components. In this context, it is important that local innovations are 'verified' by other farmers as part of a process of stimulating more widespread innovation in a village (see Component 6).

Component 3: Analysing innovators and innovations

As relatively little is documented about farmer innovation in Africa, the ISWC 2 and PFI programmes tried to gain a better picture of the characteristics of outstanding innovators and their motivations for innovating. As pointed out by Yohannes (see Chapter 16), every farmer has to be an innovator to some degree, but some farmers are known in the community as being particularly innovative. What triggers these farmers to develop new ideas or try out something new, while others operating under similar conditions do not? Is it because they are particularly rich or particularly poor in productive resources? Are most of the outstanding innovators old or young, educated or illiterate, men or women? Do innovations by women differ from those by men? Do the more innovative farmers have certain characteristics that could be 'cultivated' through programme activities? In which domains do farmers innovate? What is the potential of local innovations to be adopted or adapted by other farmers? These questions are addressed in Chapter 7 from the viewpoint of the ISWC 2 and PFI programmes, and in Chapters 8 and 16 from the viewpoint of other research on the topic.

As will be seen in Chapter 7, many of the outstanding innovators are relatively rich. Discussions among programme partners often revolved around the question of whether there was a tendency to work with richer farmers who could develop more spectacular innovations or whether outstanding innovators were richer because they were so innovative. There was a concern that the smaller improvements introduced by farmers with fewer resources were being overlooked. When country programmes such as those in Tunisia and Ethiopia began especially to seek out women's innovations, they found that these tended to be low-cost, primarily making use of local resources. Such innovations are particularly promising for resource-poor farmers (not only women) and have good potential for dissemination, but most of these innovators have less confidence and means to make their ideas more widely known than do richer, male farmers. Extension agents can play an important role in encouraging farmer-to-farmer communication about the low-external-input options developed by the less wealthy local innovators (see Component 8).

Component 4: Setting up monitoring and evaluation systems

When all is said and done, it must be possible to show all stakeholders in agricultural development whether or not the programmes have enhanced the capacity of farmers to experiment and innovate, and whether this increased capacity and the emerging innovations contribute to the well-being of farm families and to the improved management of natural resources. It is important to produce numbers, especially for donor agencies and political decision-makers, but it is just as important to analyse the process of promoting local innovation as a source of learning and guidance for future work. It is therefore necessary to document *who* has done *what*, *where*, *how* and *why*.

If the monitoring and evaluation (M&E) of programme activities is to be done well and be useful, then all partners must be able and willing to do the work. Therefore, the observations made and recorded have to be limited to the smallest possible number of key indicators of interest to those involved, and the questions need to be readily understandable. Only in this way can the learning aims be achieved.

Members of the ISWC Consortium proposed questions and forms to facilitate the monitoring of the process and outcome of the various programme activities, and to assess their impact. The proposals included more conventional aspects of M&E – collecting information defined by outsiders for analysis by outsiders – and some participatory aspects of M&E, so that the farmers and national partners could record and assess data according to their own criteria, using methods they chose or developed themselves. It was up to each country programme to adapt the questions and forms to the local conditions as well as to the local capacities to handle and analyse the information generated.

The Consortium initially thought that many farmers would perceive data recording as a burden. As it turned out, farmers who were involved in deciding what data to collect were keen to keep the records. They regarded this as a useful skill for managing their farms better because it helps them to keep track of and analyse levels of inputs and outputs. All the farmer innovators who were involved in joint experimentation in Burkina Faso insisted on keeping the records themselves. Some farmers in Uganda seemed more interested in improving their capacities to keep their own records than in improving their yields through the particular technologies they were testing. Many farmer innovators also kept visitors' books to record who came to their farms over the years.

ISWC-Tanzania made a point of regarding the M&E of innovations as an integral part of the joint experimentation and learning process, and not as a stand-alone activity. The partners in the experimentation agreed on the criteria for assessing specific innovations and experiments. The assessment itself was often done in a workshop situation, during which the development agents and scientists documented the farmers' comments. However, the families of some farmer experimenters also kept written notes (see Chapter 25).

Component 5: Exchange visits and study tours for innovators

Both the ISWC 2 and the PFI programmes gave innovators an opportunity to visit each other's farms and to exchange information and experience. Promoting Farmer Innovation in Rainfed Agriculture also took innovators on study tours to other areas and research stations. The exchange visits and tours were very effective in making the farmers aware of local innovations and the principal ideas behind them. After returning home, many participants immediately started informal experiments with new techniques they had seen. For example, after exchange visits some farmer innovators in Burkina Faso started trying out other cultivation tools, rainfed rice production, different varieties of sorghum and millet and different heights of cutting the stalks at harvest (Hilhorst, 1999).

It is important to give attention to both preparing for and following up the visits and tours. Farmer-to-farmer communication is more effective if both the visitors and the hosts are well prepared. What do both sides hope and expect to gain from the visits? An example of such preparation is given in Chapter 18. The learning effect is also greater if, after the visits, the visitors and hosts review what they found useful about the exchange and deliberate on how and to whom they will report their impressions. It was observed, for instance, in Zimbabwe that women are more likely than men to share their impressions with others, although not necessarily in a formal way, and that women tend to share with women and men with men (Edward Chuma, pers comm, February 2000). Field agents can play a role here in helping to widen the impact of the visits by creating more opportunities for feedback to others.

Another role for field agents is to arrange additional technical training or other forms of support that may be required as a follow-up to a visit or tour. In Tunisia, some farmer innovators trained other farmers, for example in grafting fruit trees, immediately during the visits, but often such training will be possible only afterwards.

The PFI programme encourages innovators to organize themselves into clusters of usually eight farmers (Critchley et al, 1999, p15). Experiences are exchanged within and between these clusters (see Chapter 17). One advantage of such small groups in one locality is that the members can meet easily and at little expense. The ISWC programme in Uganda follows the PFI approach and includes 32 farmer innovators in four networks of eight innovators each. An impact assessment showed that, after intra-network visits, each member tried out several innovations seen in the fields of other members (see Chapter 19).

In the other ISWC 2 countries, it was left up to the innovators to decide if and how they wanted to organize themselves. In Tanzania, for example, three cases were recorded in which farmer innovators, after returning home from exchange visits, started up a local group of one or more innovators and a larger number of neighbouring farmers. Not only did they want to share what they had learnt about techniques, but participation in the visits had also encouraged them to undertake joint development activities (Laurens van Veldhuizen, pers comm, January 2001).

Component 6: Farmers' evaluation of local innovations

This component can be part of the process of selecting ideas for farmer-led experimentation as well as part of farmer-to-farmer exchange on innovations. In most cases in the two programmes, 'outsiders' made an initial assessment of the innovations. Then the farmers themselves – both the outstanding innovators and their neighbours – were given opportunities to assess the innovations jointly. This was not limited to comparing yields and other technical parameters. Yohannes (see Chapter 16) found that the values and experience of the communities strongly influenced their acceptance of local innovations.

Village workshops have proved to be a useful tool for farmers' assessment of innovations. In Tigray, for example, the Bureau of Agriculture and Natural Resources (BoANR) organizes Farmers' Fora (Chapter 18) during which innovators and their neighbours examine both endogenous and exogenous innovations and discuss what is useful for local farmers. Such workshops at village or district level can also help to dispel the isolation that many outstanding innovators feel because they are doing something different from the norm. In Zimbabwe, for example, farmer innovators were often regarded as 'off-types' because they did not follow extension instructions. The emphasis of ISWC-Zimbabwe on group formation and joint experimentation served to enhance the social standing and confidence of innovators and strengthen their position vis-à-vis government services. A group approach also reduces the risk that farmer innovation programmes focus on only a few privileged farmers.

Innovation assessment workshops at village level can be starting points for designing experiments implemented by farmers selected by the community. This embeds the experimentation by individuals within a wider social process. In Cameroon, for instance, the priorities for experimentation were set during a workshop involving scientists, extensionists, farmer innovators, other male and female villagers and traditional authorities (see Chapter 21). This was done to avoid isolation of the innovators and their experiments from the rest of the community and to ensure that the techniques being tested met community criteria in technical, sociocultural and economic terms.

Component 7: Stimulating and supporting joint experimentation

The original assumption in the ISWC 2 programme document was that joint experiments by scientists, extensionists and farmers would start by the end of Year 1. This was far too optimistic. The entire process of gaining a common understanding of concepts, training in PRA/PTD, identifying and understanding farmers' innovations and informal experiments, organizing exchange visits and building up rapport between farmers, extensionists and scientists required much time. In most countries, joint experiments started only in Year 3.

If the joint experiments had started earlier, there would have been a greater risk that the scientists defined them and imposed them on the farmers. Ideally, the innovators and other interested farmers should set the research agenda and it should be based on their priorities. Scientists have a role in:

1 proposing elements for testing which the farmers might like to include;
2 advising farmers on how to design simple experiments so that both farmers and scientists have a firmer basis for evaluating the results; and
3 explaining reasons for the farmers' findings and thus helping farmers to understand better some of the principles and less visible factors influencing the outcome of their experiments.

In addition, the scientists can assist in generating 'hard' data to validate the findings in conventional scientific terms in order to convince other scientists, policy-makers and donor agencies. A further role of scientists, especially in the initial stages of collaboration with farmers and extensionists, is to observe and analyse the process in order to develop the farmer innovation methodology and to describe it clearly to others who want to engage in it.

The actual process of setting the joint research agenda differed from country to country. ISWC-Tanzania organized a series of workshops for the 'experts' (scientists and extensionists) to build up an appreciation of the important role of farmers in agricultural research and development. Parallel workshops and exchange visits were organized for farmers to build up their confidence and assertiveness so that they would be better able to argue for their interests. Once these two processes had matured, the experts and farmers were brought together in a workshop to set priorities for joint experimentation (see Chapter 5). Similarly, in Cameroon farmer innovators clearly formulated their interests and insisted on giving them priority over scientists' research interests. The scientists agreed and a process of PTD commenced (see Chapter 21).

At the start of the ISWC 2 programme, the country with the most experience in stimulating and supporting processes of joint experimentation was Zimbabwe. Here, the PTD activities were inextricably linked with the development of local institutions and were carried out with groups of farmers defined by the communities. Thus, PTD became an approach not only for technology development but also for community development (Murwira et al, 2000).

In some cases, the joint experiments focused on scientific validation of local innovations. Here, the procedures were more conventional and it was important to collect data that could be subjected to statistical analysis. In other cases, the process of experimentation was designed to allow the farmers to assess the innovations and/or to investigate farmers' and scientists' ideas to improve them further. This had the additional aim of strengthening local capacities to test and assess new ideas. Here, it was more important that the procedures and measurements could be managed by farmers, and they could eventually do it on their own. Farmers were encouraged to learn about experimentation by analysing their own experiences and mistakes.

The workshops for farmer experimenters organized by ISWC-Tanzania reflect this latter approach: with the support of scientists and extensionists, the farmers examined each other's experiments and recognized where improvements should be made (eg Malley and Mruma, 2000).

Component 8: Farmer-to-farmer dissemination of innovations

One aim of both ISWC 2 and PFI is to disseminate useful innovations. Various forms of farmer-to-farmer communication are facilitated, primarily through visits and workshops. The PFI programme organizes three types of exchange visits: farmer innovators visit other innovators to learn from each other, 'ordinary' farmers visit farmer innovators to learn from their experience, and farmer innovators visit other farmers to inform them and to train them in the use of improved technologies. The first type, involving only learning by farmer innovators, comes under Component 5, while the second and third types are examples of wider farmer-to-farmer extension.

In the ISWC 2 programme, each country has developed its own variations. In Ethiopia, Components 5 and 8 were combined in the travelling seminars, during which innovators visit other innovators and, at each site visited, meet with a wider group of people in that community. In Zimbabwe, mixed groups of farmer innovators and other villagers visit other sites of innovation, and the villagers rather than the programme choose the farmers to be involved. In Uganda, exchange visits were made between the farmer innovators working with the PFI programme in the north and those working with the ISWC programme in the south. A large number of farmers have been involved in the visits organized under the PFI programme: in Kenya alone, over 2500 farmers – about 1600 men and 900 women – have visited farmer innovators (Will Critchley, pers comm, January 2001).

The original idea was to disseminate farmers' innovations that had been scientifically validated, but soon it became obvious that farmers do not wait until a technology has been given a stamp of approval by scientists. After exchange visits, farmers in Burkina Faso who recognized potentials in certain innovations already started to apply them (Hilhorst, 1999 and see Chapter 25).

In Tanzania, this component is closely linked with Component 6: the village workshops for evaluating local innovations automatically led to sharing of ideas between farmers. However, most of the sharing probably took place outside of the activities supported directly by the programme. This included the increased opportunities for farmer-to-farmer exchange within the regular research and extension activities of the partner organizations. For instance, during the regular farmer field days, farmer innovators are asked to explain what they are doing. Information about local innovations and innovators has also been spread via the existing journal of the farmer network, Mviwata (Laurens van Veldhuizen, pers comm, January 2001).

Sharing of knowledge also includes informal farmer-to-farmer extension, such as:

- when farmers discuss new ideas during market visits or at small social meetings, for example Ethiopian women who meet for the traditional coffee ceremony;
- when farmers on their own initiative visit other farmers who are doing something that appears to be interesting; and
- when farmers report what they have seen to a larger group, for example during a village meeting held for another main purpose (see Chapter 18).

In Burkina Faso, outstanding personalities have innovated in developing farmer-to-farmer extension approaches: one innovator initiated an annual market day for exchanging information about improved traditional planting pits (*zaï*); another created a kind of 'field school' during which farmers learn by jointly treating a piece of degraded land with *zaï*; and a third innovator coaches a number of farmers in neighbouring villages by working together with them in their fields (see Chapter 20).

A fundamental issue that will determine the success of farmer-to-farmer dissemination of local innovations is whether innovators are willing to share their knowledge and experience with others. A woman innovator in Tunisia began to tell other farmers about how she uses decomposing manure to hatch eggs only after she was invited to present her innovation on the radio (see Chapter 12). Some individuals are not at all willing to inform others about their innovations, and some will do so only for payment (farmers can hire their services as consultants). In most cases, however, the ISWC 2 and PFI programmes found that farmers were eager to share their innovations. In some cases, they did so because this gave them public recognition and social esteem (see Chapter 20); in others, because they felt it was their duty to their community (see Chapter 14). Some farmers reportedly shared their knowledge with others in their village in order to avoid jealousy and possible ensuing difficulties, such as witchcraft.

Component 9: Raising awareness and lobbying for policy change

The eight countries participating in the two programmes have pursued various paths to raise awareness about farmer innovation and to influence policy in its favour. Much attention has been given to documentation and publication in the form of working papers, project reports, workshop proceedings, papers for conferences and articles in newsletters and journals. The programmes in Cameroon and Ethiopia started up local newsletters on farmer innovation. All the national ISWC 2 teams have gained access to radio, television and the press. As soon as ISWC-Ethiopia had identified some outstanding farmer innovators, it arranged radio interviews that were broadcast in the local language (Tigrigna). ISWC-Tunisia launched a two-hour programme on agriculture and innovation, which is broadcast to most of central and southern Tunisia (see Chapter 27). The Tunisia experience stimulated ISWC-Cameroon to appoint a media adviser to maximize access to mass media and stimulated ISWC-Burkina to start, in late 2000, a radio programme in the Yatenga region using the station La Voix du Paysan (Voice of the Farmer).

Several ISWC 2 teams have produced one or more video films about farmer innovation in their country and, in some cases, at least excerpts have been shown on national television. The PFI has produced a broadcast-quality video on farmer innovation in East Africa which has been aired on television in all three countries involved in that programme (PFI, 2000). It has also published a book outlining the theory and practice of its approach (Critchley et al, 1999), and more than half of the 10,000 copies had been distributed by the end of 2000.

In a strategy to enhance policy dialogue, policy-makers have been included in the ISWC steering committees and taken on 'exposure tours' to farmer innovators. In Ethiopia, for example, the head of the BoANR in Tigray plays a very active role in the steering committee, and the federal Ministers of Agriculture and Education have visited farmer innovators in the field. Also in the other countries, the national ISWC coordinators have organized meetings with, and field visits by, Ministers of Agriculture or with senior staff responsible for national programmes of research and extension. The president and vice-president of Uganda visited farmer innovators under the PFI programme. Upon the invitation of the country programmes, important policy-makers opened national or international workshops on farmer innovation.

Important events for raising awareness about the farmer innovation approach were the two conferences on farmer innovation held in November 1999 in Cameroon for francophone Africa (Tchawa and Diop, 2000) and in February 2000 in Ethiopia for anglophone Africa (Mitiku et al, 2000). These meetings gave an opportunity for the countries in the two programmes to exchange experiences, but also broadened the participation to include other projects and, above all, policy-makers who could have a positive influence on institutionalizing the approach. Also for this purpose, one day of the annual review meeting in the third and fourth years of the ISWC 2 programme was opened up to policy-makers from the host countries, Tanzania and Tunisia.

To initiate an approach to PTD through identifying and promoting farmer innovation, ISWC 2 deliberately commenced work below the national level (ie in only certain provinces or regions of a country) in order to develop successful cases of proven local innovation capacity as this is essential for effective lobbying work. Within three years, the country programmes collected substantial evidence that the farmer innovation approach produces results of interest to smallholders and supporting organizations, and that it can be an alternative to the conventional ToT model of agricultural research and extension. They are providing the accumulated evidence in various forms to decision-makers in their countries, and are exposing them to examples of farmer-led experimentation and innovation.

Component 10: Institutionalizing the farmer innovation approach

Ultimately, the ISWC 2 and PFI programmes are working towards incorporating the farmer innovation approach into the regular activities of agricultural research, extension and education in the countries where they are operating. Only in this way will it be possible to ensure the long-term sustainability of the approach. This requires a shift away from conventional ToT towards promoting farmer-led experimentation based on both endogenous and exogenous ideas. Both programmes seek to strengthen the capacity of scientists and extensionists to recognize the problems being addressed by local innovators in their informal experimentation and to participate in and support farmer-led research, as outlined in Component 7. In this collaboration, one role of extensionists will still be to make information, from whatever sources, more easily available to farmers. However, this will be with the aim of enriching farmers' experimentation, ie feeding farmers' hunger for useful new ideas to try out, rather than prescribing introduced 'improvements' which may not prove to be such. As Nielsen points out in Chapter 8, one of the major constraints to farmer experimentation identified by the farmers themselves is a lack of ideas or knowledge of new things with which to experiment.

The farmer innovation approach can be scaled up only when the concept and methods of promoting farmer innovation are included in agricultural education. Mekelle University, the lead agency of ISWC-Ethiopia, is changing its curriculum accordingly, as is the Cooperative College Moshi (CCM), the lead agency of ISWC-Tanzania (see Chapter 31). ISWC-Cameroon recently trained staff of the National Agricultural Extension and Research Programme (PNVRA) funded by the World Bank and the International Fund for Agricultural Development (IFAD), and the farmer innovation programme will become a semi-autonomous project under PNVRA. Beyond Africa, the involvement of graduate students from Wageningen Agricultural University in The Netherlands has led to an increased understanding of and interest in the farmer innovation approach among staff of the Department for Communication and Innovation Studies. These are encouraging steps forward, but there is still a long road to travel to institutionalize the farmer innovation approach in these and other countries in Africa and Europe and to make the approach less dependent on external funding and technical support.

MAJOR STRENGTHS

The two farmer innovation programmes, ISWC 2 and PFI, have sought to stimulate the enthusiasm of formal research and extension to promote local innovation in ways that combine indigenous and appropriate external knowledge. They have sought to enable institutions of formal education to prepare future researchers and extensionists to engage in this process.

The major strength of the methods used in these programmes is that they are grounded in field realities and experience (Chambers, 2000). The programmes have shown that the farmer innovation approach can be used in different agro-ecological and socioeconomic settings and is replicable in that sense. However, this does not mean that all the ten components outlined above can or should be applied in the same way or in the same order everywhere. The approach and the methods are not cast in stone. There is always a need to adapt, to refine, to learn and to improve. To quote Robert Chambers (2000): 'The challenge is to explore and invent further, and especially to embed these methodologies in the everyday practice of farmers, extensionists and scientists.'

A key feature of both programmes is flexibility. Each of the eight African countries involved has been able to place its unique accents within the overall framework of these ten components. This allowed the partners in the different counties not only to learn as they worked in their own way, building on their own history and experience, but also to learn from each other's experience. In this way, each country is strengthening its own capacity to innovate in agricultural research and development.

ACKNOWLEDGEMENTS

We would like to thank Wolfgang Bayer, Will Critchley, Jean Marie Diop, Fetien Abay, Thea Hilhorst and Laurens van Veldhuizen for their constructive comments on, and additions to, this chapter.

NOTES

1   PRA is a growing family of approaches and methods designed to enable local people to share, enhance and analyse their knowledge of life and conditions, to plan, act, monitor and evaluate (Chambers, 1997, p103)
2   Advisory support is provided by a European Consortium composed of the International Cooperation Centre (CDCS), Vrije Universiteit, Amsterdam, The Netherlands; the Drylands Programme, IIED, Edinburgh, UK, IDS, University of Sussex, Brighton, UK and ETC Ecoculture, Leusden, The Netherlands. CDCS coordinates the programme and provides SWC expertise; IIED and IDS provide support in PRA methods and policy analysis; ETC Ecoculture gives training and advice in PTD. The ISWC programme is funded by the Environment Programme of the DGIS of The Netherlands Ministry of Foreign Affairs. Each of the seven countries manages its own fund for in-country activities

REFERENCES

Bauer, E, Hoffmann, V and Keller, P (1998) 'La vulgarisation agricole: un aperçu historique', *Agriculture + Développement Rural*, vol 5, no 1, pp3–6
Bunch, R (1985) *Two ears of corn: a guide to people-centered agricultural improvement*, World Neighbors, Oklahoma City
Chambers, R (1983) *Rural development: putting the last first*, Longman, London
Chambers, R (1997) *Whose reality counts? Putting the first last*, Intermediate Technology Publications, London
Chambers, R (2000) 'Reflections on the workshop and the future', in Mitiku et al, (eds) *Farmer innovation in land husbandry*, pp146–153
Chambers, R, Pacey, A and Thrupp, L A (1989) *Farmer first: farmer innovation and agricultural research*, Intermediate Technology Publications, London
Critchley, W, Cooke, R, Jallow, T, Lafleur, S, Laman, M, Njoroge, J, Nyagah, V and Saint-Firmin, E (1999) *Promoting Farming Innovation: harnessing local environmental knowledge in East Africa*, RELMA/UNDP, Nairobi
CTA (1996) *Agricultural extension in Africa: proceedings of an international workshop, Yaoundé, Cameroon*, CTA, Wageningen
den Biggelaar, C (1996) *Farmer experimentation and innovation: a case study of knowledge generation processes in agroforestry systems in Rwanda*, FAO, Rome
Haverkort, B, van der Kamp, J and Waters-Bayer, A (eds) (1991) *Joining farmers' experiments: experiences in participatory technology development*, Intermediate Technology Publications, London
Hilhorst, T (1999) *Rapport de mission CES II*, 12 au 22 novembre 1999, IIED, Edinburgh

Howard, J, Mulat, D, Kelly, V, Mywish, M and Stepanek, J (1998) *Can the momentum be sustained? An economic analysis of the Ministry of Agriculture/Sasakawa Global 2000's experiment with improved cereals technology in Ethiopia*, in T Hilhorst and C Toulmin, *Integrated soil fertility management*, Ministry of Foreign Affairs, The Hague

Johnson, A W (1972) 'Individuality and experimentation in traditional agriculture', *Human Ecology*, vol 1, no 2, pp149–159

Malley, Z J U and Mruma, A O (2000) 'Report on farmers' reflection workshop at ADP Mbozi Farmers Training Centre, Ukwile, 12–14 September 2000', ISWC-Tanzania, Moshi

Mitiku H (2000) 'Farmer innovator awards', in Mitiku et al (eds) *Farmer Innovation in Land Husbandry*, p166

Mitiku H, Waters-Bayer A, Mamusha L, Mengistu H, Berhan G A, Fetien A and Yohannes G M (eds) (2000) 'Farmer innovation in land husbandry, proceedings of the anglophone regional workshop, 6–11 February 2000', ISWC-Ethiopia, Mekelle

Murwira, K, Chuma, E and Chirunga, F (2000) 'The experience of group approaches to research and extension in Zimbabwe', in Mitiku et al (eds) *Farmer innovation in land husbandry*, pp28–31

PFI (2000) *Promoting Farmer Innovation*, video (VHS) in English, French and Kiswahili versions, Regional Land Management Unit (RELMA), Nairobi

Reijntjes, C and Hiemstra, W (1989) 'Farmer experimentation and communication', *ILEIA Newsletter*, vol 5, no 1, pp3–6

Reij, C, Scoones, I and Toulmin, C (eds) (1996) *Sustaining the soil: indigenous soil and water conservation in Africa*, Earthscan, London

Rhoades, R E and Bebbington, A (1995) 'Farmers who experiment: an untapped resource for agricultural research and development', in D M Warren, L J Slikkerveer and D Brokensha (eds) *The cultural dimension of development: indigenous knowledge systems,* Intermediate Technology Publications, London

Richards, P (1985) *Indigenous African revolution: ecology and food production in West Africa*, Westview Press, Boulder

Richards, P (1986) *Coping with hunger: hazard and experiment in an African rice-farming system*, Allen and Unwin, London

Röling, N and Brouwers, J (1999) 'Living local knowledge for sustainable development', in G Prain, S Fujisaka and M D Warren (eds) *Biological and cultural diversity: the role of indigenous agricultural experimentation in development*, Intermediate Technology Publications, London, pp147–157

Röling, N and van de Fliert, E (1994) 'Transforming extension for sustainable agriculture: the case of Integrated Pest Management in rice in Indonesia', *Agriculture and Human Values*, vol 11, nos 2–3, pp96–108

Tchawa, P and Diop, J M (2000) 'Paysans innovateurs en gestion durable des ressources: atelier régional francophone du programme CES 2, Bamenda, Cameroun', 29 novembre–2 décembre 1999, SNV/CES 2, Yaoundé

van Veldhuizen, L, Waters-Bayer, A and de Zeeuw, H (1997) *Developing technology with farmers: a trainer's guide for participatory learning*, ZED Books, London

2

The career and influence of Barthelémy Kameni Djambou in Cameroon

*Paul Tchawa, Noubissié Tchiagam Jean-Baptiste
and Yossa Bonneau**

*A former taxi driver returned to his home village and decided to devote his
energies completely to farming. Within 12 years, he became a well-known and
respected farmer innovator, and students from the University of Dschang
regularly take training from him in agroforestry. How did this farmer manage
to make such a remarkable career?*

Barthelémy Kameni Djambou is in his mid-40s and lives in the village of
Babone in West Cameroon. He used to be a taxi driver in Douala, Cameroon's
biggest city, but returned home in 1988, when his father died. As the eldest
son, he became the head of the family. He married and his household now
consists of himself, his wife, his sister-in-law and three small children. From
his father, he inherited about 3ha of poor-quality land situated on a 20 per
cent gradient. This is about the average size of farms in the area, but he did
not regard it as enough to make a decent living for his family. As agriculture
was his only possible source of living in the village, he decided to look for
ways to modernize his farming practices and to increase production.

* Paul Tchawa is a geographer and senior lecturer at the University of Yaoundé I and
coordinator of ISWC-Cameroon; Noubissié Tchiagam Jean-Baptiste is a plant
geneticist at the University of Ngaoundéré, Cameroon, and Yossa Bonneau heads
the NGO Centre de Développement des Communautés Villageoises (Centre for
Village Community Development) in Bafang, West Cameroon

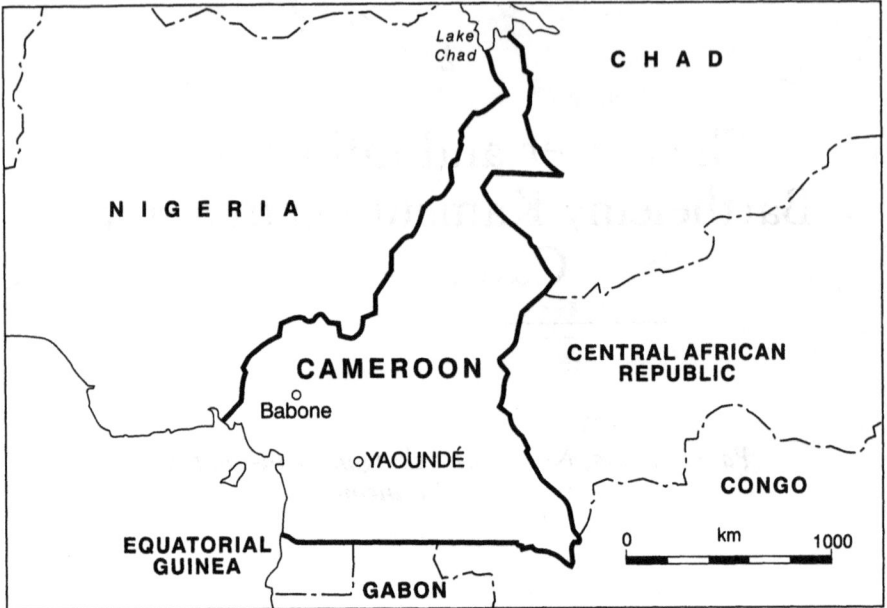

Map 2.1 *Cameroon*

SEEKING NEW IDEAS

First Djambou tried to generate cash income by growing tomatoes in the modern way recommended by the extension service. He planned to sell the tomatoes on the local market. He was the first farmer in his village to do this, but the results were disappointing and he could not even cover the costs of the chemical fertilizers. Nevertheless, he continued to try and sought additional information from the extension services, thinking that he simply was not doing it quite right yet.

At the same time, he contacted two local NGOs to see what ideas they had to offer. Through them, he received training in techniques of contour bunding and facilitating farmer-to-farmer learning. He was also given the opportunity to take part in farmer-exchange visits to other parts of Cameroon. It was during these visits that he acquired information and ideas about agroforestry and about the importance of nitrogen-fixing plants for maintaining soil fertility. He started to apply these techniques on his own farm and met with success. He felt that his experience would be a good basis for teaching other farmers.

In 1992, Djambou decided that he would try to obtain formal training as an extension officer. Although he did not have the right qualifications in terms of formal education to be admitted to this training, he was so insistent that the staff of the Cameroon Ministry of Agriculture (MoA) decided to admit him in 1993, so he would no longer bother them with his requests. In 1995, he successfully completed the 16-month course in agricultural extension. Before

Credit: Chris Reij

Plate 2.1 *Barthelémy Djambou, a key innovator in West Cameroon*

and during this formal training, Djambou continued to try to grow tomatoes and other vegetables using chemical fertilizers. He applied what he had been taught, but the results continued to be disappointing.

The turning point came in 1995, when Djambou received further training in agroforestry through the UNDP Africa 2000 Network. This triggered a complete change in the way that he farmed. He stopped using chemical fertilizers and abandoned the maintenance of terraces on his slopes as he realized that SWC could be achieved in ways that required less strenuous work. He began to devote himself to developing agroforestry practices on his farm. First, he planted rows of fast-growing nitrogen-fixing tree species (*Leucaena leucocephala* and *Calliandra calothyrsus*) on the contours in his fields. Between these rows he introduced a short-term improved fallow using the nitrogen-fixing woody species *Tephrosia vogelii, Cajanus cajan, Crotalaria funcea* and *Sesbania sesban*. After this short fallow, he sowed maize.

SHARING WITH OTHER FARMERS

In 1996, Djambou formed a group of farmers in his village with the intention of stimulating and facilitating exchange of experience with agroforestry practices. He started training the group members in agroforestry by working together with them first on his farm and then on their farms. From time to time, he passes by to visit the group members and to see how they are progressing.

In 1997, he inherited from an uncle a piece of land in a waterlogged valley bottom and he started to reclaim it for productive purposes. He mobilized the services of a retired administrator who has become a specialist in aquaculture. They had met at a farmers' workshop organized some years previously by the NGO Service d'Appui aux Initiatives de Développement (Support Service to Development Initiatives). Djambou paid for the man's travel and gave him food and board during his stay in the village. The members of the farmers' group helped Djambou to dig two fishponds. Immediately next to these ponds he built some sheds for pigs and started to use the wastes from the pigs to feed the fish.

Djambou is coordinator of the farmers' group and there is a good atmosphere of partnership among the members and a great willingness to help each other. Djambou is well integrated in the community because the people know that he readily shares with others. For example, when the fishpond is harvested, all members of the group receive some fish.

Even while he was in the process of trying out various ways to integrate crops, trees, pigs and aquaculture, Djambou maintained his interest in attending training courses in order to gain still more ideas. In 1997, he was trained in beekeeping and he immediately introduced this activity on his farm. He constructed the beehives himself and bought the honey-harvesting equipment from farmers in the area of North-west Cameroon where he had received the training. In 1998, he was trained in techniques of rearing small livestock and immediately started to plant Guatemala grass (*Tripsacum laxum*) near his homestead, already with the plan to feed to rabbits. In 1999, he added the cultivation of medicinal plants and the keeping of rabbits to his widening range of income-generating activities. He bought the rabbits with a loan from a local credit organization.

After ISWC-Cameroon commenced in 1997, Djambou joined the farmer innovator network in his region: the Réseau des Paysans Innovateurs du Haut-Nkam (REPIH). As a member of REPIH, he had the opportunity to visit other farmer innovators in West and North-west Cameroon. In this way, he gained access to seeds of a new maize variety K525 created by another farmer innovator (see Chapter 23). He tried the new variety out in his own fields and, even though his initial results were rather poor, he continued to experiment with this and other maize varieties. During the innovator exchange visits, Djambou also collected slips of vetiver grass (*Vetiveria zizanioides*) which he planted around his fishponds. Ever since 1995, he been experimenting with new plants, particularly nitrogen-fixing ones. In this way, he has built up a vast practical knowledge about how these various plants can be fitted into his farming system.

TEACHING IN HIS FIELDS

Djambou's farm in Babone has become a training ground not only for the members of the local farmers' group, but also for hundreds of visitors (roughly

Credit: Chris Reij

Plate 2.2 *Outsiders listen to and learn from Barthelémy Djambou (Cameroon)*

200 per year), including farmers from other villages and on exchange visits, national and international NGOs, MoA staff and donor agencies. In 1997, with the financial support of the Africa 2000 Network, Djambou built a classroom on his fields. Here he gives training in agroforestry and other land management practices to farmers and farmer groups from different parts of Cameroon and to agronomy students from the University of Dschang. In addition, students from agricultural schools do two to three months of practical training on his farm and live with his family. Djambou is not only a strong personality who knows what he wants, he is also an excellent speaker with an impressive command of French, a skill that he developed while driving a taxi in Douala which helps to make him a good communicator also in interacting with policy-makers. The dynamic big-city taxi driver turned farmer has now become a widely recognized innovator and teacher in agricultural development.

3

Ayelech Fikre: an outstanding woman farmer in Amhara Region, Ethiopia

*Million Alemayehu**

A woman with tremendous energy and perseverance, Ayelech Fikre has spent several decades perfecting and combining indigenous practices of SWC to build up an intensively cultivated hillside farm in the highlands of North Shewa Zone of Amhara National Regional State in Ethiopia. This chapter highlights some of her techniques and her philosophy of land husbandry

Ayelech Fikre is a 63-year-old widow who lives in Ankober district in North Shewa zone. She has one adopted son who is now married and has two children. Ayelech is the head of this family. Over 35 years ago when her father died, she inherited from him about 1ha of farmland situated in a *weyna dega* zone (medium highland, according to the local agroclimatic classification system) on a hillside over 2200m above sea level. Average annual rainfall is about 870mm, with two rainy seasons: the short (*belg*) rains from January to April and the long (*meher*) rains from July to September. Ayelech grows mainly sorghum, teff, maize, wheat, horsebean and barley. She has two oxen, one cow, one donkey and three sheep. On her own initiative, without any schooling or formal training,[1] she has applied various techniques of land improvement based on local indigenous knowledge and on her careful consideration of what she observed and what she had available to her. These techniques include the construction of stone bunds, soil fertility management and rainwater harvesting.

---

* Million Alemayehu, a SWC specialist with the Bureau of Agriculture in Amhara region, is currently a post-graduate student at Alemaya University of Agriculture, Ethiopia

Map 3.1 *Ethiopia (Northern Shewa action area)*

CONSTRUCTION OF STONE BUNDS

After her father's death, Ayelech took over responsibility for managing the farm. It was only then that she began to notice the damage done to the farmland by heavy rainstorms. One day she saw how the flood from the hillside washed away the soil and, with it, the seed she had just sown. She also observed that small gullies were forming on the land. It was then that she started thinking about what measures she could take to prevent this damage. She came up with the ideas of improving the cut-off drain that her father had previously dug into the slope above the farm and constructing stone bunds on the cropland itself. Then she set to work, together with her son, to implement these ideas on the ground, tackling first the cut-off drain by widening and deepening it, and then starting to build stone bunds.

She and her son worked on this gradually over the years, whenever they could find some time in between all their other work. Her husband was seldom there to help. They started at the lower boundary of the farm and worked their way up the slope. Before piling up the stones for each bund, they dug a foundation trench about a metre deep and a metre wide in order to make the bund very stable.

After some years, Ayelech felt that the work was going too slowly. It demanded a great deal of time and energy, and too few people were doing the work. Therefore, in order to improve her land more quickly, she asked her neighbours to assist her. In line with the local tradition of labour cooperation known as *debo*, men came to work on her farm – up to 40–50 at one time.

Since food must be served for all the people participating in *debo*, Ayelech prepared the food at night. This meant that she could then work throughout the day on designing, supervising and working together with the men in making the stone bunds. She states with pride that she designed the layout of the bunds completely by herself. Although she did not use any land-surveying instruments (eg line level), she and the men helping her managed to build the bunds along the contours.

Now all of her farmland has been treated with stone bunds. They vary greatly in size: the height varies from 0.5m to 3m and the width from 0.3m to 2m. Instead of stretching continuously across the slope from one edge of the farm to the other, the bunds are in a staggered arrangement. This was done intentionally to facilitate ploughing by oxen: the animals can move from one level to the next in a zigzag manner, passing through an interruption in one level of bunds that lies above a continuous piece of bund at the next level.

MANAGEMENT OF SOIL FERTILITY

Ayelech began to notice that the soil fertility immediately below the bunds was lower than in the area immediately above them. After considering various possibilities to improve the fertility of the soil below the bunds, she started to apply composted manure to the terraces and to plant croton (*Croton macrostachyus*) in a line immediately below each stone bund. This line of plants also helps to stabilize the bunds. She chops the croton leaves and spreads them over the less fertile parts of her farmland. After the leaves have dried, they are ploughed in during land preparation. She prepares the compost every year by putting animal manure, vegetation (mainly croton leaves) and wastes from the household and from animal feed into a pit in her backyard.

Ayelech has been practising intercropping and crop rotation ever since she started farming on her own. She intercrops sorghum with soybean and maize with horsebean. She says that her main objective in intercropping is to maximize the total crop yield that she can obtain from her one hectare of land.

HARVESTING OF RAINWATER

Ayelech produces coffee on a small plot in her backyard. Since there is no source of water (spring or river) nearby, she could not irrigate the coffee. She considered how she might solve this problem and eventually decided to try collecting run-off from her upslope farmland during the rains and diverting it to her coffee plot. She dug temporary ditches below two stone bunds in order to collect run-off from the areas between the terraces and to guide it into the coffee plot. However, after a particularly heavy rainfall one day, she saw that her coffee plot was flooded with run-off full of sediment. This made her realize that, despite the improved cut-off drain and the stone bunds, some soil was still being washed away by the rain, especially during its onset.

She then dug a trench between the coffee rows in order to harvest more run-off and, with it, her fertile soil coming from above. Moreover, she dug two pits to store the rainwater. After each trench has been filled, she diverts the run-off to the artificial waterway which eventually joins a natural waterway. When the run-off becomes clean (after the sediment load has decreased), she diverts it to the two pits. She continues this process until both pits are filled with water. Then she seals the pits. They are impermeable, so the water is retained for a long time – more than three months, according to Ayelech. She uses the water collected in the two pits mainly for watering the coffee during the dry periods. In this way, she manages to obtain a better yield than farmers in the area who are growing coffee on irrigated land.

She also collects rainwater from the roof of her house by placing a stone in a small ditch under the eaves. This blockage makes the water form a temporary 'pond' in the ditch. Immediately after the rain has stopped, she takes water from this 'pond' to grow different vegetables (peppers, tomatoes, etc) immediately behind her house. She does this during the onset of the first short rains.

Although she has tried many ways of harvesting the rainwater, she is still not satisfied and is always seeking new ideas. In her words: '*Wuha binoregn noro yemalabeklew sew chewenna nafta becha new*' (Amharic); this means, 'If I had had water, I could have grown anything except human beings, salt and diesel.'

MAINTAINING BIODIVERSITY FOR MULTIPLE PURPOSES

Apart from the different indigenous techniques applied directly on her farmland to conserve soil and water, Ayelech has also treated the steep land above her farmland by constructing hillside terraces and planting *gesho* (*Rehaminus perinoides*), also known as 'hops', which is used for local beer brewing and for which there is a high demand on the local market. She sells most of the hops she produces. In addition, she allows other indigenous tree species, such as juniper (*Juniperus procera*) and African olive (*Olea africana*) to regenerate naturally on this slope. She prunes the juniper branches so that the trees will attain quickly the right height and diameter to be sold as timber. She taught herself about the effect of pruning by leaving some trees unpruned and comparing their growth with that of the pruned trees. She uses the prunings for fuel.

Ayelech had developed considerable skills in the selection and storage of the seed of sorghum, horsebean, maize and other crops. For example, she selects sorghum seed based on the phenotype of the individual plants and threshes these separately. She mixes the seeds with pepper (a traditional means of pest control) and stores them in a cool place in her compound.

THE PROBLEM WAS HER TEACHER

Ayelech's intensively worked hillside farm has been created by her own knowledge and efforts over decades. When a visiting expert asked her 'Who taught you to do all these activities?', she laughed and replied, 'The problem.' She explained: 'People are increasing in number as time goes on, but the land remains the same. We can't multiply it; people already occupy each piece of land. Therefore, we have to take care of the land we are using, otherwise there is no way out to get land. Thus, the problem [erosion and difficulties of getting more land] taught me to do all these activities. Otherwise, I couldn't survive.'

Another of Ayelech's sayings reveals how she sees land husbandry: '*Mereten sirebat magures siberdat malbes*' (Amharic). This means: 'When the land gets hungry, feed it; and when it gets cold, cover it.' She explained this as follows: The land gets hungry when it is eroded and small gullies form, making the land open its mouth. If this has happened, it is better to feed stone, grass and whatever is convenient to that hungry land immediately. If the hungry land is left without any treatment, it will die forever. In her understanding, all sloping land becomes 'hungry', even before the gullies have formed, because the soil is washed down the slope during each rain. The land becomes 'cold' when it loses its fertility. Such land should be covered with manure, compost, croton leaves and the like which can improve its fertility. She is convinced that, even after the land has been treated with different physical measures such as stone bunds and cut-off drains, if it is used without any cover, it will become 'cold'. Her saying indicates the importance of integrating different measures (physical, biological and soil fertility management) in order to obtain higher yields in a sustainable way.

RECOGNITION OF HER ACCOMPLISHMENTS

Ayelech trained her adopted son very well in all her farming practices. Indeed, he has learned by working together with her. She is also keen to share her experiences with other farmers, as well as with experts from the Amhara Region Bureau of Agriculture (BoA), and is ready to consider any new idea that might help her to obtain better yields. She receives numerous visitors to her farm, not only other farmers and experts but also higher officials, and has also given her time to collaborate with BoA experts in making a video about her farm (North Shewa Department of Agriculture, 1999).

Her work became more widely recognized when a study of indigenous SWC practices in North Shewa zone (Million, 1998) was presented at a regional workshop. The author brought to the workshop two of the most outstanding farmers in the study area: a man, Hail Giyergis, and a woman, Ayelech Fikre. The other workshop participants were amazed at the knowledge and accomplishments of these two farmers. The following day, the President of Amhara National Regional State, who had been informed by the

Credit: Chris Reij

Plate 3.1 *Ayelech Fikre explaining her innovations to researchers and extension staff (North Shewa, Ethiopia)*

head of the BoA, invited the author and the two farmers to explain their practices to him. Ayelech was extremely pleased to have a chance to speak directly with the President and other officials about natural resource management. Shortly thereafter, the Regional Council nominated her for an award as top female conserver of natural resources in the region and she received a prize of US$500 given by the Food and Agriculture Organization of the United Nations (FAO) and the Ethiopian Ministry of Agriculture on World Food Day in October 1998.

This prize made most of the farmers in her village very proud of her and motivated them to work harder on conserving soil and water. However, a few farmers were jealous, saying: 'She didn't do anything new. We have made terraces as she did. So why has the government awarded only her?'. Ayelech does not claim to have invented anything new. However, it is now widely recognized that she has integrated various indigenous techniques of land husbandry, applied by different farmers, in a unique way on her own farm so as to make optimal use of the resources available to her. She has thus managed to conserve the land well and make it more productive. Her farm system is, in itself, a work of art, but it is also the product of hard work over many years, a work that continues as she maintains and further improves her farm system.

NOTES

1 Three years ago, Ayelech was completely illiterate. At 60 years of age, she decided that she wanted to learn to read and write. An indication of her determination – not only to develop her land but also to develop herself – is that she can now read the Amharic alphabet and write her full name easily

REFERENCES

Million, A (1998) *Indigenous conservation practices in North Shewa Zone*, Ministry of Agriculture, Addis Ababa
North Shewa Department of Agriculture (1999) *Indigenous land management systems in North Shewa Zone, Amhara Region*, video, VHS, 32 minutes, North Shewa Department of Agriculture/Ministry of Agriculture, Addis Ababa

4

Pits for trees: how farmers in semi-arid Burkina Faso increase and diversify plant biomass

*Hamado Sawadogo, Fidèle Hien, Adama Sohoro and Frédéric Kambou**

In recent years there has been a dramatic increase in the number of trees growing on farmers' fields in certain villages in the Yatenga region of Burkina Faso. This is due in part to the systematic protection of natural regeneration by individual farmers and to the use of improved traditional planting pits or zaï *for growing trees. This chapter describes the achievements of four outstanding farmers in rehabilitating degraded land and increasing the diversity of trees through the use of* zaï.

In Burkina Faso, the fight against desertification is a constant preoccupation of farmers, government agencies, NGOs and development projects. The reduction in vegetative cover has reached alarming proportions in the north of the country, leaving the soils exposed to erosion by wind and water (Rochette, 1989; Sawadogo, 1995). During the last 30 years, substantial tree-planting operations have been carried out, including the planting of village woodlots, the National Village Forestry Programme and, more recently, the campaign entitled 8000 Villages, 8000 Forests. Millions of seedlings have been planted, but survival rates have been poor. There are many reasons for this lack of success, but the main ones are the poor care of the seedlings after planting,

---

* Hamado Sawadogo and Frédéric Kambou are agronomists with INERA (Institut National d'Etudes et de Recherches Agricoles); Fidèle Hien was an ecologist with INERA until he was recently appointed Minister of Environment and Water in Burkina Faso, and Adama Sohoro is an agronomist based at INERA's regional station in Tougan

Map 4.1 *Burkina Faso (Yatenga and Zondoma action areas)*

uncontrolled grazing by livestock, cutting of trees to clear land and to obtain fuel and, in particular, the fact that farmers were not involved in the activities in ways that encouraged them to take responsibility for them (Reij, 1983; Doro, 1991).

Nevertheless, compared with the early 1980s, there has been a dramatic increase in the number of trees growing on farmers' fields in parts of the Yatenga region. Many individual farmers have protected naturally regenerating trees but some have also made considerable efforts to grow trees in improved versions of their traditional planting pits or *zaï*. An outstanding farmer innovator, Yacouba Sawadogo in the village of Gourga, originally developed the practice of growing trees in pits, but also several other farmers can be considered as pioneers in this field. They include Ousséni Zoromé in Somyaga, Ali Ouédraogo in Gourcy and Ousséni Kindo in Bogoya. Each of these farmers experimented with the technique and managed to re-establish and to protect abundant perennial woody biomass on their fields. They did this by sowing tree seeds, planting seedlings, selectively protecting the natural regenerating seedlings, and sowing and planting grasses in the pits.

LOCATION AND METHODS OF THE STUDY

These farmers live in the provinces of Yatenga, Zondoma and Lorum in north-west Burkina Faso. Rainfall is highly variable. The long-term average for the regional capital, Ouahigouya, from 1950 to 1987 was 560mm. Ouahigouya received 590mm rainfall in 1997, but it was poorly distributed over the season and the harvests failed. In 1998 rainfall was an exceptional 969mm which led to a good harvest except in low-lying areas. The average population density in this region is 55 persons/km² (1996 data); in some parts, it is as high as 100 persons/km². The grazing pressure on the natural vegetation is high; according to the 1992 national livestock census, the Yatenga region had 140,500 head of cattle, 591,500 sheep and 708,100 goats (INERA, 1994). During the dry season, the animals owned by the local farmers depend to a large part on crop residues for fodder. The traditional practice of fallowing to regenerate soil fertility has disappeared and the possibilities for expanding cultivation to new areas are extremely limited. Rehabilitation of previously cultivated and now degraded land is the only option left to farmers who want to increase production by expanding their farming area.

The experiences of 12 farmer innovators, including the four innovators mentioned above, in the field of SWC were studied using an approach that stimulated farmers to express themselves and to describe their practices and skills. Over an entire year, the authors visited the 12 farmers regularly and discussed the different techniques they were using to rehabilitate degraded land and the agroecological and socioeconomic impacts of these techniques. The researchers' observations on the rehabilitated fields also helped in assessing the impact of the farmers' activities. Farm data for the four innovators described here are given in Table 4.1.

Table 4.1 Farm data for four farmer innovators in Burkina Faso

Farmer innovator	Yacouba Sawadogo	Ousséni Zoromé	Ali Ouédraogo	Ousséni Kindo
No of persons/family	45	41	43	63
No of active persons	20	13	12	35
Land treated with *zaï* (ha)	12	11	12	14
No of cattle	4	14	4	4
No of sheep/goats	29	50	18	47
No of horses	1	0	2	0
No of donkeys	1	2	4	1
No of donkey carts	2	2	3	1
No of ploughs	2	2	2	1

The number of persons per family is well above the average of eight for the Central Plateau of Burkina Faso. This is because each of the farmer innovators heads an extended family living in one compound and is responsible for the wives and children of younger brothers who are working elsewhere, often in Ivory Coast. The average farm size in the region is 4ha.

Over a period of 15–20 years, each of the four farmer innovators has rehabilitated 11–14ha of land. Two of them, Ousséni Zoromé and Ousséni Kindo, had very few resources when they started rehabilitating degraded land in the early 1980s. Ousséni Zoromé did and does not own any land and continues to borrow all the land he farms. Ali Ouédraogo was already relatively rich when he started to work with *zaï*, whereas Yacouba Sawadogo was an average farmer, who regularly faced food deficits before he started to improve the traditional planting pits. It is striking that, in the 1980s, all four men were involved in commercial activities but, by the early 1990s, they were full-time farmers. Virtually all their fields are treated with *zaï*, and they have become relatively rich in resources, although Ousséni Zoromé has chosen to invest this in livestock rather than in land.

FROM FARMING TO REGENERATING NATURAL VEGETATION

In 1979 Yacouba Sawadogo started to use the *zaï* technique to rehabilitate land. At that time, his main aim was to produce more cereals, mainly sorghum and millet. By digging wider and deeper pits and by adding manure to them, he managed to achieve very good yields from fields that had previously been so degraded that nothing could be grown on them. His improvements allowed him to achieve food self-sufficiency for his family. In addition, Yacouba was pleasantly surprised that numerous tree species started to grow spontaneously in the planting pits. The tree seeds had been deposited in the pits by the run-off water or they were contained in the manure that had been added to the pits. He decided to protect the young trees. When he harvested the millet, he cut their stalks at a height of 50cm and the part of the stalks that remained standing served to protect the young trees. In this way, he discovered the use of pits for growing trees (*zaï forestier*). Already in the first years, the results were spectacular and highly encouraging. His next step was to start collecting the seeds of numerous useful local species of fruit and fodder trees which he introduced into the *zaï* in the next wet season. These species included sheanut, yellow plum (*Sclerocarya birrea*), grape tree (*Lannea microcarpa*) and various acacia species, but also fodder grasses such as Gamba grass (*Andropogon gayanus*) and *Pennisetum pedicellatum*.

Within a few years, the piece of barren land was gradually transformed into a 12ha forest with numerous different species. Yacouba then had to make a difficult choice because the trees and shrubs started to compete with his cereal crops. He opted for growing trees. Each year he placed the seeds of desired tree species into the *zaï*, as well as alongside the stone bunds he had constructed in his fields to prevent erosion. In the month of August, he split and replanted clumps of fodder grasses such as Gamba. In order to protect his forest from livestock, he surrounded it by cultivated fields which livestock are not allowed to enter during the growing season, according to local land-use agreements. During the dry season, he or his children protected the forest against uncontrolled grazing, woodcutting and hunting.

Another interesting innovator is Ali Ouédraogo, about 70 years old, who lives in the small town of Gourcy in Zandoma province and can be characterized as an agroforester. He started to rehabilitate degraded land around 1983.

Plate 4.1 *Yacouba Sawadogo with the forest he created in the background*

Credit: Chris Reij

Trained in the layout and construction of contour stone bunds by the Oxfam-funded Agroforestry Project, he soon discovered that trees started growing alongside the bunds in his fields because their seeds were deposited there by the run-off water. He protected this natural regeneration, but decided from 1986 onwards to stimulate the establishment of trees. When constructing stone bunds, he placed the seeds of neem (*Azadirachta indica*) trees and of certain local woody species in the furrows for the stones. During the wet season, the young seedlings helped to stabilize the bunds. Also in the *zaï*, young trees started to grow spontaneously. During weeding, Ali took care not to damage them. In this way, he managed to obtain good yields of millet, sorghum and cowpea, as well as abundant perennial vegetation. He is keen to protect the natural regeneration in his fields and is proud to say: 'All the species one can find in the bush are now growing in my fields.' He feels that: 'All barren degraded land should be under crops or trees'. Ali is constantly concerned about his trees. When he travels, his children make sure that grazing animals do not come near them.

The cases of Ousséni Zoromé and Ousséni Kindo have much in common with the cases of Yacouba and Ali described above. The major difference is that Zoromé and Kindo have much less land and this forces them to give priority to crop production while maintaining a good density of trees.

IMPACT OF *ZAÏ* AND RELATED INNOVATIONS

Impact on food security

Through the rehabilitation of degraded land using the *zaï* technique, 10 of the 12 farmer innovators studied have attained food security. As Yacouba Sawadogo explained: 'In the days before the *zaï*, I was a part-time trader and I used all the income this generated to buy cereals to feed my family. Since I started treating the land with *zaï*, I am self-sufficient in food and sometimes I sell a surplus of cereals and cowpeas to cover my financial needs.' This statement of the pioneer of *zaï* illustrates the impact that this technique can have on the food security of families that invest in it. One of the major advantages of the *zaï* technique is that it minimizes risks caused by variations in rainfall and ensures substantial yields on marginal lands (Maatman et al, 1998; Vlaar, 1992). The yields that farmers in the study areas have gained from fields cultivated using the *zaï* technique vary from 800 to 1500kg/ha, depending on rainfall, soil quality and other factors, as shown in Table 4.2.

The *zaï* have stimulated the production not only of cereals but also of leguminous crops such as cowpea (*Vigna unguiculata*). This is generally grown together with sorghum or millet as a cash crop. The substantial quantities of cowpea produced annually on the rehabilitated land contribute to the income of the farm families. Interviews with the innovators' wives revealed that the *zaï* have indirectly had a positive influence on their crops: because the men now concentrate on the *zaï* fields, the sandy soils not suitable for *zaï* have been allocated to the women who use them to grow common groundnuts (*Arachis hypogaea*) and Bambara groundnuts (*Voandzeia subterranea*).

Impact on livestock husbandry

The *zaï* also have a positive impact on livestock keeping. Many farmers stated that, before they adopted *zaï*, they had few animals. Ten of the 12 farmers mentioned that the investment in *zaï* has been paralleled by changes in their livestock husbandry practices. It is only by adding manure to the *zaï* that the farmers can obtain good yields. Farmers whose cattle were formerly managed by Fulani herders now keep their cattle at the homestead. Those farmers who keep sheep do this not only to produce fat stock for sale but also to produce

Table 4.2 Yields of sorghum and millet obtained by four innovators using the zaï *technique*

Farmer innovator	Year	Rainfall*	Soil type	Yield (kg/ha)
Ali Ouédraogo	1996	Poor	Lateritic hard-pan	870
Ousséni Zoromé	1996	Average	Lateritic	1591
Yacouba Sawadogo	1997	Poor	Gravelly	1200
Ousséni Kindo	1999	Good	Ferruginous hard-pan	1050

* This table is only indicative; the four farmers have characterized the rainfall in relative terms

manure which is either applied directly to the *zaï* or used for composting. It is now common practice among Yatenga farmers to collect the pods and fruits of specific woody species (yellow plum, acacia species, *Piliostigma reticulatum*, etc) to feed as dry-season supplement to the hay used for fattening sheep. When passing through the animal's digestive system, the seeds become softer and end up in the manure used in the pits. The seeds sprout and grow at the same time as the cereal crops, and the farmers protect them during weeding. The *zaï* have thus contributed to a stronger integration of livestock and cropping activities.

Impact on biodiversity

Meetings with the farmer innovators and visits to their rehabilitated fields revealed the existence of a great variety of ligneous and herbaceous species. When the farmers started rehabilitating the tracts of degraded land (*zipellé* in the Moré language), there were few large trees from a very limited number of species on the land. Yacouba Sawadogo counted trees of only four species: one *Balanites aegyptiaca*, one *Lannea microcarpa*, one *Guiera senegalensis* and one *Combretum micranthum* tree. Twenty years later, he has more than 60 tree species on the same land. Table 4.3 indicates the major species on his fields.

Yacouba has introduced into his forest some medicinal species which had disappeared from the region. He collected these during his travels outside the Yatenga area. When people come to visit his farm during the wet season, he asks them to dig some planting pits, plant some trees or sow some seeds that he collected.

Ousséni Zoromé counted nine trees in his degraded fields (11ha) when he started to reclaim them in 1983. By 1999 he had more than 17 species on these same fields, with a total of about 2000 trees, half of which are still young. One tree species that was at risk of disappearing about 20 years ago was the baobab (*Adansonia digitata*). Since then, the number of baobab trees has increased

Table 4.3 Some tree species in the fields of Yacouba Sawadogo

Species	Number	Species	Number
Sclerocarya birrea	+++	Sterocarpus lucens	+
Lannea microcarpa	+++	Bombax costatum	+
Guiera senegalensis	+++	Parkia biglobosa*	++
Piliostigma reticulatum	+++	Eucalyptus camaldulensis	+
Cassia siberiana	+++	Acacia seyal	++
Diospyros mespiliphormis	+++	Adansonia digitata	+
Combretum micranthum	+++	Combretum glutinosum	++
Balanites aegyptiaca	++	Anogeissus leocarpus	++
Bauhinia rufescens	++	Acacia nilotica	++
Saba senegalensis	++	Faidherbia (syn Acacia)	
Butyrospermum paradoxum*	++	albida*	+

Key: +++ abundant; ++ well represented; + small numbers
* young trees (height less than 1m)

Credit: Chris Reij

Plate 4.2 *Ousséni Zoromé has systematically protected natural regeneration in his fields (Burkina Faso)*

tremendously because the farmers have protected naturally regenerating seedlings. On Zoromé's main field, about 200 young baobabs can be found, but they are not yet productive. Another farmer innovator, Ouérémi Boudou, whose case is not treated in this chapter, has specialized in regeneration of baobabs. His three wives gain a substantial annual income from selling baobab leaves (estimated at 135,000CFA, circa US$210, in total).

Ousséni Kindo from Bogoya village started reclaiming a completely barren field around 1985. He now has 15 woody species on these fields, including about 100 baobab trees, of which 35 are already productive.

DIFFERENCES IN MOTIVES AND STRATEGIES OF THE INNOVATORS

Cereals versus trees

The farmers' motivations for regenerating the vegetation differ and depend largely on the amount of land they have. Yacouba Sawadogo owns his land and has more than enough to meet his family's subsistence needs. He aims to create a multipurpose forest of 20ha and gives priority to planting trees at the expense of producing cereals. He plans to invest more in growing medicinal woody plants and he would like to reintroduce wild fauna (small deer, hyenas, birds, etc) into his forest.

Ousséni Kindo has a large family and does not have enough land to be able feed it properly. His major objective is to produce food, while the regeneration of trees is second priority. Nevertheless, he has many trees on his fields. As soon as he feels that the tree density could reduce his cereal production, he starts cutting down the weaker trees and lops some of the remaining ones. He places the leaves of the lopped trees in the compost pit to produce fertilizer. During the wet season, he lops in particular *Sterocarpus lucens*, *Balanites aegyptiaca* and acacia species to obtain additional fodder for his animals. At this time of year, grazing land is scarce because almost all the land is under crops. Lopping allows the trees to regenerate and limits the effects of shading and, thus, competition with the cereal crops.

Ali Ouédraogo has enough land to feed his family and still has some land that could be rehabilitated. However, some members of his community dispute his ownership of certain parcels of land. For him, managing the land is one way of protecting himself from claims by others. Although the woody vegetation on his fields is abundant, he has not yet noticed a negative impact on cereal yields. Ali continues to give priority to growing cereals (although he has several years' stocks of grain), but has a vision of eventually creating a forest like that of Yacouba. He has started marking the borders of his fields and has requested a formal land title from the local administration. He will then have sufficient land security to invest more in reforestation.

Ousséni Zoromé has only usufruct rights to the land he is farming. He has had some difficulties with the landowners, but he has managed to deal with them thus far. Without doubt, his good links with the regional department of the Ministry of Agriculture, as well as the many visitors he receives, have had a positive influence on his relationships with the landowners. As a result of all his investments in SWC since 1983, his fields have a considerable amount of woody biomass. He protects this, but his major priority is food production. As he does not own his land, he hesitates to plant a live fence around the fields. This would make it easier to protect the trees against uncontrolled grazing, but, in view of the local land-use customs, planting trees around his fields could evoke negative reactions from the landowners.

Plate 4.3 *Ousséni Kindo in his rehabilitated fields (Burkina Faso)*

Credit: Chris Reij

Firewood

The lack of firewood in this part of Burkina Faso is a serious problem for the women who must often walk long distances (10–20km) to collect enough fuel for the home. As the wife of one of the farmer innovators remarked: 'To have trees on the family fields is a great richness because we can save a lot of time that we can now spend on income-generating activities.' The possibility of covering at least part of the family's firewood requirements is one reason why the farmers protect and regenerate the woody vegetation. Most of the farmers prefer local to exotic species because they are better adapted to the environment and the farmers are well aware of their multiple uses.

Medicinal products

Since the devaluation of the West African franc (CFA) in January 1994, many farmers can no longer afford to buy the commercial medicines. This has boosted an interest in medicinal plants. The farmer innovators systematically protect and introduce into their fields all the species that can be used to heal common diseases (malaria, stomach ache, jaundice, etc). The medicinal species named by the farmers include: neem (*Azadirachta indica*), grape tree (*Lannia microcarpa*), yellow plum, eucalyptus (*Eucalyptus camaldulensis*), savanna mahogany (*Khaya senegalensis*), drumstick tree (*Cassia sieberiana*) and *Guiera senegalensis*.

Yacouba Sawadogo has most strongly developed this activity. He has introduced species not previously known in his region and has focused on species that have largely disappeared because of droughts in the early 1970s and mid-

1980s. Each year he receives more than 100 visitors (mostly farmers, but also traders and office workers) who request various parts of plants (leaves, bark, roots) from him for medicinal purposes. Because of his knowledge in this field, Yacouba is in constant contact with well-known traditional healers who consider him to be their partner. The field of medicinal plants is secretive and Yacouba did not want to indicate which species he has introduced for medicinal purposes. He only indicated that he has planted species that reduce hypertension and even mental problems. Apparently, he does not ask for cash payments for his products and services, being more interested in the social esteem that he derives from this activity.

Income generation

Several farmers mentioned that they have also sold wood for the construction of roofs, sheds and the like. On an annual basis, this brought a cash income of 20,000–40,000CFA (circa US$30–60) per farmer, but the amounts can vary depending on the availability of timber in their fields and on the demand and supply on the local market (see Chapter 13 for the case of Namwaya Sawadogo). The main species for construction purposes are exotic, such as neem and eucalyptus, but certain local species are used for making chairs, mortars and pestles. The current drive to regenerate the woody vegetation is also linked to the possibility of gaining some cash income. Also the women have a stake in this: they collect leaves of the baobab, flowers of the kapok (*Bombax costatum*) and fruits of the sheanut (*Butyrospermum paradoxum* var. *parkii*) and the locust bean (*Parkia biglobosa*) trees for home consumption and to sell at local markets.

Social status

The rehabilitation of land and the reconstitution of the woody vegetation have greatly increased the social status of the innovators. Before they started to experiment and to invest, they were anonymous farmers like most others. Nowadays, their reputation extends beyond their provinces and even beyond the borders of Burkina Faso. They are in regular contact with the decentralized services of various ministries and have become focal points for improved natural resource management in their regions.

WIDER CONTEXT

It is important to place this experience with revegetation in a semi-arid region in a wider context. Around 1980, all forestry professionals and other natural resource management specialists working on the Central Plateau of Burkina Faso predicted doom and gloom. They stated that important species such as *Acacia* (syn. *Faidherbia*) *albida* were disappearing, that the stands of baobab were ageing because of overexploitation and lack of natural regeneration, and that this was also the case for perennial grasses such as Gamba grass which had retreated southwards over a distance of 200–300km in the previous 15 years (van Keulen and Breman, 1990, p182). Twenty years later, farmers are actively protecting the natural regeneration of these species and several others, and they are planting Gamba grass along the stone bunds in their fields. On many fields, more trees were found in the year 2000 than in 1980. Twenty years ago, the expanses of severely degraded land were vast and expanding. Now, thousands of hectares of this land have been successfully rehabilitated by farmers in the Yatenga region using the *zaï* technique.

This does not mean that the battle against land degradation has been completely won. Farmers involved in land rehabilitation continue to face many constraints, such as uncontrolled livestock grazing and the cutting of trees for firewood by outsiders, which are problems that can be solved only at village and intervillage level. Nevertheless, the environmental situation appears to be less gloomy now than 20 years ago because farmers have shown that something can be done.

These cases prove that, as a result of indigenous innovation and initiative, it is possible within a fairly short time span (5–10 years) to produce a considerable and diverse plant biomass that can be used for many purposes, including fodder. This facilitates the integration of livestock keeping and cropping systems, which is the basis of sustainable agricultural intensification.

REFERENCES

Doro, T (1991) *La conservation des eaux et des sols au Sahel: l'expérience de la province du Yatenga (Burkina Faso)*, CILSS, Ouagadougou

INERA (1994) *Diagnostic des contraintes et des potentialités: définition des axes de recherche pour le CREA du Nord Ouest*, INERA, Ouagadougou

Maatman, A, H Sawadogo, C Schweigman and A Ouedraogo (1998) 'Application of *zaï* and rock bunds in the northwest of Burkina Faso: study of its impact on household level by using a stochastic linear programming model', *Netherlands Journal of Agricultural Science*, no 46, pp123–136

Reij, C (1983) *L'évolution de la lutte anti-érosive en Haute Volta depuis l'indépendance: vers une plus grande participation de la population*, Institute for Environmental Studies, Vrije Universiteit, Amsterdam

Rochette, R M (ed) (1989) *Le Sahel en lutte contre la désertification: leçons d'expériences*, Margraf, Weikersheim

Sawadogo, H (1995) *La lutte anti-érosive dans la zone Nord Ouest du Burkina Faso: cas des villages de Baszaïdo et de Lankoé*, International Workshop on Food Security, Accra

van Keulen, H and Breman, H (1990) 'Agricultural development in the West African Sahelian region: a cure against land hunger?', *Agriculture, Ecosystems and Environment*, no 32, pp177–197

Vlaar, J C J (1992) *Les techniques de conservation des eaux et des sols dans les pays du Sahel*, Wageningen Agricultural University, Wageningen

Part 2

Building partnerships for innovation in land husbandry

Forging partnership between farmers, extension and research in Tanzania

*O T Kibwana**

It is generally accepted that agricultural development depends on the interaction between farmers, extension agents and researchers. What is still debated is the nature of these interactions – more specifically, the roles that these different actors should play. The ISWC programme in Tanzania has made scientists and extensionists aware of the achievements of farmer innovators, and has created situations that foster equal partnerships between research, extension and farmers in joint experimentation. This required systematic work both in the government institutions and in the field.

New ideas are the key to agricultural development. In today's dominant model, research scientists develop and test new ideas under controlled conditions, extension agents package them into 'messages' that tell farmers what to do and how to do it, and farmers are expected to adopt and apply the new ideas. A very specific status hierarchy is perceived by all the actors, with information flowing from those who know to those who do not. While the ineffectiveness of this linear model is now recognized, the question remains open as to how researchers find out whether their new technologies are relevant at field level. Mechanisms have been introduced to feed back farmers' opinions via the extension system to the scientists, but these mechanisms have done little to change the basic assumption that new technologies come from experts working at a superior level.

ISWC-Tanzania recognizes that such experts are an important source of new ideas. However, it also believes and has concrete evidence that farmers

---

* O T Kibwana is a specialist in agricultural education and extension. He is head of the Pilot Projects and Experimentation Department of Cooperative College Moshi, and coordinates the ISWC programme in Tanzania

Map 5.1 *Tanzania*

are very resourceful in generating and testing new ideas and that this ongoing process of local innovation is making a major contribution to agricultural development.

Agricultural development demands continual innovation and experimentation. All farmers innovate and experiment in their struggle to make a living from the soil. However, not all farmers innovate to the same extent. There are always those who lead the way. The challenge was to identify these farmers, to link them with scientists and extensionists and to facilitate genuine partnership.

The interactions between these actors that existed prior to ISWC 2 had led to certain attitudes, behavioural patterns and role definitions that had become taken for granted. To change these attitudes meant creating a 'new order'. The programme took a two-pronged approach, working simultaneously but separately with the institutions of research and extension and with innovative farmers, before bringing them together to plan joint experimentation.

BUILDING INSTITUTIONAL PARTNERSHIP

Partnership at national level

From the beginning, ISWC-Tanzania saw its role as facilitating the building of partnerships. The programme, coordinated by Cooperative College Moshi (CCM), decided not to establish its own action areas, but rather to work with organizations already interacting with farmers. A key selection criterion for partner organizations was their willingness to learn more about participatory approaches.

The programme commenced its work with organizations in Mbeya and Iringa regions in the Southern Highlands. This is a better watered area with relatively high potential for agriculture, in contrast to the ISWC 2 programmes in other African countries which operate in drier areas. The research partner was the Ministry of Agriculture Research and Training Institute (MARTI) Uyole, based in Mbeya, which is the designated research institute for the Southern Highlands. Coopibo, a Belgian NGO which had been working for many years in Mbeya region and had helped to set up several semi-autonomous agricultural development projects in Mbozi, Isangati and Ileje, was chosen as the extension partner in this region. In Iringa region, where the Danish-funded environmental programme HIMA (*Hifadhi Mazingira*) supports several District Agricultural and Livestock Development Offices (DALDOs), those in Njombe and Iringa rural districts were chosen as extension partners.

In Mbinga district in Ruvuma region of south-central Tanzania, the Sokoine University of Agriculture (SUA) supports the DALDO and became a research partner of ISWC-Tanzania in 1998. Expansion of the programme to northern Tanzania in 1999 brought on board the Traditional Irrigation Improvement Project (TIP) as extension partner in Arumeru and Mwanga districts in Kilimanjaro and Arusha regions.

In the meantime, in the late 1980s long before ISWC 2 began, NGO and bilateral development programmes in Tanzania had already been facilitating the formation of smallholder farmer networks. In a village, one or more groups of 10–15 farmers have formed, each with its own name, leadership and constitution. Several groups in a locality (one or several villages) make up a local network. The local networks, in turn, have come together to form a national umbrella network called Mviwata (*Mtandao wa Vikundi vya Wakulima Tanzania*, Network of Farmers Groups in Tanzania). This became the national representative of farmers as partners in the ISWC programme.

A key step in partnership development was the establishment and strengthening of a National Steering Committee (NSC), comprising all the main actor groups: farmers (Mviwata); research scientists (MARTI, Uyole, and SUA), extension (DALDOs, Coopibo and TIP) and the coordinating agency (CCM). The research and extension organizations are represented by key staff members directly involved in programme activities and the farmer network by a woman who is the chair of the Mviwata Executive Council. Her presence and active participation in the NSC has ensured that the farmers' viewpoints and interests are taken up.

Since ISWC-Tanzania started in1997, the NSC has met every three months and all except one member have attended all meetings. The member who was absent (from the first three meetings) was voted out and replaced. At the meetings during the formative stage of the NSC, the scientists tabled research proposals with a purely research focus, but these were challenged by other members in the NSC and the orientation gradually became more balanced. The farmer representatives demanded the right to understand the discussions which were therefore held in Kiswahili.

Considerable time was spent on a process of building partnership at different levels. First of all, at national level, the roles and responsibilities of each partner had to be defined clearly. The process of formulating a partnership agreement took almost six months. After ideas were collected in the NSC, the national coordinator prepared a first draft which was discussed by the committee. The comments were incorporated into a second draft which was sent out to all partners for further comments. Then a third draft was prepared, discussed in a NSC meeting and approved during the last NSC meeting of 1997, before field activities started in January 1998.

Partnership at field level

Having established the partnership at the national (policy) level, the focus was shifted to the field (operational) level. ISWC-Tanzania organized a workshop for the key researchers and extensionists who would be involved in the field activities. This workshop aimed to:

- create awareness among research and extension staff that farmers innovate and experiment in a logical way;
- enable research and extension staff to appreciate the importance of collaboration between farmers, extension and research in technology development; and
- develop a common understanding of the concepts used in PTD.

In the conventional system of research and development, the agricultural 'experts' believe they are more open to new ideas than farmers and see themselves as agents of change. The workshop helped them to recognize a new concept of 'farmer innovation' (as they were still thinking of 'innovators', 'adopters' and 'laggards' in the terminology of transfer-of-technology extension). Through well-focused field visits, they were exposed to the expertise and creativity of farmers. The workshop also helped to nurture a working relationship between research and extension as it gave participants the opportunity to understand and appreciate each other's roles and viewpoints.

In each district, two or three divisions were selected for commencing joint work. The main criteria for selection were the extension staff's evaluation of the general level of farmer innovation in the area and whether village extension officers from the division had attended the PTD workshop and understood the new concepts.

Mixed research–extension teams were formed, consisting of the Divisional Extension Officer (DEO), selected Village Extension Officers (VEOs) and a scientist from one of the two research organizations in the NSC. As only one researcher works in each region, that researcher takes part in all the divisional research–extension teams in that region. The VEOs were selected according to their interests, capabilities and disposition to regard farmers as creative. Team leaders were people from above the divisional level, known to be interested in participatory research and extension.

The teams then went to the field to identify farmer innovations. The different approaches they adopted reflected the composition and orientation of each team. In some areas, the teams asked the local VEOs to identify local innovators. Other teams asked the VEOs to convene a meeting of community leaders to discuss the general topic of farmer innovation and experimentation. Community leaders were then asked to identify local innovators.

The teams visited the farmers identified as innovative and saw and documented their work. In the case of the more technically oriented teams working through VEOs, they screened which innovations were interesting to document. Where community leaders were involved in identification, they met with the identified farmers and the research–extension team to discuss techniques and distinguish between innovations and traditional practices.

The VEOs, assisted by the researchers, created innovator profiles using a format provided by ISWC-Tanzania. Profiles covered biodata, economic status, social influence, neighbours' perceptions and motives for innovation. It was found that:

- Most innovators had responded to problems they faced during their daily work, ie their motivation was to solve problems.
- Most innovators were middle-aged men with families, but the more striking innovations were undertaken by males in their early 30s.
- Some of the older male innovators held official positions in their localities, while the younger ones were generally regarded as wayward (one was nicknamed Pwagu, a popular character in a radio play who is always trying out new ideas but with little success).
- Better-off innovators embarked on more expensive innovations requiring purchased materials and hired labour, the poorer ones on simpler, less resource-demanding innovations; however, many who started resource-poor became richer through their innovations.
- Fewer women were identified as innovators and their innovations tend to be homestead-centred (eg mixing urine with manure from stall-fed cattle).
- Most innovators claim to have been inspired by their own ideas and curiosity; few admit to having been inspired by other farmers or extension agents. Only later did it become possible to trace the origin of any particular innovation.

PREPARING INNOVATIVE FARMERS FOR PARTNERSHIP

Parallel to this process of building institutional partnership at policy and operational level, ISWC-Tanzania organized regional workshops in Iringa,

Mbeya and Ruvuma that brought together farmer innovators from several districts. A researcher, a PTD trainer and the head of the national farmers' organization made the general design of the workshops. The main objectives were to provide a forum for exchanging experiences and to stimulate networking among the innovators. This was important because innovators often felt isolated in their own communities and unappreciated by the 'experts' in research and extension services. The facilitation team for each regional workshop included a researcher, a PTD trainer, a VEO and a farmer.

The farmer innovators greatly appreciated the workshops. For many, it was the first time they had travelled across district boundaries and their first opportunity to explain to others what they were doing. They exchanged seeds and planting materials as well as ideas. During the workshops, participants examined some innovations in the field and assessed their strengths and weaknesses. New friendships were made and innovators were enthusiastic to learn more from each other.

In December 1998, cross-visits were organized in two stages. First, farmer innovators from one district visited others in the same district for three days, each group member playing host in turn. Then, a group of innovators from one district visited innovators in another district within the region. Some VEOs accompanied farmers on their intradistrict visits and the DEO went with them on interdistrict visits. After each visit, group members evaluated what they had seen and identified the ideas to try out at home.

In April/May 1999, teams of VEOs visited the farmers involved to see what they had put into practice. Farmers had been very active. The newly acquired seeds and planting materials had been tested. Some of the innovations had also been adopted, the most striking being the sowing of several maize seeds in a pit, as practised by Wilbert Mville in Njombe (see Chapter 25). Seventy-nine farmers who were trying out this technique were counted in Njombe district alone. No wonder one farmer commented: 'Learning from exchange visits is better than being visited by a VEO.'

NEGOTIATING JOINT EXPERIMENTATION

Researchers and farmers often have different ideas about what problems should be studied first. Negotiations are needed to reach consensus on the relative importance of problems. Only then can joint action start. This process requires that each stakeholder group has the capacity to express its own position. Preparatory work is needed if fair negotiations are to take place. ISWC-Tanzania tackled this on two fronts: by confronting the 'experts' and addressing the farmer innovators. The series of workshops for research and extension staff led them to appreciate the farmers' potentials. Meanwhile, the process of identifying innovators, the regional workshops and the cross-visits served to strengthen the position of the farmers who became more confident and better able to argue their interests.

Once these two parallel processes had matured, priorities could be set for joint experimentation, building on local innovations. Multidisciplinary teams consisting of agronomists, soil scientists and the VEOs visited individual farmers for discussions in the fields. Clusters of innovations were identified, for example:

- agroforestry systems;
- mixed cropping involving food crops and fruit trees;
- replenishing soil fertility with organic materials;
- testing different sowing systems;
- tapping underground water for irrigation;
- diverting waterways and managing the water;
- harvesting run-off water; and
- production of agricultural tools.

Results were summarized and presented by the researchers at a research–extension workshop for further negotiation. Finally, the proposals were reviewed by the NSC which monitors the general orientation of the action research. The woman representing the farmers' organization took very seriously her special responsibility for ensuring that the farmers' agenda was maintained.

PARTNERSHIP IN ACTION

During the first cropping season, a few farmer experimenters were identified in each action area. Research teams consisting of a farmer experimenter, the local VEO and a scientist were formed. The general framework for sharing responsibilities had already been agreed upon during the earlier workshops, but the teams still had to work out the details to fit their own situations.

Most experiments involved crops and some had been set up after the growing season had begun. In order to improve research in the coming season, ISWC-Tanzania held a workshop for the farmers, researchers and extensionists involved in the first experiments. The main aims were:

- to review the process of joint experimentation: How was it planned? How was responsibility shared? What happened?
- to derive lessons learnt so far: What went well? What problems arose? How were these dealt with? What should be done differently next time?

Generally, participants and especially the farmer experimenters were satisfied with the process. For them, the most gratifying part of the experience was that they had been treated, at long last, as partners and as equal to the 'educated elite'.

Of course, some problems were also identified. A major one was that it had been assumed that, simply by dividing responsibilities, the partners would be able to play their roles effectively. As it turned out, even in cases where the

Credit: Laurens van Veldhuizen

Plate 5.1 *Team members of ISWC-Tanzania discuss compost trials with a farmer innovator*

partners were clear about what they were supposed to do, they were not always sufficiently prepared to do it. The participants therefore requested that, in each district, practical hands-on training be given. This should focus on the tasks that the farmers, researchers and VEOs should undertake in the next cropping – hence, experimenting – season. These workshops would also serve as planning sessions for the next season – a good way to complete the reflection–action–reflection loop.

LESSONS LEARNT

The experience of the ISWC programme in Tanzania clearly shows that it is possible to break the long-established patterns of misunderstanding and often mistrust between farmers, researchers and extensionists. It is possible to nurture the development of a partnership of equals. In the Tanzanian case, this has been facilitated by the following factors:

- Like all the other country programmes in ISWC 2, the Tanzanian programme enjoys a very high degree of autonomy. This allowed for the necessary flexibility and led to a sense of ownership by the different partners.
- The NSC is composed of a combination of farmer representatives, senior staff (directly involved in the activities) from key organizations, and influ-

ential individuals such as the chair who is the registrar of SUA and is involved in policy dialogue at ministerial levels. This combination has made it possible for the NSC to make and to influence policy decisions that take into account the field conditions.

- Joint learning (including participatory training and implementation) cemented the sense of oneness among all the key actors.
- The neutral position of the coordinating agency (CCM) and the national coordinator facilitated an equitable sharing of power between research and extension staff. This has been critical in avoiding the scenario of superiority and inferiority complexes.

Nevertheless, ISWC-Tanzania realizes that it must remain vigilant. Participation, stakeholder involvement and empowerment are concepts that have gained popularity in Tanzania, as elsewhere in the developing world, but there is a danger that they become catchwords. The programme is being implemented by governmental and NGO partners who have claimed from the beginning that they believe in participation. However, experience shows that old habits die hard. Deliberate efforts have to be made to achieve a common understanding of the vision, philosophy and strategies of genuine participation. The mixed workshops and joint experimentation have been powerful tools for building mutual trust which is critical for genuine partnership. However, the NSC will have to continue to keep a close eye on the equitability of the partnership. Farmers are old hands at discerning deception.

6

Joining forces to discover and celebrate local innovation in land husbandry in Tigray, Ethiopia

*Mitiku Haile, Fetien Abay and Ann Waters-Bayer**

The ISWC programme in Ethiopia commenced work in Tigray region. It deliberately sought to include a wide array of people from agricultural research, extension and education institutions in identifying and giving due recognition to farmer innovation. This chapter describes the methods applied in discovering farmer innovators and in celebrating and stimulating their creativity.

In past approaches to promoting SWC, the emphasis has been on transferring technologies introduced from outside. In Ethiopia, these interventions often depended on mass (political) campaigns and externally financed Food-for-Work schemes. The existing practices and initiatives of local farmers in land husbandry were not taken into account when planning the campaigns, which sometimes tried to introduce techniques that were not suitable for the local agroclimatic and socioeconomic conditions. More importantly, little attention was given to what motivates farmers to improve their land husbandry systems.

Quite a different approach was taken by the ISWC programme in Ethiopia. Our starting point was the discovery and celebration of farmers' indigenous SWC techniques and innovations. From the very outset, we tried to involve as many actors as possible among the staff of research, extension, teaching and policy-making institutions and to raise their enthusiasm to support farmers'

---

* Mitiku Haile and Fetien Abay jointly coordinated ISWC-Ethiopia; the former is a soil scientist and president of Mekelle University, Ethiopia; the latter is a crop scientist with the university; Ann Waters-Bayer is an agricultural sociologist with ETC Ecoculture in Leusden, The Netherlands, and external adviser to ISWC-Ethiopia

Map 6.1 *Ethiopia (Tigray action area)*

efforts to improve their land husbandry. The 'people orientation' of our approach to SWC, working on the attitudes and motivations of actors at all levels, is just as important as the technical content. Our aim is to encourage the scientists and development agents to join farmers' ongoing experimentation and their search for new ideas to try out. The scientists and DAs thus become participants in farmer-led agricultural development.

Already in structuring the ISWC-Ethiopia programme, a variety of organizations were involved. The lead organization is Mekelle University (MU) in Mekelle, the capital of Tigray region. The partner organizations include the BoANR, Mekelle Research Centre (MRC) and various NGOs and bilateral projects concerned with agricultural development in Tigray. All are represented on the ISWC-Ethiopia Steering Committee, in addition to two external members, one from the national Soil Conservation Research Programme and one with the federal Ministry of Agriculture. The government agencies and NGOs in Tigray, including MU, contribute their own facilities, equipment and time to ISWC-Ethiopia, over and above that funded by the programme.

WIDE INVOLVEMENT IN SEARCH

In order to stimulate the discovery of indigenous innovations, ISWC-Ethiopia started with a Tigray-wide contest. The initial idea was that the competitors would be DAs, senior students, university teachers and researchers. However, it was soon realized that the real winners are the innovators themselves. The competition was therefore expanded to give recognition to both the innovators and the people who discovered them. We sought wide involvement as we are convinced that discovering farmer innovators should not be an activity confined to formal researchers. Everyone with an open eye and an open mind is capable of recognizing innovation. Involving as many types of actors as possible in this activity is an important means to ensure wide support for farmer-led experimentation to improve land husbandry.

The coordinators of ISWC-Ethiopia drew up guidelines for describing farmer innovators (see Box 6.1), indicating the type of information that could be sought. This was written in the Tigrigna language and covered the following:

- the setting: climate, soils, landscape, farming systems of the area;
- socioeconomic characteristics of the innovative farmer or group;
- information about the innovation(s) developed by that farmer or group: source and development of the idea, description of the innovation in its present form, impact on well-being and sustainability from the local perspective;
- indications of the spread of the innovation; and
- possibilities proposed by farmers and observers for further support to farmer innovation.

ISWC-Ethiopia defined a farmer innovator as someone who develops or tries out new ideas without support from formal extension services. The search is not for progressive farmers who adopt recommendations from extension, but rather for innovative farmers who are independently creative in using local resources in new ways. Therefore, innovators must be sought also beyond the circle of farmers who are involved in the official extension programme. The type of innovation sought is not restricted to physical structures, such as terraces, which DAs – as a result of the mass extension campaigns – most commonly associate with SWC. Innovations in land husbandry can include:

- protection of land or water areas;
- agronomic practices, such as different types of ridging or ploughing techniques, cover cropping, short-season or drought-resistant crop varieties;
- biological techniques, such as integrating plants into soil conservation structures;
- physical techniques, such as building barriers to trap soil being washed down from higher land, diversion drains, ponds for collecting water and irrigation techniques; and
- ways of organizing access to natural resources for land husbandry.

Mekelle University has a close relationship with the BoANR and the NGOs concerned with agricultural development in Tigray. Staff members from these organizations attend summer school and in-service training offered by MU for upgrading, and senior students from MU are attached for several months to development programmes in the field (see Chapter 31). During the courses for DAs and during the briefings for students about to start their practical attachment, the participants were informed about the contest and encouraged to seek farmer innovation. Out of their own curiosity, some staff members from MU and MRC also became active during this initial phase of discovering innovators, but they were more systematically involved during the subsequent phase when the local innovators, innovations and innovation processes were examined in more depth.

In addition, ISWC-Ethiopia worked with the BoANR in organizing a series of training workshops for DAs on the topics of Indigenous Technical Knowledge (ITK) and Participatory Technology Development (PTD). During the first such workshop, the DAs were given the assignment to seek farmer innovators in their working areas and to describe the innovators and their innovations according to the guidelines. This assignment served multiple purposes:

- It led the DAs to see farmers as innovators and to gain a deeper appreciation of farmers' abilities to experiment with new ideas and to develop their farming systems on their own.
- It served as a means by which the DAs could discover local solutions to local problems, solutions that they could then communicate to other farmers as part of their regular work.
- The initial descriptions by the DAs added to the inventory of farmer innovators, from which scientists could select innovations of wide interest for deeper investigation.

In follow-up workshops, the DAs had the opportunity to exchange their experiences in discovering farmer innovation in land husbandry and could communicate farmers' good ideas to each other.

METHODS OF SEEKING INNOVATORS

In the field, the students, DAs and MU staff applied the following methods in seeking farmer innovators:

- *Observation.* They kept their eyes open when driving, riding or walking through their work area and noted anything unusual; they stopped and asked about what they had observed and they entered into discussions about it with the farmers and sometimes also with neighbours.
- *Key informants.* They asked local leaders and older inhabitants for the names of farmers known to be trying out new things or doing something different. Some of the DAs were their own 'key informants' as they had worked in the area for several years, but had not previously been asked officially to record the existing practices and initiatives of local farmers.

Box 6.1 Guidelines for seeking local innovation in land husbandry

In the area where you work, what new things are farmers doing that they have developed without outside help from government agencies, NGOs or development projects? Who are the people who are developing these new ideas? What techniques have they developed for making better use of the land and water resources in your area?

We are seeking innovations in land husbandry developed by individual farmers, by groups of farmers or by communities. *Innovations* are something new. By 'new' we mean something that has been started within the lifetime of the farmers, not something that s/he has inherited from parents or grandparents (which would be called 'traditional'). *Land husbandry* means the way the farmer, group or community is managing and improving the natural resources to gain a livelihood and to ensure a future livelihood for their children. The *farmer innovator* is not necessarily a 'model' or 'contact' farmer with whom you are already working in order to demonstrate ideas you are trying to introduce. The farmer innovator develops or tries out new things without being encouraged to do so by you or anyone else from outside the community.

How to recognize innovators:

- Are some farmers doing things in a different way from their neighbours?
- Have some farmers made changes in techniques introduced by government, NGOs or development projects?
- Are some farmers experimenting with a new idea they have dreamt up or have seen somewhere else, just to see whether it will work on their own farm?

When you describe the innovation, please give as much as possible of the following information:

General information about the area. Place (village, district, zone); agroclimatic zone in local terms; annual rainfall/months of wet season(s); soils (at least local names); landscape (eg mountainous, plateau, hilly, valley); socioeconomic information about the area, eg average size of family and landholding per family, type and number of livestock per average family, major crops grown for home use and for cash income.

Specific information about the innovator. Name of farmer or group (male or female? approximate age?). How do other people in the area characterize the innovator, group or community? Socioeconomic information about the innovator, eg family size or number of people in group, size of landholding(s), type and number of livestock, major crops grown for home use and for cash income.

Information about the innovation. What type of innovation is it (related to crops, livestock, trees, physical structures)? What materials are used and where do they come from? Who does the work involved (farmer alone, with family members, with neighbours – if so, how is it organized)? At what time of year is this work done? What is the purpose of the innovation (what is it meant to achieve)? What is its actual effect? How long ago was it started? Where did the idea come from? What do other farmers in the area think about the innovation?

Other DAs, even those who thought they knew their area well, began to discover things they had not noticed before.

- *Tracing the history of an innovation.* In some areas there were techniques, like the building of silt traps or the planting of certain tree species, that had become quite widespread long before the extension service started its work. By asking local farmers, an attempt was made to identify the persons or group who introduced or developed this technique, and where the idea came from. Sometimes the original innovator had already died, but the second or third person to work with the innovation could be traced. Even though old, this person was often continuing to improve the techniques. The changes s/he had made over time were recorded.

Another method was to seek people who did not accept an extension package as it was. The BoANR has tried to extend packages, eg a new variety together with instructions about cultivation, fertilization, etc. The interesting farmers in terms of local innovation were those who did it 'wrong', ie differently from what had been recommended. They did not accept the package as a whole; they took some but not all parts of it. For example, one farmer did not agree with the extension instructions to plant onions on only one side of the ridge. So he did his own experiment: he planted one plot according to extension instructions and one according to his own idea (planting on both sides of the ridge) and compared the results. Such farmers are the innovators who adapt and improve introduced ideas. This method of seeking those who did it 'wrong' could be applied by MU staff and senior students, but not very well by the DAs themselves.

The DAs, students and MU staff engaged in discussions (ranging from informal conversations to semi-structured interviews) with the farmers and tried to cover as many aspects as possible mentioned in the guidelines. They went with the farmers to their fields to observe and discuss the land-based innovations on the spot. They wrote notes during each discussion and drew up reports on each farmer or group, according to the structure of the guidelines. Drawings were made of a few innovations, sometimes by the innovators themselves, in order to explain the technologies better.

Are there non-innovators?

A methodological problem arose in trying to distinguish innovators from non-innovators. In his studies in Amhara region of Ethiopia, Yohannes (1998, see also Chapter 16) found that every farmer must innovate to some degree because of the differences between farms with respect to household and plot characteristics. Some site-specific modification of a technique is always needed. Moreover, because conditions are constantly changing, farmers have to modify their farming techniques over time. Yohannes experienced difficulties in finding non-innovators: when he asked farmers if they knew of anyone who was doing something different from their parents or different from their neighbours or different from what they had done a few years back, but not something

Credit: Ann Waters-Bayer

Plate 6.1 *Researchers, development agents and farmers in Tigray (Ethiopia) jointly discussing innovations*

promoted by the extension service, he frequently received the answer: 'I do – come and look at this!'

Also in Tigray region, the ISWC-Ethiopia programme found that every farmer innovates to some degree. In this area of extreme land forms, with high plateaux (2500m elevation) and lowlands (below 1500m) separated by steep slopes and escarpments, farmers must deal with diverse agroecological conditions, even within one farm. The political upheavals and movement of large groups of people have meant frequent socioeconomic change. In order to cope with these complex and changing conditions, the farmers are constantly obliged to experiment and to adapt their techniques and strategies. They actively seek information and new ways of doing things that could improve their livelihoods. In distinguishing innovators, ISWC-Ethiopia focuses on the most striking activities that the local communities perceive to be new relative to their experience and important according to their criteria.

Seeking the women

Special attention was given to discovering women farmers who innovate in land husbandry (see Chapter 15). The search was facilitated by the fact that one of the two initial coordinators of ISWC-Ethiopia was a woman who had been working for some years with both male and female farmers in on-farm experiments (participatory variety selection). Moreover, several of the DAs and students are women. Most of the female innovators (or husband–wife innovator teams) have been identified by women.

Most of the female innovators are household heads with access to very few resources. Their informal experiments involve small, low-cost changes in local farming practices, such as digging infiltration pits in the backyard garden or finding cheaper alternatives, such as using a donkey instead of an ox for ploughing. As some of these innovations go against local tradition, the women are not so forthcoming as the men in announcing what they are doing. However, the sharp observation of some local leaders and DAs who understand the women's plight has helped to make their accomplishments known.

COMPILING AND ANALYSING THE FINDINGS

Within a few weeks, reports started coming in about a wide variety of indigenous technologies in land husbandry. The technologies include:

- trapping silt and water to create new land;
- planting local grasses to serve as 'gabion wire' to reinforce terrace walls;
- river diversion and construction of riverside terraces;
- infiltration dikes for garden irrigation;
- water infiltration furrows in field crops, with planted grasses shading the water;
- distributing manure to plots through diversion canals in indigenous irrigation systems;
- collecting the dung of wild animals (hyrax) to fertilize plots;
- revegetation of slopes with indigenous tree species;
- home-made wheelbarrow to transport stones;
- harness for ploughing with one ox; and
- ploughing by women, including ploughing with a donkey.

Database of innovators and innovations

The initial descriptions of farmer-developed technologies in land husbandry, which had been written down by the DAs, students and MU staff, were entered by a graduate assistant into a computer database. These entries included many innovations by individual male farmers, some innovations by women and a few innovations by groups of farmers, eg in improving their indigenous irrigation systems. In the database, it proved impossible to distinguish between indigenous practices and indigenous innovations as a technology long practised in one area may have been only recently tried and adapted by farmers in another area.

From this collection of descriptions, graduate assistants at MU and the ISWC-Ethiopia coordinators selected particularly promising innovations, ie those that could bring substantial benefits to other smallholder farmers and seemed to be applicable in a wider area with similar agroecological conditions.

They visited the farmers personally to see their innovations and to probe more deeply into the reasons for developing them, their advantages and disadvantages, questions that needed further investigation by formal researchers, and how support could be given to farmers' experimentation to build on these innovations.

The inventories of innovations were circulated to researchers in MU and MRC who were encouraged to take up contact with the farmers and/or farmer groups and, together with them, to work out proposals for joint research based on the local innovations. These proposals were then submitted to ISWC-Ethiopia and to other sources of funding.

The coordinators of ISWC-Ethiopia approached scientists whom they felt to be particularly open to ITK and encouraged them to make more detailed studies of the innovations in order to validate the results in scientific terms and to find questions of common interest for joint research with the farmers. As some of the innovations were integrated into complex farm systems (eg involving simultaneously agroforestry, SWC, pest management and beekeeping), teams of three to four scientists from different disciplines were sometimes involved in these studies.

In addition, workshops on PTD were organized for scientists and included a day in the field, discussing farmer innovations with various local stakeholders. These raised the interest of additional scientists to take a closer look at farmer-developed technologies. In those cases where further studies of certain innovations were made, the new findings were added to the original entries in the database.

Ten researchers from both the technical and the social sciences in MU and MRC have been involved in a variety of studies on, eg indigenous practices of managing soil fertility and crop pests, community-based irrigation systems, interactions between farmers' SWC practices and their cropping strategies, and farmers' views on land-use policies. These researchers have met periodically to compare methods and results. Part of their agreement with ISWC-Ethiopia is that they feed back their findings to the farmers for discussion, farmer validation or correction of the results and deepening of the analysis so that the discussions can lead into planning of joint experiments by farmers and scientists.

Delving into sociocultural aspects of innovation

Studies were also made of the innovation processes and the conditions that were conducive to developing and disseminating innovations. We assumed that the social position and relations of innovative farmers would influence the extent to which their innovations were locally valued and likely to spread. Investigations were therefore made of the social position of the innovators within their communities according to local criteria, and their communication links within and beyond their communities.

Through training and advisory sessions, ISWC-Ethiopia encouraged the researchers to use participatory techniques, such as Venn diagrams and the mapping of communication networks by men and women farmers and by community leaders. Investigating the processes and conditions of innovation in a participatory way helps the local people to identify directions that they can take themselves in order to promote the development and spread of useful ideas.

Certain innovations challenge the local culture. Through participatory methods of examining the merits and demerits of these innovations and by seeking alliances with respected leaders in local institutions, the programme tried to stimulate cultural change, particularly with respect to the role of women (see Chapter 15).

IDENTIFYING INNOVATIONS APPLICABLE BY THE POOR

As the vast majority of farmers in Tigray have few resources at their disposal, apart from their own knowledge and labour, we paid particular attention to the resource base that was needed to develop and adopt indigenous innovations. The innovators were classified according to resource endowment and the innovations were examined with respect to resource requirements.

In order to understand local perceptions and criteria of wealth and well-being, farmer innovators were ranked in relation to others in the community. The names of the innovators identified within a community were written on small cards and local persons other than the innovators were invited to rank their relative position by sorting the cards into different socioeconomic groups according to criteria that were important to the community.

In order to identify innovations that can be adopted and adapted by the poorer farmers, the innovators were asked to assess their own innovations according to:

- Effectiveness: what had they hoped to achieve with their innovations and to what degree did they feel they had achieved these objectives?
- Wider applicability: how appropriate did they think their innovations were for different wealth classes in the community?
- Impact: how did they think that their innovations affected their own farm and family, their neighbours and the community as a whole?

In village workshops (see Chapter 18) the other community members also had a chance to assess the local innovations in these terms.

CELEBRATING LOCAL KNOWLEDGE AND CREATIVITY

In Tigray, farmer innovation and experimentation is being encouraged by celebrating local knowledge in order to give it public recognition and raise its social esteem. The ideas generated by farmers are disseminated in various ways and other farmers are encouraged to experiment with them. Indigenous practices and innovations in land husbandry are being made more widely known to other farmers, DAs, scientists, policy-makers and the general public through the following means.

Awards to top innovators

The head of the Tigray BoANR asked his staff to organize meetings in all village areas to honour local farmers who have developed outstanding innovations in integrated land management leading to significant yield improvements. The top innovators at district level, selected from those identified in each village, were awarded prizes. As part of the ceremony, local people visited the winners' farms and saw their innovations. A similar ceremony was held to honour the three top innovators in each of the four zones of Tigray. At a regional ceremony in Mekelle, the zonal winners were invited to describe their innovations, and to explain what they had done and how they disseminate their new ideas to others. This meeting was attended by regional policy-makers, agricultural researchers, the ISWC-Ethiopia coordinators and Steering Committee, and the Dutch coordinator of the ISWC programme. The prizes provided by ISWC-Ethiopia consisted of a certificate as outstanding innovator, plus a sum of money sufficient to buy an ox.

Awards to innovative women

Additional prizes were awarded to women innovators. In this case, the Steering Committee stressed that innovators could also include women who improved their livelihoods by going against social norms and doing their own ploughing instead of sharecropping with men. Animal traction is an indigenous practice in Ethiopia, but has always been the domain of men. The awards are meant to encourage and give public recognition to women who innovate by challenging this tradition, in addition to recognizing other agricultural innovations by women.

Audiovisual and written media

The ISWC-Ethiopia coordinators contacted radio and television reporters and took them to the field to interview the innovators; the broadcasts were made in Tigrigna. The ceremonies for awarding prizes to farmer innovators were also covered by local news media. A newsletter of local innovations is published in Tigrigna, as well as in English for dissemination in other parts of Ethiopia. A series of research reports in English gives formal researchers an opportunity to publish on farmer innovation and PTD, and thus to gain some recognition themselves. This is an important source of motivation for them. Publications in international journals and congress proceedings (eg Fetien et al, 1998, 1999; Mitiku et al, 1998; various articles in ILEIA, 2000) also raise the status of local innovation in the eyes of scientists, DAs and farmers in Ethiopia.

Exchange visits of farmer innovators

A travelling seminar for farmer innovators was held in August 1998. It brought together farmer innovators working on similar themes, eg water control, gully reclamation, selection of species for newly created micro-environments, integration of beekeeping and bee forage species into land husbandry. The visits were meant to stimulate the cross-fertilization of ideas. Further details about these exchange visits and their impact are given in Chapter 18.

Workshops and seminars at all levels

ISWC-Ethiopia arranged numerous small workshops of 15–30 people each, with the aim of raising awareness of farmers' capacity to experiment and innovate: workshops for middle-level management staff of BoANR, MRC and NGOs and workshops for field agents and for farmers. Out of its own conviction that this should be its task, BoANR assumed responsibility for organizing the village workshops to bring together farmer innovators and their neighbours to take a closer look at the innovations, to discuss what is useful for whom in the community and to consider how the opportunities can be developed further. The farmers' innovations are presented – often by the innovators themselves – at regional and national meetings of educational, research and development organizations in Ethiopia. Members of the ISWC-Ethiopia team have also presented the process and findings at international symposia.

Seeking the collaboration of community leaders

In our investigations of farmer innovation, we noted the important role played by enlightened community leaders in giving recognition to farmers who follow new and promising paths to development. For example, some community leaders and DAs in Tigray give public recognition to women who do their own ploughing. ISWC-Ethiopia deliberately sought the collaboration of community leaders, including religious leaders, to gain their support in promoting local innovation, particularly by women.

Stimulating the interest of other development agencies

ISWC-Ethiopia aims to generate interest in farmer innovation and PTD in as many institutions as possible. We therefore visit other development support agencies operating in Ethiopia in order to explain the programme and to invite people from these agencies to visit farmer innovators or to participate in workshops. We encourage scientists to approach these agencies for support in financing studies of ISWC and farmer-led experimentation and technology development in land husbandry.

ENHANCING FARMER EXPERIMENTATION

The ISWC 2 programme assumes that farmer innovation is based largely on informal experimentation. The role of outsiders is to recognize these experiments and how they reflect farmers' assessment of local problems and possibilities, and to help farmers build on these ideas. The collaboration between farmers and the formal research and extension sector includes what Loevinsohn (1990) calls 'feeding farmer innovation'. By providing new ideas and linkages with sources of information (other farmers or formal researchers), ISWC-Ethiopia stimulates farmers to innovate further. The DAs play a key role in encouraging farmers to experiment with new ideas. They can help farmers to find options for testing and to help to evaluate the results together with farmers, rather than trying to transfer ready-made technologies that may not suit the local preferences or agroecological conditions. The knowledge generated through the farmers' experiments leads not only to the creation of site-appropriate technologies; it also increases farmers' capacity to adapt to changing conditions.

Farmer innovators can be entry points into a process of further development of technology through participatory experimentation. Farmers' informal experiments are defined, controlled, implemented and assessed by the farmers themselves, using their own inputs and doing their own observations and recording. Participatory experiments are defined, implemented and assessed jointly by farmers and scientists and/or DAs in such a way that all partners in the research process learn from them. Our approach puts less emphasis on situation analysis than on the 'classical' PTD approach (van Veldhuizen et al, 1997) as the farmers in the midst of their own experimentation have already analysed (although perhaps not systematically) the local situation. However, some PRA tools, such as mapping and ranking, have proved useful when innovators, other farmers and outsiders jointly analyse the context of the local innovations and decide what is worth pursuing in participatory experimentation. The mapping of bioresource flows not only shows how the innovations function within the existing farming system, but can also lead to the recognition of additional constraints and opportunities that the farmers have not yet addressed.

Before entering into such a partnership in experimentation, scientists could spend many years making profound studies of the behaviour and methods of farmers in their informal experimentation. However, our interest is in strengthening the existing farmer experimentation as quickly as possible. This means that research into how farmers do their experiments has to accompany rather than precede the PTD process. We use the following methods to stimulate DAs and scientists to join farmers' experiments:

- *Development agents learning about farmer experimentation.* In a series of learning workshops, we ask DAs to identify and describe farmers' experimentation. This gives a better idea of the types of innovations the farmers

are seeking, the technical and socioeconomic constraints they are trying to address, and how this process of farmer experimentation can be fed with appropriate information and materials.
- *Participatory experimental design workshops.* At the same time, we encourage scientists to investigate local innovations more deeply and to consider how they could support the technology development process already under way at farm or community level. We involve the scientists in PTD training, that includes experimental design workshops, together with groups of farmers who are eager to look into questions of common interest.

FAVOURABLE CONDITIONS FOR PROMOTING LOCAL INNOVATION

This approach to the development of land husbandry systems through farmer innovation and farmer-led experimentation calls for innovation within the formal research and development system within Tigray. Fortunately, the conditions for this type of innovation have been favourable. These include:

- *Spirit of renewal.* After the victory of the liberation forces over the Derg regime, Tigray entered a period of reconstruction and revival. There was an openness for taking new paths. This enthusiasm was not muted by the international border conflict (1998–2000), although some collaboration with farmers had to be postponed in areas invaded by the Eritreans.
- *Strong support of the extension service.* Keen interest has been shown and strong support given to the programme by the head of the BoANR who is a member of the Steering Committee. He recognized similarities between the farmer innovation approach of ISWC 2 and the approach of disseminating best local farming practices that was taken in Tigray during the war of liberation (see Chapter 29). ISWC-Ethiopia could thus build on a strongly felt tradition.
- *Existing linkages between government agencies and NGOs.* The programme also benefited from the tradition of open joint evaluation (*gumgum*) of government agencies and NGOs during meetings at least twice a year. This meant that there was already communication between them which needed only to be refocused on jointly supporting farmer experimentation.
- *Relevant experience of the NGOs.* Already before ISWC-Ethiopia started, the Irish-supported Eastern Tigray Development Programme had begun to introduce participatory methods of improving microwatershed management, and the UK-supported Community-Oriented Development Project had gained experience in farmer participatory research (see Chapter 22). Thus, scientists and extensionists who were sceptical of farmers' abilities to do research could visit convincing examples. The farmer experimenters also presented their methods and results at workshops for research and extension staff.

- *Practical orientation of university teaching.* Mekelle University has a strong practical orientation in its teaching and research, and aims to produce graduates who can contribute to development in Ethiopia (see Chapter 31). Staff and students had already undertaken research into ITK. This was one reason why MU was chosen as the lead organization for ISWC-Ethiopia.
- *Close links to policy-makers.* One of the ISWC-Ethiopia coordinators is involved in various regional and national bodies concerned with agricultural research, education and extension, and thus can take advantage of numerous opportunities to make the ISWC approach more widely known and accepted in formal circles. He also promotes farmer-led approaches to dryland development during his informal meetings with policy-makers.

INCLUSIVE APPROACH

From the very start, ISWC-Ethiopia has involved people from all levels within research, development and teaching institutions. This widely inclusive rather than narrowly exclusive approach was chosen as a path towards institutionalizing support of farmer innovation and experimentation in the formal research and development system. We did not want ISWC-Ethiopia to be a pilot PTD project in a small area that would operate for several years in splendid isolation and then face a problem of scaling up. Instead, we decided to encourage as many people and organizations as possible to take small steps: starting with recognizing farmer innovation, leading to identifying ways of feeding farmer experimentation, and gradually progressing into more systematic collaboration of farmers and scientists in further development of the existing systems of land husbandry. Our entry activities into this PTD process were therefore focused on celebrating and promoting a spirit of innovation, experimentation and collaboration.

The programme had the great advantage that it did not have to start from scratch. Past experience during the civil war had forced a strong degree of independence. This already ploughed the ground for creativity in local technology development. Tigray is still far from reaping the full benefits of this approach, but the enthusiasm generated among farmers, DAs and a growing number of scientists for seeking collaboration in innovation development gives promise for sustained growth of this approach.

REFERENCES

Fetien A, Mitiku H and Waters-Bayer A (1998) 'Farmers' innovations in land and water management', *ILEIA Newsletter*, vol 14, no 1, pp21–23
Fetien A, Mitiku H and Waters-Bayer A (1999) 'Dynamics in IK: innovation in land husbandry in Ethiopia', *Indigenous Knowledge and Development Monitor*, vol 7, no 2, pp14–15

ILEIA (2000) 'Grassroots innovation', *ILEIA Newsletter*, vol 16, no 2, (French version: 'Promouvoir l'innovation paysanne'; Spanish version: 'Innovación desde las bases')

Loevinsohn, M (1990) 'Feeding farmer innovation', *ILEIA Newsletter*, vol 6, no 1, pp14–15

Mitiku H, Fetien A and Waters-Bayer A (1998) 'Recognising and encouraging farmer innovation and experimentation in dryland farming in Tigray, Ethiopia', proceedings of the Association of Farming Systems Research and Extension 15th International Symposium, 29 November–4 December 1998, Pretoria

van Veldhuizen, L, Waters-Bayer, A and de Zeeuw, H (1997) *Developing technology with farmers: a trainer's guide for participatory learning*, ZED Books, London

Yohannes G M (1998) 'Farmer assessment of local innovators in northern Ethiopia', *Farmer Innovators in Land Husbandry*, nos 4/5, pp12–16

Part 3

Farmer innovation: process, evidence and analysis

An initial analysis of farmer innovators and their innovations

*Chris Reij and Ann Waters-Bayer**

Do farmer innovators have some common characteristics? What triggers them to innovate? Where do they get their ideas? In which fields are they innovating? To what extent are their innovations spreading and how? It is clear that any generalization is full of pitfalls on account of the considerable socioeconomic and agroecological differences within and between countries in Africa. What may appear to be true in the Yatenga region of Burkina Faso need not hold true for another part of the country, let alone for other countries in Africa. Here we discuss some of the characteristics of the farmer innovators who have been identified thus far through the ISWC 2 and PFI programmes. The results may say more about improvements that could be made in the methodology of the farmer innovation approach than about the traits of innovators.

WHAT DO THE IDENTIFIED INNOVATORS HAVE IN COMMON?

Most of them are men

As an average across all eight countries in the ISWC 2 and PFI programmes, about three-quarters of the identified innovators are men. Although women often do a large share of the farmwork, it is usually the men who are the household heads and represent the family in public, and are therefore most

---

* Chris Reij, a geographer at CDCS, Vrije Universiteit, Amsterdam, is international coordinator of the ISWC programme; Ann Waters-Bayer, an agricultural sociologist with ETC Ecoculture in Leusden, The Netherlands, advises the programme in participatory approaches to research and extension

likely to take credit for any changes made on their farms. This may partly explain the lower percentage of female innovators identified. However, it is also a question of how innovators are identified and by whom, and what is considered to be an 'innovation'. In those countries where women were involved in the identification process, it generally proved easier to find female innovators. Also, where the focus of the search was not limited to SWC in a narrow sense but included livestock husbandry, gardening and processing of farm products – fields in which women usually have substantial decision-making power – a higher number of female innovators were documented. For both these reasons, about 50 per cent of the innovators identified in Tunisia are women (see Chapter 12). After recognizing a bias in its methodology, the PFI programme introduced training in gender awareness; this led to an increase in the number of female innovators identified (see Chapter 10).

The name of the ISWC 2 programme probably led the partner organizations to focus primarily on techniques of physical SWC which are usually the domain of men and are implemented only by very exceptional women, such as Ayelech Fikre (see Chapter 3). In contrast, many women innovate in collecting and sowing selected grasses, protecting the spontaneous growth of tree seedlings and conserving local plant biodiversity for medicinal, cosmetic or other purposes (see Chapter 15). These less spectacular innovations are likely to be overlooked at the initial stages in a farmer innovation programme, before the partners have developed a keen sense of the myriad of small changes that, in combination, can eventually make a large difference.

In areas like Western Kenya, where the percentage of female-headed households is high because of male labour migration to help support the family, there was no significant difference in the number of innovations carried out by male and female heads of household. This was despite the fact that the women were sometimes constrained by having to wait for the husband's approval before starting up new initiatives on the farm (see Chapter 8). It is possibly because this study deliberately tried to encompass all innovations in farming, rather than only in SWC, that a gender balance was found in terms of innovativeness. Moreover, the study was based on a random sample rather than a selected (and therefore probably biased) one.

In many cases, it would probably be more appropriate to refer to 'family innovations' rather than make a distinction between male and female innovators because families often work very closely together in building up their farms (see Chapter 10). For example, Wilbert and Emelita Mville in southern Tanzania collaborate in integrating their innovations in pit planting, manuring and irrigation (see Chapter 25). In Ethiopia, innovations by husband–wife teams were also documented (see Chapter 6). Moreover, most innovators will need support from the rest of the family as a new technique may require extra labour, divert resources and involve some risk and therefore, at least in some cases, require consultation within the family. On the whole, insufficient attention has been given thus far to innovation from a family perspective.

Many of them are strong personalities

The successful innovators who have been identified tend to be strong personalities capable of withstanding considerable social pressure. Some of them have faced hostile treatment because members of their community felt that they were not respecting cultural traditions. For example, when Tensue Gebremedhin, a widow in Tigray, started to plough with an ox and a donkey, she broke two taboos: a woman behind the plough and a donkey in front of it. Only when the local council and extension agent began to give her moral support did the other community members gradually start to accept her (Mamusha et al, 2000).

However, this impression of strong personalities may be biased by the fact that, in the initial search for innovators, those who were 'found' were mostly successful people already known to outsiders or individuals practising very conspicuous innovations. It was more difficult to discover those farmers who are making small and less visible but nevertheless important improvements in their farms, and perhaps do not even regard themselves as innovators.

Most of them are relatively old and experienced

The average age of farmer innovators involved in the PFI programme is 44 years (Critchley et al, 1999). The ISWC-Tunisia team found that most of the innovators with whom it is working started to experiment when they were between 30 and 50 years old, ie after they had married and assumed responsibility for the family and its land, and could finally realize their own ideas about improving the farm (see Chapter 11). In a random sample of farmers interviewed in East Africa, Nielsen (see Chapter 8) found that the level of innovation dropped off only when the farmers are fairly old (in their late 60s). In northern Ethiopia, Yohannes (see Chapter 16) found that most innovators identified by the community were over 50 years of age. The improvements in the innovators' farms were due to the long-term effects of many incremental changes over the years. Those who have come to harvest the fruits of their experimentation and to be recognized as successful innovators were therefore older than average. What has been recorded is the sum of a series of innovations or the maturity of a gradual process of developing innovations over several decades (eg Zigta GebreMedhin in Chapter 14).

However, not all innovators are old. In Cameroon, for example, a recently formed network of 15 farmer innovators in Western province has four members in the age group of 23–32 years. These are young men who, after losing their jobs in urban centres during the economic crisis in Cameroon in the 1990s, returned to their home villages and took up farming. For example, Martin Nkegne (32 years) used to be an electrician and is now experimenting with sprinkler irrigation and trying out several new techniques that he observed in other areas. There may be a bias in the identification of these innovators: extensionists are more likely to notice the techniques that have been brought in from elsewhere as these are also likely to be the types of techniques being promoted by the extension services.

Another complicating factor is that, in many rural African societies, it is not readily accepted that younger generations do something different from their elders. This cultural factor may constrain innovation by young people as found among the Luo in Western Kenya (see Chapter 8), or at least constrain their willingness to admit that they are trying something new.

Most of the widely recognized innovators are relatively rich

Although the situation varies from country to country, the general finding has been that the established, widely recognized and successful innovators are richer than average. This may be because they are older (see above) and have had time to develop their farm. It is enlightening to look at the history of these wealthier innovators. In Burkina Faso, for example, many started off 15–20 years ago as relatively poor people, but their investment of time and energy in improving their land allowed them gradually to expand and diversify their production and to improve their yields. Especially in dry years, such as 1997 and 2000, the innovators who were applying a wide range of water harvesting and soil management practices had higher cereals yields than did their neighbours. This suggests that, through their creativity and initiative, they had decreased their vulnerability to drought and improved the food security of their families. Their innovation had an incremental character. Thousands of farmers in Burkina Faso, Niger and Mali have invested in making improved traditional planting pits, but at a slow rate, usually managing to rehabilitate only about 0.2ha of land per year (Hassane et al, 2000). They did not expect miracles overnight. Nor did the farmer innovator from southern Zimbabwe, Cleopas Banda, who told the ISWC 2 coordinator during a field visit in November 1998: 'I started my experiments in 1987 and I felt in 1994 that I really benefited from them' (see Box 7.1).

In Tanzania, two students from Wageningen Agricultural University studied the socioeconomic background of the farmer innovators identified through the ISWC 2 programme. The innovators included a roughly equal number of rich and poor farmers. The former were innovating because they were curious and enterprising, had the resources to experiment and could afford to fail. The latter were innovating because they were highly motivated to improve their situation and, having no resources to apply solutions offered from 'outside' (the extension service), used their creativity to find their own solutions. For example, because they could not afford to buy chemical fertilizer, they found ways of combining manure, urine and crop residues to maintain soil fertility for maize production. The study revealed no direct relationship between innovativeness and economic status (Verhoeven and van der Kroon, 1999).

Similarly, Nielsen (see Chapter 8) found no correlation between innovativeness and farm size. This may be because his study did not confine itself to only the conspicuous innovators and because, in the densely populated areas in East Africa, the innovativeness was linked with the intensification of farming rather than with the expansion of farm area.

Wealth is, of course, relative. In Ethiopia, for example, most of the innovators identified through the ISWC 2 programme would be classified as 'resource-poor'. In the highlands of Tigray, population density is high and few farmers (also few farmer innovators) have more than 1ha of land to cultivate – additionally this is often divided over three or four plots. Even here, innovative men and women have managed to increase their food security, eg by cultivating land they had reclaimed from gully erosion or by doing their own ploughing and therefore not having to practise sharecropping (see Chapter 15).

Exposure to other areas stimulated innovation

Many innovators have been exposed to other areas, usually through labour migration or military service. They picked up ideas while in other parts of the country or abroad and, in some cases, made earnings that they could invest in agriculture (equipment, livestock, etc). For example, one innovator saved money while working for years as a watchmaker in Tunis and invested in building a small concrete dam in the arid foothills near his home village – a dam that was unique for the region. The father of another innovator gained knowledge about fruits and fruit trees while working as a cook for the *Bey* (king) of Tunis and then introduced fruit trees into his village in southern Tunisia. Some outstanding innovators in Ethiopia were ex-soldiers who experimented at home with techniques they had seen while in service (see Chapters 14 and 16). Especially in Cameroon, several cases of innovation after urban-to-rural migration were recorded, such as the taxi driver turned agroforestry teacher Barthelémy Djambou (Chapter 2) or the above-mentioned electrician turned sprinkler designer Martin Nkegne.

Many of the women innovators did not have the same opportunities for travel as did the men. Especially in Tunisia, women are less likely to migrate to the towns and have little exposure to other places than their home villages or those into which they marry. In the sub-Saharan countries participating in the ISWC 2 and PFI programmes, there are somewhat fewer cultural obstacles to travel by women than in Tunisia but, on the whole, the women innovators identified thus far have seen less of the country than the men. An exception is in situations of unrest and war, when people move to other areas as fighters or refugees and observe different farming techniques. This was the case for both female and male innovators in Tigray, such as the late Hailay Hagos who experimented with water-management techniques he had seen as a refugee in Sudan (Fetien et al, 1998).

Most innovators are full-time farmers

In most African countries, the majority of farm families derive their livelihoods not only from crop and livestock production but also from a range of activities outside of agriculture, particularly in the dry season. According to a recent study (Bryceson, 1999), farmers in sub-Saharan Africa derive 60–80 per cent of their income from non-farming activities. In seven of the eight countries in the two farmer innovation programmes, it was found that most of the innovators devote most of their working time to farming. They are often in their fields, digging pits, constructing bunds, planting and protecting trees, caring for their livestock, producing compost, carting compost, and so on. In essence, innovators are good farmers and hard workers.

On the one hand, it appears that the more innovative farmers can produce enough from their land and therefore need not seek off-farm sources of income to the same extent as do the less innovative farmers. Indeed, some innovators have abandoned commercial activities entirely in order to concentrate on farming. Instead of diversifying into off-farm activities, they have chosen to diversify their agricultural activities (eg the four agroforestry innovators in Chapter 4).

On the other hand, many farmers who devote most of their time to agriculture still have outside sources of income and it may be these additional sources that give them some flexibility to experiment. This is especially obvious in Tunisia, where almost half of the male innovators identified through the ISWC 2 programme stated that their major occupation is outside agriculture. Among them are teachers, drivers, masons and tailors. They argue that their farms are too small in relation to the sparse rainfall to produce enough food and cash to cover the financial needs of their families.

Farmers in the sub-Saharan countries might have the same argument as those in Tunisia, but less opportunity to gain the skills and positions to earn these types of off-farm income. In some remote areas, as described in Chapters 14 and 16, it is not clear whether the farmers are investing huge amounts of labour in innovation and intensification in agriculture because they love the land and the work, or because they lack alternatives with better returns to labour. In Tanzania, some farmers claimed they were experimenting with organic farming practices because they loved the soil and did not want to spoil it (Laurens van Veldhuizen, pers comm, January 2001). Better insights are still needed into the relationships between innovation and intensification on the one hand, and such factors as diversification in sources of off-farm income, the possibilities for investment in agriculture, and attitude to the land and to farming, on the other hand.

Creativity and formal education are not correlated

The level of formal education does not appear to be a determining factor with respect to farmers' creativity and propensity to experiment. In Ethiopia, Yohannes (see Chapter 16) found that there was no significant correlation between the level of formal education and the innovativeness of farmers. Some of the most remarkable and communicative innovators identified through the ISWC 2 and PFI programmes are virtually illiterate, and this is also the case in Tunisia which has the most widespread system of formal education of all the eight countries involved. Most of the farmers in the Yatenga region of Burkina Faso who are improving the traditional *zaï* technique for rehabilitating degraded land are illiterate (see Chapter 4), yet they have made an extraordinarily important contribution to the development of sustainable agriculture in the Sahel.

Innovative farmers tend to develop integrated farm systems

It has proved difficult to classify innovators according to the type of innovation or even to separate out specific innovations, as many are interconnected and mutually reinforcing. Extremely innovative farmers have developed highly integrated farming systems.

Barthelémy Djambou in West Cameroon (see Chapter 2) combines cereal growing, agroforestry, aquaculture and the raising of rabbits, poultry and pigs. The farmers in Upper Babanki in North-west Cameroon (see Chapter 9) have improved soil fertility management, developed a local gravity-based irrigation system and designed a simple tool to facilitate harvesting. One innovation stimulated the next. Yacouba Sawadogo in Burkina Faso started with improving traditional planting pits for growing cereals, then began using them for growing trees, created a seed bank to collect all crop varieties cultivated in the planting pits and introduced a 'market' model for promoting and sharing experience with planting pits (see Chapters 4 and 20).

These are only a few of the many examples of integrated innovation and chains of innovation that challenge all attempts to classify farmer innovators into tidy categories.

MOTIVATION FOR INNOVATION

Population pressure on a limited natural resource base appears to be an important incentive for innovating and investing in agricultural diversification and intensification. Where farmers have their 'backs against the wall' and few options left, experimentation and innovation find 'fertile ground'. Farmer innovators frequently recount that they were driven by the need to feed their families. For example, when Yacouba Sawadogo was confronted with frequent harvest failures provoked by droughts in the 1970s and many villagers migrated to other regions, he decided to stay on the land of his ancestors and find solutions to this problem. Many similar stories could be told.

The Special Adjustment Programmes of the World Bank also triggered a spate of innovations. After subsidies on inorganic fertilizers were removed, their prices sky-rocketed and, in many countries, smallholders could no longer afford to buy them. Farmers then started to revive traditional soil fertility management practices that had previously been abandoned (personal communication during a field visit in Tanzania in March 1997) and traditional crop varieties that grow better than the 'high-yielding' varieties under low-external-input conditions (see Chapter 8).

Farmers are keen to experiment with technologies that promise to create win–win situations, substantially increasing production and, at the same time, maintaining or improving the environment. Higher yields are important not only because they improve food security at household level, but also because more agricultural products can be sold to generate cash for other expenditures

BOX 7.1 THE INTEGRATED FARM SYSTEM OF CLEOPAS BANDA, FARMER INNOVATOR IN ZIMBABWE

Cleopas Banda is a farmer in his late 30s who lives in Zvishavane district of southern Zimbabwe where annual average rainfall is about 600mm. He has one wife and seven children. His main source of income is his farm of 2ha plus five head of cattle, ten goats and a few chickens. His innovative energies have been devoted to trapping water and circulating it in his fields. He started doing this in 1990 and says it is his own idea, although he gained some inspiration from the work of the renowned farmer innovator from the same district, Maseko Phiri.

Credit: Chris Reij

Plate 7.1 *Cleopas Banda with mango tree on a raised bed with buried clay pots*

Banda began by digging large ditches alongside his contour bunds in order to harvest and conserve as much rainfall and run-off as possible. At the height of the wet season, the ditches are full of water which slowly infiltrates into the rest of his land. In 1997 he built a dam to create a pond at a lower corner of his farm to catch any water that spilled from the infiltration ditches. He rears different types of fish in the pond. He established a vegetable garden below the dam and uses gravity irrigation to water it. He uses liquid manure from his cattle pens to fertilize the garden. He has also planted mango trees on a raised bed and covered it with pebbles to reduce evaporation. In this raised bed, he buried clay pots which he fills with water in the dry season. In this way, he economizes on water use in this semi-arid region. He explained that growing the trees on a raised bed also reduces termite attack.

It took many years before he began to derive benefits from this integrated system of water and soil fertility management, livestock integration, aquaculture and horticulture. His success was due to good local knowledge, close observation and reflection, and commitment and hard work by both himself and members of his family. Banda says he is still experimenting and observing and trying to understand how his system works.

In the last couple of years, he has received an average of about 150 people during field days and visits by individuals, and he keeps a register of visitors. Government extension agents and development projects often refer people to him. He explains his innovations so well that he is now known among both farmers and extensionists as The Professor.

Source: Edward Chuma and Kudakwashe Murwira, coordinators of ISWC-Zimbabwe

(Hassane et al, 2000). When farmer innovators in East Africa were asked why they had innovated, the main reasons were, in order of importance, to provide food for home consumption, to increase household income and to maintain or increase soil fertility (see Chapter 8). Likewise, Critchley et al (1999) found that farmers innovated in order to improve their crop yields and to increase their cash income.

SOURCES OF INSPIRATION FOR INNOVATION

Where do farmer innovators get their ideas? Is it their own creativity? Have they been influenced by extension agents or other outsiders? Have they observed the technique elsewhere, such as during visits to another region? It is often difficult to obtain reliable answers to these questions because both the interviewers and the interviewees may have various definitions of 'innovation' and the farmers also have their pride. It is not unusual for farmers, like others in a society, to believe (or want others to believe) that they were 'the first'.

When Critchley et al (1999, p50) interviewed 74 farmer innovators in East Africa, 22 claimed that the innovation was their own idea, 16 said they were influenced by training and advice from extension, 11 reported a mixed influ-

Credit: Will Critchley

Plate 7.2 *Sustainable increases in yields are an important motivating factor*

ence, 10 mentioned having seen the idea already somewhere else far from home, 10 reported that the 'innovation' was a family tradition and 5 mentioned other sources of inspiration.

A study by Sumberg and Okali (1997) of farmer experimentation in Kenya, Zimbabwe and Ghana revealed that the main source of the ideas that farmers were testing differed greatly between the countries. In Ward 21 of Chivi district in southern Zimbabwe (which is also an action area of ISWC-Zimbabwe), almost three-quarters of the 29 experimenting farmers interviewed said that they were trying out what they had observed elsewhere or had been suggested to them or actively promoted, while the rest said that their experiments were based on their own ideas. The other extreme was in the Eastern region of Ghana, where 65 per cent of the farmer experimenters stated that they were building on their own ideas. A partial explanation for these differences could be that, during the colonial period in Zimbabwe, a uniform extension message was imposed on the farmers and traditional farming practices were strongly discouraged (see Chapter 28). This policy destroyed much of the indigenous knowledge and made independent initiative more risky.

ONE OF MANY POSSIBLE TYPOLOGIES OF LOCAL INNOVATIONS IN SWC

As mentioned above, it is often difficult to categorize local innovations because they are closely integrated with each other. Moreover, they often serve multiple purposes. Nevertheless, some general fields in which farmers are developing new technologies can be discerned (Critchley et al, 1999). As both ISWC 2 and PFI focused on land husbandry in the sense of managing soil and water, most of the identified innovations are related to this, although some country programmes deliberately cast the search net more widely. For this reason, innovations related to biodiversity management, crop varietal selection, pest control, sowing density and other cultural practices, implement design and food processing have also been found, and are described in several chapters of this book.

Looking at land husbandry in a more narrow sense and looking primarily at the accomplishments of the exceptional innovators rather than the 'everyday' innovations of a larger number of smallholders, the local innovations identified through the ISWC 2 and PFI programmes can be divided into the following fields.

Improved soil fertility management

With the exception of southern Tunisia where farmers cultivate fertile loess soils (see Chapter 11) and the Irob farmers in similarly dry areas of Ethiopia who likewise harvest water and silt (see Chapter 14), most of the innovators were actively seeking ways of improving soil fertility. These include the night paddock manuring system and farmers' experiments with liquid manure in North-west Cameroon (see Chapter 21). Farmers on the Central Plateau of Burkina Faso who invest considerable labour in rehabilitating degraded land with improved planting pits have started to produce more and better compost to put into the pits, because only then will they draw optimum benefits from their investment in the land (see Chapter 24). An analysis of innovations in southern Tanzania mentions a wide range of soil fertility management practices, including soil covering of trash lines in coffee farms, the use of liquid manure, multiple maize stands in pits with manure (see Chapter 25), ridging to incorporate organic matter and spot application of manure (Temu and Malley, 1999).

Most of these practices of organic matter management are well known. What is new is the way the farmers combine them and try to optimize the use of scarce resources to maintain fertility while conserving soil moisture.

Harvesting and economizing on water

Innovations in this field include the *zaï* in Burkina Faso (see Chapter 4), the infiltration ditches and trenches in southern Zimbabwe (Box 7.1) and the stone silt-traps that Irob farmers in northern Ethiopia have been building up over the last 40 years (see Chapter 14). Outstanding examples of economizing

Plate 7.3 *Improved soil fertility management leads to better harvests (Burkina Faso)*

on the use of water were found in the arid areas of Tunisia, such as the use of upturned plastic bottles to irrigate melons, an idea developed by the elderly woman innovator Rgaya Zammouri (see Chapter 12). As women in many societies are expected to carry water, it is not surprising that they have become very inventive in terms of water economy.

Gully control and rehabilitation

The PFI programme in East Africa documented several cases of treating gullies using trash barriers, stone barriers, earthworks, live barriers and pits (Critchley et al, 1999, pp52–53, 57–58). One example of a person who excels in gully rehabilitation is Grace Bura in Tanzania (see Box 7.2). Also numerous farmers in mountainous areas of Ethiopia are investing considerable efforts in reclaiming land in gullies (Mitiku, 2000).

Agroforestry practices

Several examples of local innovation in agroforestry are documented in this book. They include the use of planting pits for growing trees in Burkina Faso (see Chapter 4), the grafting of fruit trees in Tunisia (see Chapter 11), protecting spontaneous growth of tree seedlings by Tigrayan women (see Chapter 15) and stabilizing terraces with local trees of economic importance in Ethiopia (see Chapter 18).

Any typology of innovations will depend on who is doing the classification and for what reasons. The typology above, made according to technical categories, is likely to be of greater interest to agricultural scientists than to

---

Box 7.2 Grace Bura: a local expert in gully rehabilitation

Grace Bura, a Tanzanian woman in her 50s, is the main farmer in the family. Her husband, a retired teacher, tends only a small plot of rain-fed grapevines. In 1982 Grace acquired and decided to reclaim some severely gullied land. She packed the gullies with checkdam 'sandwiches' of trash and soil in alternate layers and planted *mikayeba* (tree cassava) cuttings on top of the dams. Gradually, the gullies healed. She extended the lines of trash and *mikayeba* until they formed contour lines across her land. The *mikayeba* is not only a living structural support for her soil; its leaves are also a source of fresh vegetables. Grace fallows the land for up to two years, then digs in the vegetation during the growing season as a green manure and, late in the season, plants a catch crop of maize. If it produces cobs, she harvests them; if not, she feeds the plants to her stall-fed dairy cows (one 'grade cow' and one of a local breed). Grace continues to rehabilitate land. Where the gullies are more severe, she uses cuttings of *Commiphera* which form a stronger barrier than *mikayeba* when they establish. Closer to home, where the cattle are kept, are fields intercropped with pigeon pea, manured with waste from the cattle and house compound, and protected with trash lines. Other farmers have copied Grace's gully-control system, but she does not know how many.

Source: Critchley et al (1999)

---

extensionists or farmers. These may be more interested in the resources (labour, land, capital) needed to implement the different innovations as they could then identify which innovations could be applied by farm families with different amounts of these resources. Alternately, in areas with considerable differences in agroecological conditions, such as the mountainous areas of Ethiopia, it may be more important to draw up a typology according to agroclimatic zone or soil type. Scientists are more likely to be interested in classifying and studying outstanding original innovations, whereas farmers may be interested simply in hearing about everything that is new and potentially useful for their area.

The Suitability of Innovations for easy Dissemination

An assumption on which both ISWC 2 and PFI were based was that innovations developed by farmers would be low cost and more acceptable to smallholders than innovations developed by scientists and transferred to farmers through the conventional extension system. There are good examples of simple, low-cost innovations that spread quickly and fairly spontaneously. For example, 14 farmers from the Illéla district in Niger visited the Yatenga region of Burkina Faso in 1989, observed the improved *zaï* and, upon their return, started to try out these planting pits in their own fields, initially covering a total of 4ha.

Already in 1990, 75ha of land in Illéla district had been treated with *zaï* and, when drought occurred in that year, farmers had a good harvest only on those 75ha. This sparked a process of even more rapid spread of the *zaï* technique and led to the emergence of a market for severely degraded land that could be rehabilitated with *zaï* (Hassane et al, 2000). Although the spreading of *zaï* in Niger was largely spontaneous, it did require some external support to initiate it. This consisted of a project funding a study visit by the farmers from Niger to Burkina Faso and subsequent provision of funds for some tools and training.

In some cases, if the innovator's land is close to a major travel route, such as the footpath passing Yohannes Tesfaye's riverside plot with the devil's tie (see Chapter 14) or the barren land rehabilitated by a '*zaï* school' next to a tarmac road (see Chapter 20), a recognizable innovation can spread without any special efforts by the innovator or anyone else to make the idea known. The devil's tie was the result of Yohannes' own lateral thinking and 'clicked' immediately with other farmers who were used to working with stones but had not thought of placing them vertically instead of horizontally to withstand the pressure of the water. They could immediately grasp the principle. This is a prime example of an innovation that could spread spontaneously because it was favourably located, used exclusively local resources and was easy to comprehend.

However, some of the more spectacular innovations by farmers or farming communities are not simple to learn or to do, are not low cost and are not likely to spread easily and spontaneously. For example, gully control and rehabilitation require considerable technical skills that not all farmers possess and high investment of labour – up to several hundred person–days – depending on the size of the gully, the number and size of checkdams needed, the type of revegetation and the materials used. To many farmers, as impressed as they may be with the accomplishments of extremely energetic farmers and/or well-organized groups that have indeed reclaimed severely degraded land, this is a hard act to follow. Unless external support is provided, only the relatively rich farmers who can afford to hire labour are likely to carry out this kind of conservation work and, even then, only if they expect some economic gain in the end.

Especially in Tunisia, some of the initiatives taken by men are expensive (checkdams in combination with water storage in cisterns) and will spread only if government subsidies are available (see Chapter 11). Such subsidies are common in Tunisia, but not in the other countries in the farmer innovation programmes. In these countries, more emphasis must be put on influencing policy to encourage investment in the land in ways that require lower material costs, although considerable labour. An example is the effort made by farmer innovators in Tigray to ensure that farmers who reclaim gullies or create new land behind silt-traps can retain the right to use the land (Mitiku, 2000).

In Uganda, it was found that the farmers (more often women than men) who tried out local innovations about which they had heard tended to be poorer than the original innovators (see Chapter 19). The reasons behind this are not yet clear. It may be that the wealthier farmers are in a better position to take the risks involved in innovating, while the poorer ones are prepared to adjust their practices when they see good alternatives that suit their needs.

However, this will apply only where the richer farmers' innovations can also be realized with the limited resources available to the poorer ones.

Why do certain innovations spread whereas others do not? It is not only a question of access to labour and other resources. An innovation such as using plastic bottles for very localized irrigation makes use of readily available and inexpensive materials and is easy to apply, yet it started spreading only after it was made known via the radio. The spreading of innovations is influenced by a number of factors, including the costs and benefits of the innovation as perceived by the farmers, and whether the farmers are aware of the innovation. Some of the innovations by women in particular, although often simple and low cost, have not spread quickly because the women have not had the same opportunities as the men to share new ideas. The spread of local innovations to other smallholders who might find them useful will often depend on the deliberate efforts made by farmer innovators, development projects, NGOs and the extension services to make these innovations more widely known. This is a strong argument for building better linkages between these various actors in agricultural development, as well as linkages with mass communication media as channels for disseminating information about local innovation.

REFERENCES

Bryceson, D (1999) *Sub-Saharan Africa betwixt and between: rural livelihood practices and policies*, Working Paper 43, Africa Studies Centre, Leiden

Critchley, W R S, Cooke, R, Jallow, T, Lafleur, S, Laman, M, Njoroge, J, Nyagah, V and Saint-Firmin, E (1999) *Promoting Farmer Innovation: harnessing local environmental knowledge in East Africa*, RELMA/UNDP, Nairobi

Fetien A, Waters-Bayer A and Mitiku H (1998) 'Traditional practices and farmers' innovations in land husbandry: some examples from Tigray, Ethiopia', *Farmer Innovators in Land Husbandry 1*, ISWC-Ethiopia, Mekelle

Hassane, A, Martin, P and Reij, C (2000) *Water harvesting, land rehabilitation and household food security in Niger: IFAD's Soil and Water Conservation Project in Illéla District*, International Fund for Agricultural Development (IFAD), Rome/CDCS, Vrije Universiteit Amsterdam

Mamusha L, Fetien A and Waters-Bayer, A (2000) 'Women challenge cultural norms', *ILEIA Newsletter*, vol 16, no 2, pp40

Mitiku H (2000) 'Lobbying for policy support to local innovation' *ILEIA Newsletter*, vol 16, no 2, pp23–24

Sumberg, J and Okali, C (1997) *Farmers' experiments: creating local knowledge*, Lynne Rienner Publishers, Boulder

Temu, A E M and Malley, Z U J (1999) *Report on the analysis of farmer innovators and innovations for the development of themes for participatory research*, ISWC-Tanzania, MARTI Uyole

Verhoeven, M and van der Kroon, S (1999) *Are innovative farmers innovative in spreading the news? A research after innovative and non-innovative farmers in the Kilolo and Njombe Divisions of South Tanzania*, MSc thesis, Wageningen Agricultural University, The Netherlands

8

Why do farmers innovate and why don't they innovate more? Insights from a study in East Africa

Flemming Nielsen[*]

Why are some farmers more innovative than others? Is it because they have access to more resources than other farmers or because a lack of resources forces them to be innovative? Are some farmers more curious than others? Do some simply enjoy playing around with new things? Does gender make a difference: are female-headed households less innovative than male-headed ones? Answers to these and many more questions about farmer innovators and innovation processes are often only anecdotal or based on only a small number of case studies. This chapter reports on a survey that was designed to gain a wider overview of farmer innovation in East Africa.

There are many reasons for seeking to find out why farmers innovate. The answers can provide academic insight into the how and the why of development. But it could also be argued that we have a moral obligation to bring about a long overdue recognition of the role of small-scale farmers in development. From a practitioner's point of view, the answers can guide interventions to support innovation by farmers.

Before trying to explain the variation in farmers' innovativeness, it would make sense to see to what extent this actually varies between farmers. Maybe the variation is so small that it is not really an issue. I say 'maybe' because, despite an impressive number of publications on farmer innovation, basic

---

[*] Flemming Nielsen is an agricultural geographer from Denmark, currently based in The Netherlands, where he is part-time post-doctoral researcher with Wageningen Agricultural University and part-time senior researcher with ILEIA (Information Centre for Low-External-Input and Sustainable Agriculture)

Map 8.1 *Research sites in Kenya and Tanzania*

information on such questions is still missing. A few other authors have recognized this peculiar situation; for example, Sumberg and Okali (1997) point out that the same anecdotal evidence is cited again and again without sufficient basic research being done.

The prominence of anecdotal evidence can be explained easily. First of all, the realization that farmers innovate stems from individual contacts, eg a project staff member or a field researcher comes to know an innovative farmer and then documents his or her activities. This evidence is, by its very nature, anecdotal if the contact is not a result of random sampling. Secondly, a resistance to acknowledging farmers' innovative capacity has meant that hitherto it has been more important to show that farmers do innovate than to collect statistically sound data about these activities. A problem with this way of collecting information is that farmers who really stand out from the majority are easily noticed. This leads to the risk that documentation of farmers' innovativeness paints a misleading picture.

The word 'anecdotal' has a negative connotation. It may be better to speak of case studies. If one wants to understand the inner workings of the innovation process, case studies have many advantages, but problems arise if one wants to extrapolate the findings. For instance, what does a case study of a farmer's innovation in irrigation tell about the innovative activities in the entire area? Nothing really, unless we know something about the activities of a representative sample of other farmers in the area.

SURVEY OF FARMER INNOVATION

In order to gain a more general overview of farmer innovation in East Africa, basic information about this was collected in a survey of over 500 farmers. Each one was interviewed for one hour and some farmers were visited several times. The results of this study can serve as a good complement to case studies of farmer innovation, such as those found throughout this book, but cannot be a replacement for case studies.

Collecting basic information on farmer innovation turned out to be more difficult than it at first appeared. The problem is that basic concepts like the definition of an innovation are neither clear nor widely agreed upon. Each researcher must therefore come up with his or her own working definitions. The consequence of this is that the research methodology has a major impact on the findings – not an unusual situation but too often underestimated. To allow for a better understanding of the findings presented here, and some of the limitations of the study, important aspects of the research methodology are discussed in the next section.

What is an innovation?

A prerequisite for collecting information on farmer innovation is knowledge about what an agricultural innovation is. No generally accepted definition is used by scientists and extensionists involved in agricultural research and development. Also the legislation on patents and copyright provides no rigid or universal definition of an innovation. Besides that, how do farmers define innovation? In each of the four language areas covered in this study, I could find no single word in the everyday language that could be used to stand consistently for 'innovation'.

For the purpose of this study, I therefore chose to define an agricultural innovation as 'a new thing or method used in farming'. To establish a common reference point for discussing innovation with farmers in the four language areas, I worked out an 'introductory speech'. After testing and several revisions, this is what the interviewers explained to each farmer at the commencement of the interview:

> *An innovation is something new. For example, it can be a new maize variety, composting, use of new tools, line planting instead of broadcasting or a new combination of crops. Some innovations come from outside, like chemical fertilizer, while others are developed by farmers themselves, like herbicide made from local plants. We are interested in both innovations that you have made yourself and innovations that you got from elsewhere. We also call it an innovation if you try new planting times or change the spacing of crops compared to what you used to do. So an innovation is anything new you are doing in your farm.*

This explanation was followed by several guide questions, such as: 'What have you done to improve your farm over the last year?' 'Did you try new ways of preparing the land?' 'Did you try new times of planting?'

The field-testing included a translation from English to the local language by one translator and a back-translation by another translator. In this way, many distortions of meaning caused by words that have almost, but not completely, the same meaning in different languages could be captured.

Who identifies an innovation?

Defining an innovation as 'something new' leads to another question, namely: 'New to whom?'. For instance, a farmer may experiment with early planting without knowing that other farmers in the area have done similar experiments. As this study was focused on innovation from a farmer's perspective, what was new to the farmer being interviewed qualified as an innovation.

THREE STUDY SITES

The research was carried out from 1998 to 2000 in collaboration with the African Highlands Initiative (AHI).[1] This meant that the initial choice of sites was limited to the nine AHI benchmark sites in East Africa. These had been selected so that, taken together, they would represent the natural and socio-economic conditions found in the major parts of the Central and East African highlands. Fortunately, each benchmark site covers an area that is large enough to prevent researchers from stepping on each other's toes but small enough to make comparison of data meaningful.

I chose one primary site for in-depth research and two secondary sites for testing to what extent findings from the primary research site were valid elsewhere. The primary research site was in Western Kenya (Vihiga, Siaya, Kakamega and Butere/Mumias districts) and the secondary sites were in central Kenya (Embu district) and in the Usambara Mountains (Lushoto district) in Tanzania.

Western Kenya was traditionally the bread-basket of the nation, but high population growth has turned the area into a labour reserve. Typically, only women, children and old people live permanently on a farm while the able-bodied males work in towns elsewhere in the country. Agricultural practices, crops and animals are often traditional and the use of modern agricultural inputs is minimal. Market access and the availability of agricultural services is poor. Using a grid sample plan, 221 farmers were selected for interviews in a 20km^2 area that covers two major ethnic groups, the Luo and the Luhya. A further 203 farmers were sampled from five villages, one at each corner of the 20km^2 area and one village in the centre of the area.

The site in central Kenya is on the slopes of Mount Kenya. The area has high agricultural potential with fertile soils, sufficient water – partly from streams – and farms large enough to support the families living there.

Compared to Western Kenya, the use of modern inputs in Embu district is widespread, market access is good and the standard of living is high. Here, 40 farmers were interviewed.

In contrast to the sites in Kenya, the one in the Usambara Mountains of Tanzania is relatively isolated. Soil degradation is a problem, and the standard of living is low. Despite this, a pattern of male out-migration, as in Western Kenya, has not emerged. At this site, 41 farmers in one village were interviewed.

The data in this chapter are based on findings from all three sites, with the exception of those referring to ethnic group and innovation, for which only the data set from Western Kenya was comprehensive enough to draw conclusions.

VARIETY OF INNOVATIONS AND VARIATIONS IN INNOVATIVENESS

The 505 farmers interviewed listed a total of 1614 innovations that they had initiated during the previous 12 months (see Table 8.1). Growing new crops – often new varieties of crops already cultivated by the farmers – was the most common type of innovation by a large margin (44.3 per cent of all innovations). It is interesting to note that, particularly in Western Kenya, what farmers called 'new crop varieties' were, in some cases, traditional varieties that were being revived. A case in point is maize: high-yielding varieties (HYVs) of maize have been popular for many years, but the declining economy and removal of subsidies on external agricultural inputs have forced many farmers to move back to low-external-input systems in which the traditional varieties perform better than the HYVs.

Table 8.1 *Types of innovation during the previous 12 months on 505 farms in Kenya and Tanzania*

Innovation category (n=1614)	% of n
New crops	44.3
Soil fertility	18.3
Land preparation and planting	10.9
Pest and disease control in animals	6.5
Erosion control	5.5
Storage	5.1
New fodder	3.2
Crop pest, disease and weed control	3.1
New livestock species or breeds	2.0
Water control	0.4
Harvest and post-harvest processing	0.3
Other innovations	0.4
Total	100.0

Measures to improve soil fertility, such as the application of manure and compost, accounted for less than 20 per cent of the innovations. Trading in manure was observed at the three sites and, particularly in Tanzania, the importance of manure for maintaining soil fertility was mentioned often.

All the innovations mentioned by the farmers were, in fact, minor changes to existing farming systems and practices. The farmers who introduced new animal species into their farms for the first time probably diverted most greatly from their past practices, particularly in those cases where a farmer who had previously kept no cattle acquired a cross-bred dairy cow and kept it under a zero-grazing regime. However, since such farmers were copying the practices of neighbours to a large extent, this can hardly be termed a radical innovation for the area. Making minor changes to what is already known is a wise strategy for avoiding excessive risks and, over time, hundreds or thousands of minor changes can lead to a considerable overall change.

Some development practitioners and theorists assume that a small minority of farmers are actively innovating and that the majority just copy the successful innovations made by this minority. If that were the case, then one would expect that most of the 1614 innovations recorded in this study would be carried out by a fraction of the 505 households. However, as is clear from the distribution of innovations shown in Figure 8.1, practically every household is innovating. Thus, the variation between households is much smaller than many observers would expect. Nevertheless, there is some variation, as the number of innovations per household during the previous 12 months ranged from zero to ten.

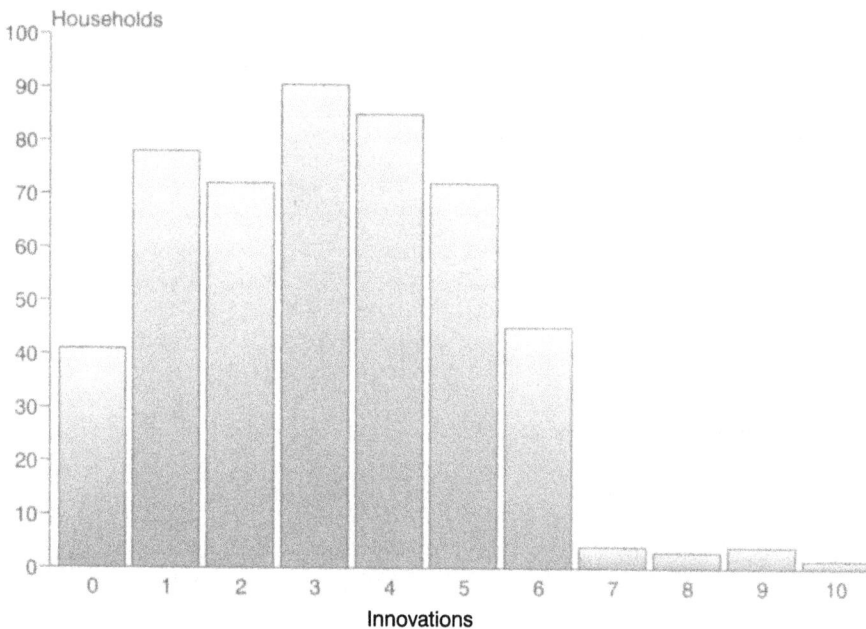

Figure 8.1 *Number of innovations by number of households over the previous 12 months*

EXPLAINING INNOVATION: THE EMIC VIEW

In anthropology, a distinction is made between the view from outside – the etic view – and the view from within – the emic view. Case studies usually capture the actors' point of view, ie they have an emic approach. Surveys are often made to gain an etic view. For instance, a survey may be made to measure the statistical correlation between education and the use of computers without asking the surveyed people their opinions about this relationship. Both approaches have their merits and each can be used to reveal different types of information.

In this study, only an emic approach could reveal what motivates farmers to innovate and what they see as major obstacles to innovation. However, an etic approach may reveal the importance of factors that the individual farmer cannot easily observe, such as the relationship between innovativeness and gender, age, distance to services, etc. This section of the chapter presents an emic view, ie farmers' views on the reasons for innovation and obstacles to innovation.

Why innovate?

For each innovation listed by the farmers, they were asked about the reason(s) for the innovation. The 2028 answers recorded are categorized in Table 8.2. The main reason why farmers innovated was to provide food for their family's own consumption, closely followed by innovations to increase the household income and those aimed at maintaining or increasing soil fertility.

Table 8.2 *Farmers' reasons for their innovations over the previous 12 months*

Reason for innovation (n=2028)	% of n
Own consumption/food security	17.2
For sale	13.4
Increase/maintain yield	12.9
Increase/maintain soil fertility	9.3
Adapt to soils	5.3
Prevent/cure animal diseases	5.1
Fast growth	4.7
Control erosion	4.5
Preservation during storage	3.6
Provide fodder	3.4
Pest/disease/weed control in plants	3.1
Easy weeding	2.7
Improve/maintain milk production	1.4
Lack of land/improve use of land	1.4
Greater drought resistance	1.3
Better taste	0.9
Other reasons	9.5
To test its performance/to try out	0.3
Total	100.0

Only six innovations were undertaken out of curiosity without any particular goal in mind. In other words, curiosity experiments do not appear to be very common among these farmers. All the innovations tried out of curiosity were concerned with testing new varieties of maize and beans. Thus, they were not of a different type than innovations undertaken for other reasons than curiosity.

Why not innovate more?

The farmers were asked to list all the obstacles they faced in trying to innovate and then to rank them using a traditional Bao game. The game board has two rows of holes. For each obstacle the farmer mentioned, a card was made and placed next to a hole. When all the obstacles were listed, farmers placed beans in the holes according to how important s/he found the obstacle. The scores are weighted so that each farmer has an equal say in the combined scores shown in Table 8.3.

Many farmers explained that they had particular innovations in mind with which they wanted to experiment, but lack of money prevented them from going ahead. Other farmers stated that they had the money to buy what they needed for the innovations but lacked the land for experimentation. Farmers at all three sites regarded theft as a problem. It was not unusual to hear about a farmer who had planted new seeds or seedlings only to find them stolen or destroyed during the night. In Western Kenya, where it is common that women manage the farms while the men work elsewhere, some of the women said they were keen to try out various innovations but could not begin before their husbands had approved. It is interesting to note that over 10 per cent of the farmers interviewed mentioned a lack of ideas or knowledge of new things with which to experiment.

Table 8.3 *The obstacles that farmers see to innovations*

Obstacles to innovations (scoring by farmers, n=502)	Importance (average score in % of n)
Lack of money	39.1
Lack of land	11.6
Theft	10.6
Lack of new things/ideas to try	10.4
Lack of labour	8.8
Wild animals and pests	7.5
Drought	6.6
Lack of time	2.9
Absent husband has to approve	1.4
Lack of working tools	1.1
Total	100.0

EXPLAINING INNOVATION: THE ETIC VIEW

Several factors that might influence the number of innovations at household level were included in this study, such as level of education, size of household, amount of land available, age of household head and degree of contact with other areas. Interestingly, most of these factors were not mentioned by farmers. This indicates that an etic approach can give additional insights that an emic approach does not capture. Not all the factors covered in the study have been analysed thus far. This section focuses on the relationship between innovativeness and gender, farm size, off-farm income, age and ethnic group.

Gender

In total, 211 of the 505 households surveyed were headed by women. The household was considered to be female-headed if the man was working elsewhere and the woman was the main day-to-day decision-maker. Female-headed households are particularly common in Western Kenya where many men work outside the area and return home only once or twice a year.

In this study, no statistically significant difference was found in the number of innovations carried out by male- versus female-headed households. Actually, the female-headed households had made an average of 3.3 innovations during the previous 12 months, compared to 3.1 innovations by male-headed households, but the difference is far from significant (*t*-test probability of 0.297959).

Size of farm

A correlation between farm size and innovativeness could have been expected for a number of reasons. For instance, owners of big farms are often rich, have access to more resources, including information, and can better afford failed experiments. However, no correlation between farm size and innovativeness was found (Pearson $r=0.04$, $p=0.391798$).

Off-farm income

Off-farm income plays a role for most of the households surveyed. In some cases, household members worked elsewhere and sent money home, or people living on the farm worked seasonally for other farmers or on plantations. It was difficult to predict what effect the availability of off-farm income would have on innovativeness by farmers. On the one hand, farmers who engage in off-farm work have less time for farming and may consider farming less important but, on the other hand, they probably have more access to information and resources than people who work full-time on their farms.

In this study, the households with off-farm activities as the main source of income were found to be more innovative than those depending mainly on income from the farm (4.1 versus 3.6 innovations over the previous 12 months, *t*-test, $p=0.025232$).

Table 8.4 *Number of innovations and innovative households according to ethnic group*

Ethnic group	No of innovations in previous 12 months	No of households
Luo	2.6	269
Luhya	4.7	155

Age

As had been expected, the study revealed that the level of innovation was lower among older heads of household, but the drop was marked only among farmers in their late 60s. The peak in innovativeness was found among farmers in the age bracket of 35–40 years.

Ethnic group

The most controversial finding is the strong correlation between ethnic group and innovativeness found in Western Kenya, the primary study site that covered 20km² and included areas inhabited by the Luo and Luhya ethnic groups.

As can be seen in Table 8.4, the Luhya appear to be much more innovative than the Luo. How can this big difference be explained? First, it must be examined whether methodological problems led to the difference. Because the languages of the two ethnic groups differ greatly, different field assistants were needed to work in the two areas and they may not have been equally thorough in their work. Several field assistants were used in both areas, and we could find no differences in the overall findings within the Luo and Luhya areas respectively that correlated with the different field assistants. Therefore, this could be ruled out as an influencing factor.

A mapping exercise showed that areas with 3.5 and less innovations per household are those settled by Luo farmers and the areas with more than 3.5 innovations per household are those settled by Luhya farmers. The correlation between settlement area and number of innovations is evident. These findings were fed back and discussed with Luo and Luhya farmers and with extensionists and researchers working in the area. Everyone agreed that the findings were in line with their own observations.

Particularly farmers and extension staff gave explanations focused on the remnants of the traditional organization according to age groups in the Luo society. Each age group had certain rights and obligations. Today, this cultural system survives to the extent that young family members are discouraged from doing things that older brothers or relatives have not done. Many stories were told of young Luo farmers who wanted to develop their farms but did not dare to advance further than their older brothers. Other people told of Luo men who had become rich in Nairobi or elsewhere and wanted to develop farms back home in Luoland but could not do so because they would surpass their older brother's achievements. Some of them have bought large farms instead in the Rift Valley or elsewhere, while they maintain their small traditional farm in Luoland. The Luhya farmers do not experience similar restrictions.

Other explanations focused on attitudes to agriculture. Many Luo farmers told how they consider farming to be a last option of which they were not very proud. Education and work in other areas is the preferred option. Only if this failed would they return to farming. In contrast, Luhya farmers were proud of their farms and did not consider farming to be an inferior occupation.

CONCLUSIONS

Many case studies, such as those in this book, have shown that some farmers do indeed experiment and innovate in very creative ways. The study presented here tries to go further in order to see the general picture, ie what is the typical innovative behaviour, instead of focusing on a few outstanding experimenting farmers.

It was found that people in almost every household innovate and there was no group of farmers that stood out in terms of their experimentation. This is an important finding for interventions aimed at promoting and supporting innovation by farmers. There is no need to make great efforts to identify a small exclusive group of farmers who, as is often assumed in development theory, are experimenting and showing the way to the rest of the community. Such an exclusive group does not appear to exist.

Numerous innovations were found but all reflected minor changes and attempts to optimize existing practices. It would be too hasty to conclude that farmers are interested only in minor innovations. They follow a low-risk approach out of necessity and, given circumstances where they can experiment without bearing the full risk, they might be interested in trying out radically new things.

Innovations made out of curiosity were indeed taking place, but were very rare. Innovations were generally undertaken with the goal of improving the standard of living or well-being in a particular way. This utilitarian attitude may make it more difficult for scientists to involve farmers in more basic research as opposed to adaptive research.

Some farmers indeed felt that lack of money or lack of land constrained their capacity to innovate, but a major problem in their eyes was also a lack of information about new techniques with which to experiment. This is already widely acknowledged by extension services, and most development interventions are aimed at providing farmers with new information. However, too often information is passed down to farmers as rigid prescriptions, disregarding farmers' desire to adapt, combine and modify inputs to suit their needs. The farmers need a buffet of choices, not a standard meal.

Theft is another problem that almost every innovator seemed to be facing. In the morning, some of them found their new crops uprooted or their new animals stolen. This problem is so widely reported in the study areas that any programme to promote farmer innovation should consider how to deal with it. One way, for example, might be to encourage community instead of individual ownership of local experiments with new technologies, or at least group planning and evaluation of experiments to create a sense of ownership.

Female-headed rural households are very common today, especially in Kenya, as men often work elsewhere because farming can no longer sustain the family. Some women reported that they find it constraining to have to wait for their husband's approval for new initiatives if he visits the farm only once or twice a year. Nevertheless, female-headed households were found to be just as innovative as the male-headed ones, so it appears that many women are finding ways to get around this constraint.

In this study, the size of the farms did not correlate with the number of innovations, a finding that may surprise some people who link farm size with wealth and this, in turn, with the ability to innovate. Poor people may not have the resources to buy, for instance, a cross-bred dairy cow, but this does not mean that they do not innovate at all. The study indicates that such farmers are simply undertaking other innovations within their ability and resource limits.

It was also interesting to note that the level of innovation appears to be fairly constant through most of the productive life of farmers and that they become less innovative only when they become quite old.

The finding that ethnic group was the factor that showed the highest correlation with level of innovativeness highlights how important cultural factors are for issues that outsiders often reduce to a technology-related problem, such as lack of capital or information. In this case, making credit facilities available or improving the delivery of information about techniques is unlikely to have much impact unless some of the fundamental cultural issues are also addressed. In the case of the Luo, for example, interventions may have to focus more on working with people from older age groups, either to convince them to undertake innovations themselves or to allow them to retain control of testing new ideas by delegating the task to younger people in the community.

NOTES

1 The African Highlands Initiative is an agricultural research programme in East Africa that involves several international agricultural research centres, regional research institutions, commodity networks, universities and NGOs, and receives technical support from ICRAF (International Centre for Research on Agroforestry)

REFERENCES

Sumberg, J and Okali, C (1997) *Farmers' experiments: creating local knowledge*, Lynne Rienner Publishers, Boulder

9

Chain of innovations by farmers in Cameroon

*Paul Tchawa**

In the highlands of North-west Cameroon, one farmer tackled soil fertility problems by developing a system of keeping cattle overnight on fields to be cultivated. This set off a chain of innovations by other farmers, such as a new tool to handle the increased harvests, an irrigation system to take fuller advantage of the fertilized areas and growing fodder for the cattle. It also improved relations between farmers and herders.

NIGHT PADDOCK MANURING

Samuel Toh's farm in Upper Babanki is almost 2000m above sea level and receives about 1500mm average annual rainfall from May to September. The population density is about 150 persons/km². The 'grassfields' where Mbororo Fulani pastoralists keep their cattle lie above the farming areas. In the early 1980s, Toh saw that his soils were becoming poorer and that, with population growth, there was less space for the traditional long fallow to replenish soil fertility. In the surrounding hills grazed by Mbororo cattle, he began to collect manure and transport it in jute sacks to his field. As this was strenuous work, he decided to ask a Mbororo herder to bring his cattle to spend the night over three to four weeks in a field around which Toh built a wooden fence. Toh then cultivated the fertilized area. The bumper harvest showed that his new system of night paddock manuring worked.

Over time, Toh improved his system. For example, he noticed that the animals tended to concentrate in one corner of the field and the manure was

---

* Paul Tchawa, a geographer and senior lecturer at the University of Yaoundé I, is coordinator of ISWC-Cameroon based at SNV in Yaoundé

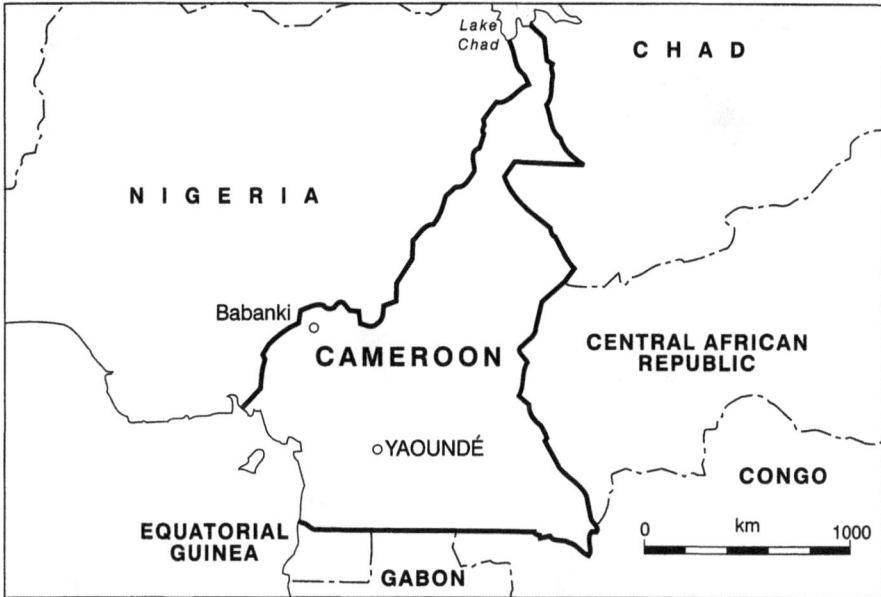

Map 9.1 *Cameroon*

not well distributed. He subdivided the paddock and the cattle were moved each night to different subdivisions.

Toh's innovation met with extraordinary success. One plant in particular is regularly grown after manuring: black nightshade (*Solanum nigrum*), the leaves of which are eaten like spinach and are highly appreciated in North-west Cameroon and in the cities of Yaoundé (Central province), Kumba (South-west) and Douala (Coastal). Almost all farmers in Upper Babanki (more than 500 families) have adopted the night paddock manuring system, and a stream of traders in 'bush taxis' weave through the villages to collect the leaves and take them to the city markets. Usually, the farmers grow nightshade for two years and then maize for another two years, after which the cattle come back to manure the fields again.

Besides bringing direct benefits in terms of income, the innovation has borne other fruits; other innovations that would not have seen the light of day had Toh not developed the night paddock manuring system.

NEW HARVESTING TOOL

With better soil fertility, farmers were faced with about five times as many nightshade leaves to harvest several times per season. The work of breaking off the stems by hand had become very hard. In the early 1990s, another local farmer, Philip Ndong, tried to harvest with a knife, but was not satisfied as it was not sharp enough. Moreover, because the women and children joined in harvesting, several knives were needed and these were costly.

Plate 9.1 *Philip Ndong invented a simple and low-cost harvesting tool (Cameroon)*

Ndong then tried using a razor blade held directly with the fingers. This cut the stems better, but also often cut his fingers. He therefore took a piece of bamboo about 20cm long and attached the razor blade to the end. After trying out several types of blade, he settled on one with three holes which could be fixed well with thread to the bamboo. With this tool, which costs less than 25CFA (circa FF0.25 or US$0.04), the price of a razor blade, the nightshade leaves can be cut quickly and efficiently and, because the stems are not damaged, leaf regrowth is stimulated. Neighbours were sceptical at first, but now all nightshade growers in the area use Ndong's innovation. It spread spontaneously.

IRRIGATION IN RESPONSE TO MARKET DEMAND

At this point, another farmer in Babanki, Christopher Vitsuh, noticed that the market demand for nightshade leaves was not satisfied in the dry season, when the price is three times higher than in the wet season. This inspired him to conceive a system of irrigation so that he could produce nightshade leaves in the off-season.

Since the 1960s, small canals are dug in the Babanki area to conduct water towards brickmaking yards. In 1986, Vitsuh thought of using the same technique to lead water to his farm. The night paddock manuring system had greatly increased nightshade leaf production in the wet season and the fertility could still be used in the dry season, but water was the missing factor. Vitsuh started a small gravity irrigation system which expanded as additional families wanted to be connected to it. In 1999, the system was irrigating more than 10ha and over 40 farm families were benefiting from it.

When Vitsuh had first conceived his idea, he contacted some advisers in water engineering. After examining the site, these experts estimated that it would cost six million CFA (circa FF60,000) to set up the system. As Vitsuh could not afford this, he had the choice of giving up the idea or working out something himself. He did the latter, and his initial network of 5km of canals cost him only 110,000 CFA (circa FF1100).

To start the process of making canals, Vitsuh identified streams that could be diverted. Depending on the location of the plots of farmers involved in the work, the most appropriate routes for the canals were chosen. As the land is prone to erosion and the canal sides could cave in, the farmers planted live hedges to stabilize them. When they had to cross a deep gorge or major watercourse, they used hollowed-out logs as pipes to link the two steep banks.

SOCIAL INNOVATION AND MUTUAL INSPIRATION

The new technology also led to social innovation. The community set up a committee to manage the distribution of water to the different plots and to resolve possible conflicts. Water is distributed on the basis of strict rules that the farmers set themselves; non-respect of the rules is punished by a fine. A farmer who has not contributed to digging the canals must give the management committee 20 litres of palm wine, a basket of maize flour and a cock in order to gain the right to irrigate his or her plot.

This innovation is characterized by the gathering of people around a common problem, the simplicity of the means used and its great potential to improve income. There was no outside intervention in building and managing this new irrigation system. Farmers in other parts of the village still seek the innovator's support to be linked to the network of canals. Vitsuh conveyed this request to the ISWC-Cameroon programme. As a result, for the first time a land surveyor has joined Vitsuh to survey the entire system and to help him to improve and extend it.

This case shows that, as isolated as some farmers' innovations may seem at first glance, there may be close and logical connections between them. In Babanki, one innovation triggered the next and the next (see Box 9.1). The explosion in nightshade production led to a high demand for cattle manure and a more than twofold increase in the number of cattle kept in the area over the past 20 years. To reduce costs (in terms of materials and time) of enclosing the animals overnight, some farmers have begun to experiment with live fencing. Under the contract with the herders, the farmers have to feed the cattle during one month and some have started to plant fodder grasses.

The relationships between the sedentary farmers and the mobile Mbororo herders used to be tense because the cattle sometimes damaged the crops and

---

BOX 9.1 CHAIN OF INNOVATIONS BY BABANKI FARMERS

- Night paddock manuring system
- Contracts between farmers and herders
- New harvesting tool
- Irrigation system
- Live fences for paddocks
- Growing fodder grasses

---

the farmers expanded their fields into grazing areas. The contracts between the Babanki farmers and the Mbororo herders for enclosing the cattle overnight on farmers' fields for a month of each year has improved the relationships between the two groups.

It is also interesting to note that the links between innovations also link the innovators, who admire and respect each other. The development and mastery of an innovation by one person stimulates others. As a farmer in Babanki said: 'After fertilizing a patch of ground, you lose a lot if water cannot reach it.' The farmers obviously do not regard these innovations separately. It is not surprising, therefore, that Samuel Toh, Phillip Ndong and Christopher Vitsuh have supported each other actively in developing their innovations in fertilizing, harvesting, irrigation and still more.

SPREADING INNOVATION

Relationships between Mbororo pastoralists and sedentary farmers in other parts of North-west Cameroon are often tense. With the help of CIPCRE (Cercle International pour la Promotion de la Création), study visits to Upper Babanki have been organized for several villages where the situation is particularly tense. For example, in July 1998, a delegation from Mbiame village made a 'look and learn' visit to Upper Babanki. The group of 14 persons was made up of representatives of the Mbiame administration, the sedentary farmers and the Mbororo. They heard explanations made by the Mbororo from Babanki and Samuel Toh about the benefits of their collaboration. Once back in their village, they organized a feedback meeting for the rest of the village farmers and pastoralists who attended in great numbers. The chief of the Mbiame Mbororo, who had taken part in the visit to Upper Babanki, immediately decided to put a herd of 150 head of cattle at the disposal of the Mbiame farmers. During the meeting, the responsibilities of each party were defined and several groups immediately started to fulfil the conditions for receiving cattle for night paddocking (Mbafor, in Tchawa, 1999). This is a promising step towards a better understanding between herders and farmers.

REFERENCES

Tchawa, P (ed) (1999) *Rapport annuel CES 2 Cameroun*, SNV/CES 2, Yaoundé

Women and innovation: experiences from Promoting Farmer Innovation in East Africa

*Milcah Ong'ayo, Janet Njoroge and Will Critchley**

The primary aim of PFI in East Africa is to increase diffusion of appropriate techniques of SWC and other forms of natural resource management in order to enhance agricultural production on a sustainable basis. The approach is intended to benefit rural women and men equally but, at the beginning, it proved difficult to involve women. This chapter explains how gender analysis studies and sensitization workshops helped the project to come closer to a gender balance.

Promoting Farmer Innovation has operated since 1997 in Kenya, Tanzania and Uganda. The project document states that, by the end of three years, 500 farmers in each country will have adopted improved SWC techniques. It specifies, furthermore, that half of these farmers should be women. This is for both ethical and practical reasons: ethical because it acknowledges that women and men have the same development rights, and practical because it recognizes that both sexes are crucial actors in development.

However, the project team soon realized that the prevailing sociocultural and socioeconomic factors made it difficult to identify and involve an equal number of male and female innovators in the project activities. This was partly due to the staff structure in the partner organizations (both governmental

* Milcah Ong'ayo is a gender specialist working as a freelance consultant with a Nairobi base; Janet Njoroge is based at the UNSO/UNDP office in Nairobi and assists in managing PFI, with a particular brief for gender issues; Will Critchley is an agriculturalist with CDCS, Vrije Universiteit, Amsterdam, and advises the PFI and ISWC-Uganda programmes

Map 10.1 *PFI action areas in Kenya, Tanzania and Uganda*

extension services and development NGOs). It was particularly difficult to find female extension agents at divisional and village level. Three challenges emerged that had not been anticipated adequately during project planning:

1 to ensure that women were represented adequately among the innovators identified and among the end beneficiaries of the project;
2 to identify innovations that were attractive to women; and
3 to build capacity to deal with gender issues among both female and male staff of the partner organizations.

EARLY DAYS: GENDER IMBALANCE

The initial focal areas of PFI were in Mwingi district in Kenya, in Dodoma Region in Tanzania, and in Soroti, Kumi and Katakwi districts in Uganda. The workshops on PFI methodology, which were held in each country at the start of the project, dealt with the topic of identifying farmer innovators (FIs) in some detail but in an essentially gender-neutral manner. When PFI commenced field activities at the end of 1997, it was immediately clear that many more men than women were being identified as FIs. This was true for each of the three countries, although the Tanzania team came closest to a reasonable balance because of its somewhat more gender-sensitive approach. Overall, only 19 per cent of all FIs identified were women. The predominance of male

FIs could have meant simply that more men were innovating than women, but reflection within the project team led to the feeling that female FIs were being overlooked. The critical questions therefore were:

- Was there a general gender bias in the project approach in favour of male farmers and innovators?
- If there was indeed a bias, how could this be corrected?

PFI has focused on identifying innovators in land husbandry. We should not assume that women innovate exclusively (or even mainly) in the household and that therefore few innovations by women in land husbandry will be identified. We hope to help reduce women's workload in the home, but we should not make simplistic assumptions about what activities women will, or will not find of interest – that should be their choice. Therefore, the project felt it should give women the opportunity to show and express their capacities and interest in activities related to land husbandry.

GETTING TO THE ROOT OF THE PROBLEM: GENDER ANALYSIS

During the course of the first year, PFI decided to hire consultants to carry out gender analysis studies in each country. Table 10.1 presents the schedules of the gender studies and subsequent gender sensitization workshops and the names of the lead consultant in each country.

The two main aims of the gender studies were:

1   to establish the relative roles of women and men in the rural economies in the focal areas; and
2   to recommend how PFI could become more sensitive to gender issues.

With respect to the first objective, the studies did not reveal any great surprises. While both women and men are involved in agriculture and both play an important role in SWC and land management, the decisions about these activities are made basically by men. Under the traditional division of labour, women and girls have a much heavier general workload than do men and boys because of the additional responsibilities for family maintenance. It was also

Table 10.1 *Gender studies and sensitization workshops under PFI*

Activity	Kenya	Tanzania	Uganda
Gender analysis study	October 1998 by Milcah Ong'ayo	November 1998 by Pendo Nyanda	March 1999 by Daisy Eresu
Gender sensitization workshop	March 1999 by Milcah Ong'ayo	August 1999 by Pendo Nyanda	July 1999 by Daisy Eresu

observed that the women's contributions to land husbandry were not fully recognized by either the male farmers or the (primarily male) extension staff. This may be one reason why women were being overlooked when FIs were identified.

With respect to the second objective, a number of recommendations were made both for addressing gender-related constraints at the household level and for improving the capacity of PFI and its partners to deal with gender issues. The findings and recommendations can be divided into inherent problems and those associated with the project's activities, as shown in Table 10.2. This table also includes gender-related findings and recommendations of an external review of PFI carried out in August 1999 by Dr Donald Thomas and Dr Bancy Mati.

MOVING FORWARD: GENDER SENSITIZATION WORKSHOPS

Some of the recommendations were very general and some (eg seeking ways to lighten the women's general workload) were clearly outside the capacity of a small project like PFI. However, other problems were quite specific and practical remedies could be found. Some of these problems were addressed in the sensitization workshops that were subsequently held in each of the three countries. These workshops were targeted specifically at staff members from government agencies and NGOs who were directly or indirectly involved in implementing PFI. In Uganda and Tanzania, FIs were also involved (three men and two women in Uganda, four men and two women in Tanzania). The project team was surprised to hear that few of the participants had received gender training previously.

The simplest way of seeking evidence of the impact of these workshops is to look at the change in gender balance of FIs identified in the three countries. A policy of affirmative action was put into place: women innovators were specifically sought. There has indeed been a significant improvement: the proportion of female FIs increased on average from 19 per cent (mid-1998) to 33 per cent (end of 1999). Details are given in Table 10.3. It should be noted that the 'before' numbers differ slightly from those presented in Critchley et al, (1999) because of further identification and screening after that publication was prepared.

Three things should be noted when looking at these figures. First, it is culturally more difficult in some areas to gain access to women, whether they are innovators or not. In Uganda, for example, this was much more of a problem for PFI in the north-east (Soroti, Katakwi and Kumi districts) than for the ISWC 2 programme in the south-west (Kabale district). Secondly, the true split between male and female FIs is still unknown. It certainly should not be automatically assumed that it is 50:50. To some degree, the gender studies highlighted the fact that men have more means at their disposal to innovate

Table 10.2 *Findings and recommendations of gender studies and external review of PFI*

Findings	Recommendations
Inherent problems	
Unequal division of labour and heavy workload of women and girls, who therefore: • have little time to innovate • cannot take part in other strategic community activities	Seek ways to lighten workload; create awareness about benefits of sharing work within the family; promote labour and time-saving innovations
Women have limited control over resources and the benefits of their labour; this has a negative effect on: • their motivation to innovate • food security at family level	Create gender awareness in the community; promote innovations most relevant to women
Men dominate decision-making at family and community level; this implies: • sociocultural barriers to women's participation in project activities • lack of women's exposure to/experience with new ideas • lack of women's authority to decide	Target women for training; create gender awareness in the community; work with women's groups to strengthen women's capacity
The position of women is subordinate; they are generally: • less mobile than men • less self-confident • perceived negatively by men	Identify more women FIs; target women for training and tours; create awareness among men of women's contribution to development; build women's confidence through training and participation in innovator groups
Project-associated problems	
Implementing staff members lack gender sensitivity	Conduct gender sensitization workshops for staff of partner organizations
Low level of participation of women in PFI activities	Identify and work with women's groups; increase number of women FIs involved; organize training at appropriate times and venues
Inadequate focus on female-headed households	Identify and work with female-headed households as innovators and adopters
Limited number of female extension staff	Identify female contact persons in the communities
Lack of consistent gender strategy in PFI and partner organizations	Review criteria for and methods of identifying FIs; develop guidelines for PFI and partners
Inadequate focus on women's innovations	Identify and promote innovations relevant to women; help female FIs spread their innovations
Inadequate gender capacity of partner organizations of PFI	Look for gender-sensitive partners where possible; sensitize selected contact persons on gender issues

Table 10.3 *Gender ratio of farmer innovators in PFI before and after gender sensitization workshops*

Country	Number of farmer innovators identified and verified (and % women)	
	Before gender sensitization	After gender sensitization
Kenya	34 men/6 women (15%)	34 men/16 women (32%)
Tanzania	16 men/8 women (33%)	36 men/24 women (40%)
Uganda	13 men/1 woman (7%)	20 men/5 women (20%)
Total	63 men/15 women (19%)	90 men/45 women (33%)

than women, ie they control more of the family's overall resources. Thirdly, as a result of affirmative action to recruit more women, some of the women 'innovators' identified after the sensitization workshops turned out to be, in fact, adopters of innovations they have seen under PFI rather than genuine creators of new technologies. They were effectively 'potential innovators'.

It was interesting to note that, when the types of innovations in land husbandry developed by women were compared to those developed by men, they did not appear to differ greatly. Nevertheless, the women's innovations often demand considerable labour. For example, Mrs Grace Bura in Tanzania builds barriers made of earth and brush in order to block gullies; Sister Martha Mwaso, also in Tanzania, has reclaimed large sections of river-bed through labour-intensive efforts; Mrs Kalekye in Kenya excavates large quantities of earth to reclaim gullies for agricultural production. Other examples are given in the boxes in this chapter.

Signs of Change

Shift towards better gender balance

Various other changes have taken place since the sensitization workshops. These were documented in the course of a follow-up impact assessment study (Ong'ayo and Njoroge, 2001). The major new developments in each country that showed a shift in gender balance were as follows:

Kenya

- A female contact person (the District Home Economics Officer) was identified and seconded to work with PFI; this facilitated the work with women and strengthened the gender approach of the project.
- Training was decentralized to enable more women to participate and several women's groups were mobilized to visit FIs; by mid-2000, 25 women's groups had participated in training and visits under PFI.
- Many women have evidently adopted innovations they had visited, though this has yet to be quantified by an impact assessment.
- The extension staff began to use a family-oriented approach to involve all members of the farming households in the PFI activities.

BOX 10.1 KAKUNDI KITENG'U AND HER SUGARCANE PITTING TECHNIQUE IN KENYA

Kakundi Kiteng'u is about 50 years old and lives in a dry zone of Mwingi district in Eastern Kenya where the annual average rainfall is less than 500mm. Her farm of just over 20ha is larger than the average for the area and she owns several head of cattle. She is just one example of many women innovators discovered by PFI who have developed ingenious innovations and has thus managed to ensure a regular income. She gained her inspiration by observing the movement of the water table in holes dug in the banks of a sand river. Kakundi decided to plant sugarcane cuttings in wide, deep holes (about 1m deep) just above the moisture line so that the roots can reach the water. The cuttings thrive even in the driest of seasons. As the growing season progresses, the plants become taller, reaching well out of the pits. Their roots can resist temporary flooding. The holes gradually fill in with sediment. Kakundi sells the sugarcane locally where it commands a good price and provides her with an estimated income of KSh40,000 (US$570) per annum. She has been instrumental in spreading her pitting technique to several other farmers; she estimates about 50, both men and women. Despite being only partly literate, she has begun to keep simple records with the help of her husband and children. She measures the inputs and outputs related to her innovation and compares this with an adjacent control plot where she has planted cane in the locally common way.

- Participation of women in visits to FIs has increased: the latest calculation (presented at PFI Uganda's Review Workshop in Jinja, December 2000) was that, of the 2500 farmers who have made such visits, approximately 1600 are women.
- Greater attention was given to female-headed households (many of the newly recruited FIs fall into this category) and youth (35 young farmer groups have been taken to visit FIs).
- A greater percentage of women (from the mandatory 25 per cent up to 60 per cent, in some cases) were elected to soil conservation committees at village level.

Tanzania
- Husbands have reportedly allowed wives to join PFI activities.
- Extension staff began to recognize explicitly 'family' innovations.
- The dissemination activities of PFI gave emphasis to women farmer groups.
- Women were given greater general recognition as innovators: eg village governments began to work with female FIs in mass mobilization activities to promote better land husbandry.
- Leadership in FI groups and networks was restructured to allow women to participate.

BOX 10.2 FLORENCE AKOL HARVESTS WATER FOR BANANAS IN UGANDA

Florence is a 40-year-old farmer and a housewife with a family of 12 to support. She has a tiny holding of land – about 1ha – and has only a few goats. She lives in Kumi district in northern Uganda where the rainfall varies between 500mm and 750mm rainfall per annum. She is also the dynamic chairperson of a recently formed women farmer's association, the Atege Innovators' Farmers Association. Her main innovation, which she started in 1990, is a combination of harvesting water and improving soil fertility in a *matoke* (cooking) banana plantation. Florence collects water running off the road and leads it into her plantation. She has created a system of trenches to allow the water to circulate through the plantation and to flow into basins around the individual banana stools. She also mulches and plants grass barriers in the plantation. There is some doubt whether the water harvesting can be claimed as her own innovation as there are variations of this practice in several nearby farms. Nevertheless, her holistic management system is probably unique in the area. Florence has been visited by a study tour group of farmers from the ISWC-Uganda programme in Kabale. She received advice from those farmers in banana stool maintenance. She has attended various PFI-supported workshops, she herself has joined a study tour and she networks with other innovators. The advice and constructive ideas she has received from them have covered mulching, pruning and composting. Florence regards her efforts as rewarding and she points to various achievements, such as less erosion, more water for her farm and larger bunches of bananas. She has ambitious plans to improve her livelihood, such as setting up a home bakery and raising turkeys.

Credit: Will Critchley

Plate 10.1 *Women innovators in Uganda look at bananas grown in mulched plots*

Uganda
- Gender contact persons (women and men) were identified in each district.
- Extension staff began to identify and give recognition to 'family' innovations.
- Women's and men's groups formed and became active in disseminating innovations.
- Extensionists began to give more attention to the youth; for example, in a PFI-supported tour to East, Central, Southern and Western Uganda in November 1999, three male and one female youths were included.
- In polygamous families, all wives became involved in PFI activities on a rotational basis.

Changes in attitude and behaviour of extension staff

After the gender sensitization workshops, many extension agents began to show interest in gender issues and made use of available opportunities to discuss gender issues in training sessions and other meetings with farmers. During the sensitization workshops and follow-up meetings, they made positive comments about the gender concept and the role of women, such as: 'I realized gender is not only an issue for women but a concept that can help me improve my extension work.' Some staff members reported (in the same settings as noted above) that the training helped them to appreciate the contribution of women in agriculture and to become more aware of the sociocultural factors affecting their participation in the PFI activities.

Important changes in the working approach of the extension agents became evident. They gave much more attention to women as an important part of their target group. They became concerned about the constraints that hinder women from participating effectively in meetings, rather than blaming them for not participating and assuming that they were not interested. Some extension agents started to go out of their way to encourage women to attend meetings. Through discussions with male farmers, the extension agents persuaded them to allow their wives to participate. The extension agents encouraged women to present their experiences in meetings and training sessions, even when their husbands were present. This deliberate support for women demonstrated the extension staff's belief in women and paid dividends: numerous women started to express themselves more freely in mixed-gender meetings.

Changes in attitude and behaviour of farmers

The gender sensitization workshops helped to dispel some fears of male farmers (and male extension agents) about the concept of gender which were expressed at the beginning of the workshops, eg:

- 'Gender is about making women rule over men.'
- 'Gender sensitivity would make our women big-headed.'
- 'Developing women threatens marriages.'

BOX 10.3 SUSANNA SYLVESTER AND HER NOVEL COMPOSTING TECHNIQUE IN TANZANIA

Mama Susanna lives in Kondoa district of Dodoma region, Tanzania, where average rainfall is below 500mm per annum. She is elderly and heads the household that farms an area of about 6ha, about twice the average in the area. Susanna was relatively poor, like most of her neighbours, but was able to increase her income and buy land, as a result of her innovation. On her own initiative she developed a system of mixing crop wastes with fodder residues, manure and urine from her zero-grazed cow and stall-fed pig. She puts this mixture into large pits and adds ash and waste water from washing to keep the compost moist. By applying this high-quality compost, she keeps her land much more fertile than the local standard and also manages to sell a surplus of compost. Susanna cuts and carries the fodder for her cow from the nearby woodland, which is protected from grazing by government mandate. Susanna was encouraged by PFI to record the inputs and outputs in her composting system and gave her some guidance in doing this. The zonal research institute at Mpwapwa (she is the farmer representative on the institute's advisory committee) took a sample of her compost for analysis. Susanna might be able to speed up decomposition by using the three-pit system that is promoted in conventional extension, but she thinks this would demand too much extra labour for the large quantities of compost she produces. She has recorded over 30 farmers – men and women – who have followed her novel technique of making compost. While several farmers had made compost before PFI started, the number reportedly grew rapidly after farmer-to-FI visits organized by PFI, which included visits to Mama Susanna's farm.

Credit: Will Critchley

Plate 10.2 *Susanna Sylvester is a specialist in composting (Tanzania)*

Perceptions of male farmers towards women gradually started to change. A male teacher in Mwingi district, Kenya, whose wife is an active innovator farmer, publicly expressed his strong feeling (during a follow-up meeting) that, if women are given responsibility, they can manage the family resources as well as men. This change in men's attitudes was also demonstrated by the fact that some men have allowed their wives to travel to workshops and tours, although this has not yet been quantified. Women were even elected into decision-making committees at village level. For example, one of the female innovators working with PFI, Grace Lunyonga in Tanzania, was elected Village Government member in 1999. Some male farmers have invited female innovators to demonstrate certain techniques – eg groundnut mounding (Grace Lunyonga, Tanzania); gully control (Grace Bura, Tanzania); harvesting water for growing bananas (Florence Akol, Uganda) – to help them and colleagues to learn.

As a result of the training, tours and exchange visits, women farmers acquired knowledge and skills and gained greater exposure to new ideas. This, together with the recognition and support given by extension staff and some male farmers, led to an improved self-image among the women. They were given opportunities to explain their innovations in meetings of farmer networks and groups and in other public meetings. In Kenya, a training-of-farmers session (a modified form of training-of-trainers) was organized for FIs to enhance their presentation skills. The participation of all 16 women innovators in this training helped to increase their confidence.

The changes that became apparent in the attitude of some women towards their role in public inspired other women to come forward and express their views and experiences more freely. Among some men (both farmers and extension agents), it also inspired more respect for women. The men explained in one follow-up workshop that, 'Women too can express themselves as well as, and sometimes even better than, men' and that 'women can be as innovative as men'.

Women also expressed their realization that they, too, have useful knowledge and ideas and that they can present these just as well as men can. This process of discovering the capacities of oneself and of others is perhaps the most appropriate way to develop gender sensitivity and to bring about meaningful change in this culturally sensitive sphere.

CONCLUSION

The explicit equal-benefit objective of PFI led the project to address a gender bias in identifying innovators that was recognized at an early stage of the project. The systematic approach of conducting gender studies and sensitization workshops in each of the three countries disclosed some problems that are inherent to any rural development project working in such areas (eg the place and power of women in rural society). However, it also showed up problems that were specific to PFI (eg how to identify female innovators and ensure their recognition) and that could be addressed by the project. These recommendations were acted upon. In this chapter, we have touched on some of the main impacts of the gender sensitization process. Some headway has obviously been made. However, achieving as much gender equality as is reasonably possible under the prevailing sociocultural conditions will require continuous consciousness of the issues and of the opportunities. As we pointed out at the outset, this is not just an ethical question, but also a pragmatic one. Unless women are fully involved alongside men, the value of projects such as PFI will be diminished both morally and tangibly.

REFERENCES

Critchley, W R S, Cooke, R, Jallow, T, Lafleur, S, Laman, M, Njoroge, J, Nyagah, V and Saint-Firmin, E (1999) *Promoting Farmer Innovation: harnessing local environmental knowledge in East Africa*, RELMA/UNDP, Nairobi

Ong'ayo, M and Njoroge, J (eds) (in preparation) *Gender and innovation*, UNSO/UNDP, Nairobi

11

Innovators in land husbandry in arid areas of Tunisia[1]

Noureddine Nasr, Bellachheb Chahbani and Chris Reij[*]

Major socioeconomic changes during the last four decades have led to a decline in farming in the mountainous parts of central and southern Tunisia and gradual abandonment of traditional techniques of harvesting rainwater, such as the jessour. *In these arid areas, farmers and scientists are carrying out joint experiments designed to reduce the labour required to maintain the* jessours. *In the process, they are stimulating each other to innovate to increase the productivity of rain-fed agriculture.*

HARVESTING WATER FOR AGRICULTURE

In nearly two-thirds of Tunisia, in mainly the centre and south of the country, the average annual rainfall is less than 200mm. Here, except in the irrigation schemes and oases, agriculture would be impossible without water harvesting. One of the most widely used traditional techniques to harvest water is the *jessour* (El Amami et al, 1980), an ancient system of collecting run-off from long slopes. This is still used in agriculture in mountainous regions to this day. Farmers have built earthen dams (*tabias*) across the valley floors to trap the run-off water and silt. Although the entire system is called *jessour*, the word refers in fact only to the cultivated valley floor.

Two major rural livelihood systems have traditionally coexisted in southern Tunisia. One involves sedentary farming by agropastoralists in the

---

* Noureddine Nasr is an agronomist and geographer with the Institut des Régions Arides (IRA) and PTD trainer in ISWC-Tunisia; Bellachheb Chahbani is a SWC specialist with IRA and coordinator of ISWC-Tunisia, and Chris Reij is a geographer with the CDCS, Vrije Universiteit, Amsterdam, and international coordinator of ISWC 2

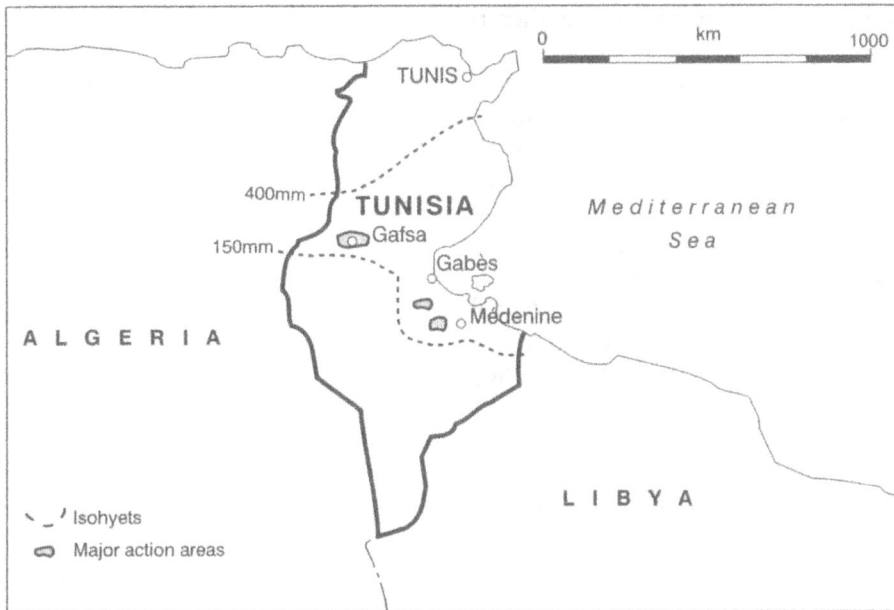

Map 11.1 *ISWC action areas in Tunisia*

Matmata mountain range, and the other nomadic and transhumant herding of camels, sheep and goats in the adjacent plains between the mountains and the Mediterranean Sea. Sedentary farming in the mountains is based on olive, fig and palm trees growing in the *jessours*, in combination with cereal and legume crops sown in years of good rainfall. Agriculture in this arid region involves high risks on account of the very low amount and high variability of rainfall (Nasr, 1993). The infrequent but heavy rainstorms cause considerable damage to the *tabias* and much labour must be invested to repair them.

Three major socioeconomic changes during the last four decades have had a profound impact on the rural livelihood systems both in the mountains and in the plains. The first is that many men migrated to urban centres in northern Tunisia or Europe to seek employment. This increased and diversified sources of income for the families in rural Tunisia (Nasr, 1998). The second is the enormous boom of the tourist sector along the coast of central and southern Tunisia which generated demand not only for labour but also for fresh vegetables and fruits. The third is the descent of sedentary farmers into the plains. The government invested heavily in water-harvesting systems in the plains, not only to reduce the risk of damage by floodwater to infrastructure in the coastal zones, but also to replenish the groundwater tables. At the same time, this created opportunities for agriculture based on the harvested water. The descent into the plains had a double effect: it led to the abandonment of *jessours* in the most isolated mountain valleys and it reduced grazing resources that were available to the pastoralists (Nasr et al, 1998). Communal grazing lands were increasingly transformed into private cultivated land. Parallel to this, livestock numbers grew, increasing further the pressure on the dwindling grazing resources.

Because of the high labour inputs, low productivity and high risk associated with farming in the mountainous areas, many young men have abandoned it. They prefer jobs in trade and commerce or the tourist sector. Under these circumstances, it is a major challenge to make farming more remunerative and attractive to young people. The decline in interest in farming is most strongly felt in the immediate vicinity of tourist centres, such as Matmata, and in isolated valleys. Because the extended families have disintegrated into smaller nuclear families, less labour is available at household level. The use of machinery in mountainous terrain is difficult and costly. The challenge is to reduce the maintenance requirements of the *jessours* and to increase the productivity of farming based on rainwater harvesting.

Technological change could reduce the workload and improve the image of farming. A small number of scientists began to take a closer look at the innovations developed or introduced by the people who continue to practise agriculture in the region.

A PARTICIPATORY APPROACH TO THE DEVELOPMENT OF RAINFED FARMING

The ISWC programme in Tunisia effectively started in August 1997. The lead agency of ISWC-Tunisia is the Institut des Régions Arides (Institute for Arid Regions). In October 1997, IRA organized a national training workshop of PRA and PTD for researchers, development agents and staff of agricultural training centres. Some participants came from the national Ministry of Agriculture, some from regional government services and some from NGOs. The main aim of this workshop was to change the attitudes and behaviour of scientists and development agents who had little practical experience with participatory approaches to research and extension.

The second major activity of the programme was to create awareness at the level of the five Regional Centres for Agricultural Development (CRDAs) of the Ministry of Agriculture in central and southern Tunisia. Workshops were held at all the CRDAs to explain the farmer innovation approach to PTD. During these workshops, several participants voiced their scepticism, expressing the view that 'only researchers can develop technologies'. The ISWC-Tunisia team in IRA argued strongly that farmers also innovate. However, at such an early stage in the programme, there were few cases to support this claim. Nevertheless, some participants actually identified innovative farmers during these workshops.

The third major activity of ISWC-Tunisia was to start, in a more systematic way, to identify local innovators and to analyse their innovations. Several male innovators were identified fairly quickly. However, for cultural reasons, it is difficult for male extension agents and researchers to interview rural women. The programme therefore decided to select and train professional women (mainly teachers) working in town who return to their villages for the long summer holidays, to recognize and document women's innovations related to agriculture and processing agricultural products (see Chapter 12).

Examples of farmers' innovations

Some male farmers were found who had begun to diversify the species of fruit trees planted in the *jessours* in response to the growing demand from the urban and tourist sector for apples, pears, peaches, plums, apricots, grapes, almonds, etc. Some farmers now have more than ten species of fruit trees in their fields – a radical change from the traditional olives, figs and palms. It is not unusual to find several varieties of each species (early maturing to late maturing varieties) chosen by farmers with a view to spreading the risk of harvest failure. Some farmers are very skilled in grafting fruit trees, even grafting different species on one tree. For instance, a combination of apples and pears or of peaches and plums can be found. Farmers also graft on to the roots of a tree because this allows the young plant to grow in the shade. The greatest surprise to the scientists and extension staff was that some farmers grafted fruit trees on to the roots of the jujubier (*Ziziphus lotus*). Development agents used to regard this as a 'useless' species and the small trees were systematically uprooted in the plains of central Tunisia. The farmer innovators who graft on the roots of the jujubier regard this plant as an indicator for reasonable levels of soil moisture and good soil fertility.

Another interesting innovation was the construction of a concrete dam by a farmer in a region where such dams had never been built before. He also constructed small sediment traps in the catchment to reduce the silting of his dam in which he had invested considerable money. He used the water behind the dam for supplementary irrigation to grow a wide range of fruit trees and some vegetables.

Many innovators tend to practise more than one innovation. For instance, a farmer in Gasr Jaouamaa village (Médenine), Béchir Nasri Jamii, has introduced various species of fruit trees, is very skilled in grafting, gives supplementary irrigation to his trees using water stored in a cistern, has adopted and adapted a water-saving technique that was being tested by an IRA scientist in a neighbouring farmer's field and changed the design of local beehives, thus increasing the honey production substantially. Béchir's father was the first person in the village to introduce new species of fruit trees. He worked as a cook for the *Bey* (king) of Tunis and, during his visits to his village in the 1940s and 1950s, he brought home a wide range of seedlings. In those days, the villagers reportedly thought it was ridiculous to grow these types of trees. The son, who worked as a painter in France for ten years, has continued to build on his father's knowledge and skills. The growing demand for fresh fruit in the cities has led to a veritable explosion in fruit-tree diversification in the village, from an innovation originally introduced by a sole farmer. The first peaches to arrive on the Médenine market are from Gasr Jaouamaa and they fetch a good price.

Plate 11.1 *Abbès Sandi grows a wide range of fruit trees in a pre-desert region, and irrigates them from a small dam he constructed (Tunisia)*

Characteristics of innovators

In the mountains of Matmata, the farms tend to be relatively small for such a dry area (10ha or less) but, in the regions of Gafsa and Sidi Bouzid, they are larger (20ha or more). The owners of small farms cannot survive on plant production alone. Many owners of larger farms keep flocks of sheep. About 40 per cent of the 41 innovators identified by ISWC-Tunisia by the end of 1998 stated that their major income came from non-farm activities. Among these innovators are drivers (of private pick-up trucks or in the agricultural services), masons, a tailor, a shop owner, a teacher and a health worker. The innovator who constructed the above-mentioned concrete dam spent several years in Tunis repairing watches before he decided to return to his village and invest in farming the money he had earned in the capital city.

The innovators identified thus far are relatively old and experienced. Of 32 innovators interviewed, 20 are over 50 years of age. Most of those older than 60 had started to innovate in the 1970s, a period that marked the end of a government policy aimed at promoting collective agriculture and the start of support to agricultural development in the dry regions.

The level of formal education of male innovators is higher than that of female innovators (see Chapter 12). Six male innovators (19 per cent) attended primary school, eight (25 per cent) reached secondary school level and 12 (37.5 per cent) attended the Koran school only. The 15 per cent of male innovators who are illiterate are all over 50 years of age.

Visits to farmer innovators

In late 1998 and early 1999, ISWC-Tunisia organized four visits to farmer innovators in Sned, Béni-Khédache and Mareth, with about 20 farmers, researchers and extension agents participating in each. These visits inspired some of the farmers to try out what they had seen on the innovators' farms. A camera team from national television accompanied one such visit. After having seen the new techniques on television, other farmers also started to try some out. A systematic study of the viewers' reactions to the television programme was not made. However, one of the innovators, Béchir Nasri, reported that farmers who had seen his improved version of the 'buried stone pockets' (described below) on television had tried to make some themselves and had invited him to come to their farms to see if they had done it well.

Joint experimentation

In the second half of 1999, a small number of scientists, extension agents and farmers started to carry out some joint experiments. Some of the scientists had already been trying for many years to find ways to reduce the maintenance requirements of traditional techniques such as *jessours*, as well as to test technologies to economize on water use. Whenever possible, they had used local techniques as starting points.

Since the ISWC-Tunisia programme started, experiments that have been carried out jointly by scientists and farmers, based on the latter's techniques, have included:

- the use of plastic bottles to irrigate individual plants (watermelons in 1999, potatoes in 2000);
- economizing on the use of water in greenhouses; and
- building cisterns to store water for supplementary irrigation (by gravity) of fruit trees and vegetables.

If the various experiments on the efficient use of irrigation water lead to good results, these could be used to take advantage of financial support from the government of Tunisia, which systematically subsidizes up to 60 per cent of the costs of technologies that economize on water use in farming. It will also be much easier for farmers to apply for credit to cover most of the remaining costs. One particularly promising experiment is the temporary storage of water in a small concrete dam in the foothills close to Gafsa (average annual rainfall 140mm) on a large piece of marginal land that the farmer had bought ten years previously. He built the dam in order to see how he could use water-harvesting techniques. Once the dam is full of water, this is pumped to a large cistern constructed downslope close to his arable fields. Storing the water in a cistern avoids evaporation. This water is then used for supplementary irrigation of olives and almonds planted behind *tabias*. The local CRDA, which is monitoring the results of this experiment, has already received requests from several other farmers interested in developing similar systems on their farms.

An experiment to economize on water use in greenhouses is being carried out in the Mareth area. The farmer is comparing irrigation by submersion with drip irrigation with a buried plastic distributor of water for individual plants. The ISWC-Tunisia coordinator from the IRA invented this third option. The first indications are that this technology reduces the water needs substantially. This is of great importance to farmers who buy piped water to grow crops in greenhouses. The scientists and development agents involved in this trial expect that this third technique will reduce the crop water requirements by two-thirds.

The joint experimentation has triggered not only cooperation but also competition between scientists and farmer innovators. The best example is the case of Béchir Nasri. During the first half year of the ISWC-Tunisia programme, he simply observed what IRA researchers and a neighbouring farmer were trying out together on an experimental plot. Then, one day, he approached the researchers to tell them that he had found a solution to their problem of pumping water out of a cistern without silt blocking the rubber hose. From then on, he produced a range of innovations, for instance, a mechanical timer to control the duration and the quantity of water use for supplementary irrigation, a tool for threshing cereals and a technique for feeding honeybees. The programme had obviously triggered Béchir's creative capacities, which he is now developing fully. In recognition for his work, he was invited to join the Tunisian delegation to the regional francophone workshop on Farmer Innovation in Land Husbandry organized in Cameroon in November 1999. He was also given the opportunity to present his innovations at an International Fair on Agricultural Technology held in Tunis in June 2000.

ROLE OF SCIENTISTS

In the ISWC 2 programme in Tunisia, IRA researchers have played a strong role in studying indigenous techniques of SWC and developing ways to improve them. The testing of these potential improvements on farmers' fields has led to joint observations by scientists and farmers and intensive discussions. In some cases, the farmers and/or their neighbours have been stimulated to improve the scientists' improvements still further. These changes have, in turn, stimulated new ideas among the scientists.

Experiments to reduce maintenance requirements of *jessours*

Traditional water-harvesting techniques such as the *jessours* face numerous technical and socioeconomic constraints. The latter have already been discussed above. A major technical constraint concerns the high ratio between the catchment and the cultivated area (at least 20:1). Large catchments guarantee adequate run-off in years of low and average rainfall, but occasional high-intensity rainfall events cause floods which can damage all the *tabias* in a valley. Because the infiltration capacity of loess soils is limited, the run-off water can stagnate for weeks in the *jessours*, causing damage to both trees and annual crops (Bonvallot, 1979; Chahbani, 1984, 1990, 1997). A researcher at IRA developed a technology to evacuate excess water. This was tested in a farmer's field in the village of Béni-Khédache. To avoid destruction of the spill-ways and of the dam during both normal and exceptional overflow, the lateral spillway was replaced by two joined tubes: one vertical and one subhorizontal. The drainage system consists of a basin and a floater. The initial results were not as positive as the scientist had hoped. The above-mentioned farmer innova-tor, Béchir Nasri who had observed the initial experiment, suggested some improvements and these are now being tested.

Experiments to make more efficient use of water in *jessours*

In farmers' fields, scientists have been testing several techniques intended to increase the efficiency of water use in the *jessours*. One of these is the 'buried stone pocket' for the localized underground irrigation of fruit trees. The origi-nal idea introduced by the IRA researcher was as follows: the bottom and lower edges of a planting pit (1 x 1 x 1m) are lined with stones (limestone, sandstone, lime crust, etc) laid in three or four layers with two or three rows of stone each. The pocket of stones is then covered on three sides with plastic sheeting to prevent soil from entering the spaces between the stones. When the pit is filled again with soil, a plastic tube with a T shape (3–7cm diameter and 80cm length) is fixed vertically between the stones near the fourth side of the pit. Water flows by gravity through a rubber hose from a cistern higher up the slope to a tap near the pits. Another rubber hose connects the tap to the plastic tube in each 'stone pocket' in order to irrigate the fruit-tree seedling planted in it. This technology leads to faster growth of the individual fruit trees, while making very limited use of water. Farmers who have tested it have observed substantial increases in fruit production.

The farmers have not simply adopted this technology; they have been active in adapting and improving it to fit their own circumstances. Their tendency has been to reduce the depth and breadth of the 'pocket' originally introduced by the scientist. At a depth of about 40cm, some farmers have laid out a small circle of stones, leaving an opening in the centre. They insert a plastic pipe vertically between the stones, cover the stones with soil, plant a tree seedling in the centre of the pit and provide water through the plastic pipe rather than submersing the soil around the tree. One farmer decided to put the plastic pipe closer to the tree so that he could continue to plough the land around it. Another farmer modified the 'buried stone pocket' technique so that it could be used for growing watermelons. The scientists are observing and learning from these farmers' experiments.

Several farmer innovators have now started on their own initiative to record the details of their experiments in a notebook. This allows them to compare their experiments with a control plot in their own fields.

INSTITUTIONALIZING THE APPROACH

Through the ISWC-Tunisia programme, researchers, development agents and policy-makers have become more aware of farmer innovators and their innovations. New links have been created between these different stakeholder groups. Development agents and even policy-makers are following with interest the process of identifying innovations, joint experimentation and spreading the results. The intention in the next phase is to intensify and expand the farmer innovation approach in central and southern Tunisia as well as to other parts of the country. One way to achieve this will be to strengthen the links that have already been made with the Presidential Pilot Project on Agricultural Extension being implemented under the responsibility of the National Farmer's Union.

NOTES

1 This is a considerably revised version of a paper submitted to the Tenth International Soil Conservation Organizations (ISCO) Conference, 23–29 May 1999, Indiana, USA

REFERENCES

Bonvallot, J (1979) 'Comportement des ouvrages de petite hydraulique dans la région de Médenine (Tunisie Sud) au cours des pluies exceptionnelles de mars 1979', *Cahiers de l'ORSTOM, Série Sciences Humaines XVI, 3*

Chahbani, B (1984) *Contribution à l'étude de l'érosion hydrique des loess de Matmata et de la destruction des 'jessours': bassin versant de l'Oued Demmer, Béni-Khédache, sud Tunisien*, doctoral thesis, 3rd cycle in applied geomorphology, University of Paris I

Chahbani, B (1990) 'Contribution à l'étude de la destruction des ouvrages de petite hydraulique dans le sud tunisien', *Revue des Régions Arides 1*, Institut des Régions Arides, Médenine

Chahbani, B (1997) 'Les acquis de recherche dans le domaine de la conservation et de la valorisation optimale des eaux de ruissellement dans les régions arides du centre et sud tunisien', in 'Actes du Séminaire International sur les Résultats Scientifiques pour le Développement des Zones Arides, décembre 1996, Jerba', *Revue des Régions Arides 8*, pp77–92

El Amami, S et al (1980) 'Les aménagements hydrauliques traditionnels (*meskats et jessours*); un moyen de lutte contre l'érosion', Séminaire National sur l'Erosion, 5–6 juin 1980, Sidi Thebet

Nasr, N (1993) *Les systèmes agraires et les organisations spatiales en milieu aride: cas d'El-Ferch et du Dahar de Chénini-Guermessa (Gouvernorat de Tataouine)*, doctoral thesis, Paul Valéry University, Montpellier III

Nasr, N (1998) 'L'impact de l'émigration à l'étranger sur les systèmes fonciers et le développement agricole en zone aride: cas de Bir Lahmar', in *Migrations interna-*

tionales entre le Maghreb et l'Europe; Actes du colloque maroco-allemand de München, 1997, LIS Verlag, Passau, pp175–180

Nasr, N, Sghaier, M and Chahbani, B (1998) 'Transfert de technologies pour optimiser les systèmes de production basés sur les techniques traditionnelles de conservation des eaux et des sols. Actes du Séminaire International sur les Résultats Scientifiques pour le Développement des Zones Arides, décembre 1996, Jerba', *Revue des Régions Arides 9*, pp456–464

12

Women's innovations in rural livelihood systems in arid areas of Tunisia[1]

Noureddine Nasr, Bellachheb Chahbani and Radhia Kamel[*]

In central and southern Tunisia, women are involved in almost all activities in both rain-fed and irrigated farming, and are also responsible for specific tasks, such as collecting firewood; managing the ovens (tabounas); fetching water; harvesting grains, fruits and vegetables; collecting traditional fodder; hoeing; weeding; irrigation and feeding and watering animals. Some women have managed to increase production and cash income by developing innovations based on their experience in these activities. Here, the outcome of a survey to identify such women innovators is presented.

IDENTIFYING WOMEN INNOVATORS

At the outset of ISWC 2 in central and southern Tunisia, training was given in PRA and PTD in different regions. This was meant to raise awareness about innovation by farmers, both men and women, and to lead to the identification of specific innovators. The trainees were researchers and staff of the Departments of Soil and Water Conservation in the regional branches of the Ministry of Agriculture. All of them were men. In addition, one-day workshops were held at the regional headquarters of the agricultural services in Médenine, Gabès, Gafsa, Sidi Bouzid and Tatouine. Some 160 staff members took part who were predominantly men (95 per cent). After these workshops, innovators were indeed identified, but most of them were, likewise, men.

---

* Noureddine Nasr is an agronomist and geographer with IRA Gabès and PTD trainer in ISWC-Tunisia; Bellachheb Chahbani is a SWC specialist with IRA Médenine and coordinator of ISWC-Tunisia, and Radhia Kamel is a sociologist with IRA Gabès

Map 12.1 *ISWC action areas in Tunisia*

In the local culture, male researchers and development agents from outside the area are usually not permitted to talk with village women. As the ISWC team at the Institut des Régions Arides (IRA) was composed at the time exclusively of men, it was decided to ask some professional women from technical agencies and local institutions, but mainly female teachers and students returning to their villages for the long summer holidays, to identify rural women's innovations. The ISWC team trained 15 women to document the role of women in farming and processing agricultural produce. Within two months, they managed to identify 31 female innovators. This identification process is continuing through a regional radio programme on agriculture and innovation (see Chapter 27) and other activities of ISWC-Tunisia and has been facilitated by the recent recruitment of a female sociologist to the team.

In the initial survey, most women innovators were found in the regions of Gafsa and Sidi Bouzid, where population density is higher and agriculture is more diversified and intensive than in the Médenine and Tataouine regions (Abaab et al, 1993; Nasr, 1993). Sidi Bouzid has the largest number of irrigated schemes, here cropping (both rain-fed and irrigated) is well integrated with livestock keeping (Abaab et al, 1993).

CHARACTERISTICS OF WOMEN INNOVATORS

Thirty of the women innovators were married. The women were between 23 and 84 years old, most being in their 30s and 40s. The older women innovators, in particular, had little formal education as they came from rural mountainous areas where, until recently, there were few opportunities, especially for girls, to go to school. One of the women had attended secondary school and eight (25 per cent) primary school. The others (72 per cent) were illiterate; these 22 women were over 40 years of age and lived in relatively remote and deprived areas. However, with the spread of electricity and education in the rural areas during the last 30 years and particularly during the last decade, the women have more contact with a new culture through radio, television and their school-going children.

SPHERES OF INNOVATION

Married women are responsible for taking care of their homesteads and families and are in charge of certain agricultural activities. Rabbits and poultry are their major sources of cash income (Chahbani and Nasr, 1999). The survey revealed that the women innovate most actively in those spheres that concern them directly. The main economic activity of all but one of the women innovators was rain-fed and/or irrigated crop production and raising sheep and goats. Most of them also practised some handicrafts. The sphere in which the largest number of women (11) was found to be innovating was in livestock keeping. Other innovations were in cropping (7 women), handicrafts (6), use of medicinal plants (3), efficient use of energy for charcoal making and improved stoves (2) and food processing, specifically the processing of milk from sheep and goats (2).

Handicrafts included making carpets and other products out of wool and weaving mats and other household items out of alfa grass (*Stipa tenacissima*). Women innovators in this sphere were found in all age groups and in all regions. Specific innovations were producing woollen mats and extracting natural dyes from leaves, roots and bark.

The innovations related to crops included fig pollination techniques and using plastic bottles for water-efficient irrigation of melons. For example, Rgaya Zammouri in Zammour village (Médenine), who is over 70 years old, uses 1.5 litre plastic bottles to irrigate watermelons and melons. She buries each bottle in the soil with the cork downwards in which she makes up to three tiny holes with a needle so that water is released directly beside the plant. She fills the bottles with water from a cistern fed by run-off rainwater. The water infiltrates slowly near the plant roots and thus escapes the high evaporation in this region. She started this innovation in the 1997–98 growing season. She used to carry the water from the cistern to the field in a bucket, but now the ISWC programme has supplied her with a water tap and a rubber hose to facilitate her work. Her innovation is simple, efficient and low in cost and therefore has the potential to spread much more widely.

Eleven women (35 per cent of those identified) have innovated in livestock keeping, specifically with sheep and goat feeding, and with keeping poultry, bees and rabbits. For example, Mbirika Chokri, a 70-year-old woman living in Sidi Aich (Gafsa), practises rain-fed farming and specializes in poultry. Her

Credit: Bellachheb Chahbani

Plate 12.1 *Mbirika Chokri incubates chicken eggs in dry cattle dung (Tunisia)*

innovation consists of incubating chicken eggs in dry cattle dung. She puts the eggs with some straw in plastic bags to preserve some humidity. Each bag contains 16–20 eggs. She puts the bags in small holes dug in the manure, covers them with a piece of cardboard to protect them against damage and covers the cardboard with a thin layer of manure. Each day, she opens the bags to check the temperature of the eggs and to turn and aerate them. From day 20 the eggs start to hatch. She puts the chicks into a box to protect them from the cold and feeds them couscous, vegetables and bread.

Mbirika started this innovation in 1995 when one of her chickens, whose eggs were about to hatch, suddenly died. She decided to put the eggs into a pile of dried cattle dung. After some days the eggs hatched, to her delight. She decided to use manure again in the same way to hatch eggs. Mbirika now masters this technique very well and produces numerous chicks. She did not share her experience with her neighbours, but she accepted the request of ISWC-Tunisia to present her innovation in the Agriculture and Innovation programme of the Gafsa regional radio and later also on television.

POTENTIAL FOR THE SPREAD OF WOMEN'S INNOVATIONS

The livelihood systems in central and southern Tunisia have changed radically in recent decades. New production systems have replaced the traditional pastoralism which had been the dominant source of livelihood in this area for centuries (Nasr, 1993, Abaab et al, 1993). There are also increasingly closer links between the countryside and urban markets, and rural women need more cash to satisfy new needs. Women innovate not only to increase their income, but also to decrease their workload. For instance, economizing on the use of water for irrigation reduces the time and energy spent on fetching water.

Several women stated that their innovations grew out of their own ideas and creativity, or were a chance discovery. The oldest innovations by women – in handicrafts and medicines – are rooted in local knowledge but adapted (in design, materials or use) to the new socioeconomic context. Generally, women's innovations, such as the above-mentioned ones involving bottles for localized irrigation or incubating eggs in manure, are simple, practical and low-cost and therefore have a good potential for spreading.

More and more Tunisian researchers and development agents, as well as policy-makers at regional and national level, are coming to recognize the innovative capacities of rural women. In 1999 and 2000, researchers and several women innovators began collaborating on experiments to develop their innovations further. The challenge is to improve and expand this approach within Tunisia and beyond.

NOTES

1   This is an expanded version of an article by Nasr et al, (2000) that appeared in the *ILEIA Newsletter* focused on Grassroots Innovation and the French version *Promouvoir l'Innovation Paysanne*

REFERENCES

Abaab, A, Nasr, N, Ben Salah, M and Sghaier, M (1993) *Caractérisation des milieux et des systèmes des zones arides et désertiques tunisiennes*, CIHEAM-IAM, Montpellier/IRA, Médenine

Chahbani, B and Nasr, N (1999) *Le développement participatif de technologies basé sur les innovations des hommes et des femmes en zones arides de la Tunisie*, IRA, Médenine/CDCS, Vrije Universiteit, Amsterdam

Nasr, N (1993) *Les systèmes agraires et les organisations spatiales en milieu aride: cas d'El-Ferch et du Dahar de Chénini-Guermessa (Gouvernerat de Tataouine)*, doctoral thesis, Paul Valéry University, Montpellier III

Nasr, N, Chahbani, B and Ben Ayed, A (2000) 'Innovation by Tunisian women in dryland farming', *ILEIA Newsletter*, vol 16, no 2, p20 ('Les innovations des femmes en Tunisia', *Bulletin d'ILEIA*, vol 16, no 2, p16)

13

Namwaya Sawadogo: the ecologist of Touroum, Burkina Faso

Jean-Baptiste Taonda, Fidèle Hien and Constant Zango[*]

The story of Namwaya Sawadogo is one of rags to riches. He started as a petty trader and, through great innovative energy and with the support of development agents who recognized his potential, he has managed to establish a highly integrated system of agrosylvopastoralism – ie cultivation, tree farming and livestock keeping. He is now a widely known innovator in Burkina Faso and has received many visitors, including the Minister of Agriculture.

INTRODUCING NAMWAYA SAWADOGO

Namwaya Sawadogo was born in 1943 in Touroum, a village located in Pissila Department in Sanmatenga province of Burkina Faso. He has three wives and 12 children, but has to feed another five relatives, which means that he supports a total of 20 persons. His farm covers 14ha, almost three times more than the average farm size in this area. He keeps four cattle, 15 sheep, nine goats, one horse, one donkey and poultry (chickens and guinea fowl) near his home and owns another 15 head of cattle which are cared for by Fulani pastoralists. Thus, one can say that he has become a rich man compared to most Burkinabé farmers, but when he started farming in this area, he had only some guinea fowl, one donkey and 1ha of land – not enough to feed a family in this semi-arid climate with only about 600mm annual rainfall.

---

[*] Jean-Baptiste Taonda is an agronomist with INERA and Head of the Saria Research Station; Fidèle Hien is an ecologist, formerly with INERA and now Minister of Environment and Water, and Constant Zango is an animal scientist with the Namentenga Integrated Regional Development Project in Burkina Faso

Map 13.1 *Burkina Faso*

Namwaya has developed or taken up several new ideas related to land husbandry, but his major innovation is that he has become a forest farmer. Over the years, he has established a eucalyptus plantation covering 4ha to produce timber for sale. The wide spacing of the trees allows him to grow groundnuts or other annual leguminous crops in between.

MOTIVATION TO INNOVATE

In the 1970s, Namwaya was a small itinerant trader in natural medicines who was constantly on the move. He often did not see his family for weeks or even months at a stretch. However, after he married and became the head of a household, he decided to settle and concentrate on agriculture, limiting his commercial activities to the village.

In 1982, a government-built dam inundated almost all the fields of the Sawadogo family which consisted of seven households ('hearths'), and there was land left for only two of them. Faced with this difficult situation, Namwaya decided to move to the periphery of the village where his uncles gave him about 1ha of reasonably fertile soil and the *chef de terre* (the man traditionally responsible for distributing land) also granted him the right to use the uncultivated poor-quality land surrounding this field if he wished to do so. He cultivated this poor land for several years, but never managed to get a good harvest from it. Even with the 1ha of good land, he could not produce enough food to feed his household. With his back against the wall, he decided to see what he could do to rehabilitate the poor land.

THE PROCESS OF INNOVATION

In 1988, Namwaya started to construct stone bunds along the contours on the poor land. He also planted a perennial grass (*Andropogon gayanus*) along the bunds. Contour bunds made of stones had become a well-known technique in Touroum which was promoted by the Association for the Development of the Kaya Region (ADRK), of which Namwaya was a member. This NGO special-izes mainly in savings and credit, but between 1986 and 1998 it was also very active in the field of SWC. Through ADRK, Namwaya bought a donkey cart on credit, primarily to be able to transport the stones. The contour stone bunds had some positive impact on yields, but still he could feed his family for only nine or ten months of each year.

The real turning point came in 1990, when government forestry agents proposed to train him in techniques of establishing and maintaining a tree nursery. He had already been trying to raise seedlings before this, but he felt that he did not master the techniques adequately, so he gladly accepted their offer and attended a short training course. At the same time, he tried to expand his cultivated area by applying mulch to part of his barren degraded land. In 1990, his fields produced enough millet and sorghum to feed his family during the entire year, for the first time since the dam had inundated the family fields in 1982.

In 1991, Namwaya established a tree nursery next to the dam and produced 3000 plants in that same year. He planted 1ha of relatively good land to a mixture of locust-bean trees (*Parkia biglobosa*) and *Faidherbia* (syn *Acacia*) *albida* and intercropped the young seedlings with millet. On three sides of his field he planted a live fence composed of several local woody species that he could use for different purposes (fodder, medicines, etc) in addition to protecting his field. This was his start as an agroforester. In 1992, however, most of the planted trees died. He decided to replace them with eucalyptus (*Eucalyptus camaldulensis*) trees. Also in this year, Namwaya dug his first compost pit. Both ADRK and the government extension service were promoting this technology. He used his donkey cart to transport water to the compost pit from the dam reservoir about 5km from his farm.

In 1993, Namwaya doubled the size of his eucalyptus plantation, growing the trees in lines about 8m apart so that he could grow crops in between. That same year, he also planted eucalyptus on 2.5ha of ancestral lands in Touroum that had been given to him by his mother's brothers. In 1994, after having been trained in animal husbandry through ADRK, he started applying what he had learned with his own livestock and used his own resources to build a shed to store fodder. Also in that year, he took part in an agricultural fair in the regional capital Kaya, where he was distinguished as a model farmer and awarded a certificate of honour. He then decided to buy a plough with some credit from ADRK.

In 1995, Namwaya participated in a study visit to the Yatenga region organized by ADRK. On his return home, he decided to dig planting pits (*zaï*)

Plate 13.1 *Namwaya Sawadogo in his eucalyptus plantation (Burkina Faso)*

in his fields, like the pits he had seen in Yatenga, even though this meant that he would no longer be able to use his newly acquired plough on this land. He started a second compost pit. In 1995, a year of exceptional demand for tree seedlings, his cash income from selling seedlings from his nursery was 600,000CFA (circa US$950). This is a remarkable income compared with the estimated average income in Burkina Faso at the time (US$230). The positive impact of the *zaï* soon became evident. Two years later, there was a serious drought and Namwaya was the only farmer in Touroum who could harvest enough to meet the food needs of the family. He provided cereals to those who requested support from him and felt that this would morally oblige them to experiment with new practices as he had done.

The gradual increase in the size of his household over the last decade has enabled Namwaya to invest in the expansion and rehabilitation of land for agroforestry. With three wives and 12 children, many of whom are of working age, he now has a considerable labour force.

WIDE RANGE OF INNOVATIONS

Namwaya has progressively integrated a number of components into his farm and has gradually developed an agrosylvopastoral system. This development was closely linked to the evolution of his own knowledge and to the level of equipment, labour and financial resources at his disposal. Whenever Namwaya started up a new activity, he always looked carefully at how it could be done efficiently in order to minimize costs and labour energy.

Innovations in crop production

Various agricultural technologies are combined in Namwaya's farm: contour stone bunds, *zaï*, barriers made of perennial grasses (*Andropogon gayanus*), mulching (with cut wild grasses and lopped leaves of the shrub *Piliostigma reticulatum*) and composting. He sows millet and sorghum in the same planting pits as a strategy to reduce cropping risks. The quantity and distribution of rains in any one season determines which of the two cereals he will harvest.

Innovations in agroforestry

Namwaya has mastered all aspects of producing tree seedlings, also of local species. His key to success was his close observation of how the seeds germinate. This helped him to develop his own technical knowledge for treating seeds. Through observation, Namwaya also identified which tree species can be multiplied through root suckers. They include *Faidherbia albida*, tamarind (*Tamarindus indica*), kapok (*Bombax costatum*), *Diospyros mespiliformis*, *Balanites aegyptiaca* and neem (*Azadirachta indica*). In his tree plantation, Namwaya prefers to sow nitrogen-fixing annual crops such as groundnuts between the lines of eucalyptus. If he does not grow a crop, he uses this space for grazing. In this case, he just lightly scarifies the surface of the soil in order to improve the growing conditions for grasses.

Innovations in animal production

In Namwaya's view, the purpose of livestock is to support crop production by providing organic matter for the soil. In his words: 'There can be no cropping without livestock.' His livestock holdings are diverse: bovines, equines and poultry. His innovations in animal keeping are of both a technological and an organizational nature. He keeps his cattle, sheep and goats in a stable during the dry season so that he can collect their manure systematically. This means that much extra work has to be invested, especially by his wives and children, in feeding and watering the animals. To be able to feed his livestock efficiently, he has built the hangar for storing fodder immediately next to the stable. Each head of cattle and his horse has its own feeding trough as his theory is that this will reduce the wastage of fodder to a minimum. The two compost pits are immediately next to the stable so that the manure can be evacuated easily into the pits. The animals' urine flows through a drain that leads from the stable directly into one of the compost pits.

Namwaya cuts whatever fodder he can find that his animals are willing to eat. His theory is that most fodder species have complementary virtues. Some are good for nutrition, whereas others are good for animal health. During the dry season he feeds his livestock large quantities of ground *Piliostigma reticu-*

latum pods. He has produced his own concoctions to treat certain animal diseases. He also produces salt-licks himself, using an extract of *Hibiscus* spp, some bark of savanna mahogany (*Khaya senegalensis*), some clay from a salty marsh and some salt (NaCl) and natron, both of which are Sahel products sold on local markets.

Conservation of biodiversity

Namwaya regards livestock as playing a central role in the life of man, family and society. Positive attitudes towards animals can be a source of divine blessing and more strongly so if the animal is not domesticated. He therefore tries to create conditions on his farm that allow wild animals to survive and reproduce. He does not allow hunters on his land and, during the dry season, he cares for birds by hanging calabashes with water in the trees. As a result, his trees abound with birds during the dry season and he is convinced that they contribute to the natural regeneration of the vegetation by spreading the seeds of various trees and bushes.

BENEFITS DERIVED FROM INNOVATION

Of great importance to Namwaya has been the training that he has been privileged to receive as this has enabled him to develop his knowledge and practical skills in several fields. During the last ten years, he has not been obliged to buy grain for his household, not even in years of drought. When he could not produce sufficient grain himself, he could cut down some trees and sell them on the market in order to raise cash for buying food. We estimate that, in a year of 'normal' rainfall, Namwaya harvests about 8000kg of millet and sorghum, which means that he normally produces a substantial surplus.

According to his own estimate, his annual cash income from his various farming and commercial activities is about 350,000CFA (circa US$550). Namwaya has a shop on the market of Pissila, about 10km from his home, where he sells medicinal plants during the weekly market day. He estimates that his net income from this is about 150,000CFA per year. Added to this are 100,000CFA from the sale of livestock,[1] 25,000CFA from the sale of cereals and the remainder from the sale of tree seedlings. This is probably a very conservative estimate; his income, from forestry activities in particular, is underestimated. His total annual cash income is more likely to be, on average over the yearly fluctuations, above 500,000CFA (circa US$800).

In addition, Namwaya has been honoured several times, most recently with a medal at an agricultural fair in Bagré in 2000. He has also been awarded cash prizes at fairs: 150,000CFA (circa US$240) in Ouagadougou in 1994 and 200,000CFA (circa US$320) in Bogandé in 1998. Both the recognition and these prizes, which are substantial in relation to the income of most farmers in Burkina Faso, have greatly motivated him to continue innovating and have allowed him to invest in both livestock and his ethnopharmaceutical business.

When assessing the impact of his innovations of his life, Namwaya speaks in terms of his gain in respectability, responsibility and popularity, but also his increased financial capacity and his ability to support those in need. He is proud of what he has achieved and enjoys the higher social status that his achievements bring. His market stall in Pissila with natural pharmaceutical products attracts many clients, even from other parts of Sanmatenga province. 'People come from far to consult me,' he states, and he does not hide the fact that this activity is lucrative. Namwaya views his trees as a form of life insurance and looks to the future with much confidence: 'When I am old, I can live from the income and the products of my plantations.'

NOTES

1 In April 2001 Namwaya sold three head of cattle for 750,000CFA.

14

Outwitters of water: outstanding Irob innovation in northern Ethiopia

*Asfaha Zigta and Ann Waters-Bayer**

In an arid mountainous corner of northern Ethiopia, remote from government services, the formerly pastoral Irob developed a very labour-intensive form of cropping on pockets of harvested soil and water. Two of the many remarkable innovators among these people are presented here: the masters of engineering in stone, Zigta Gebremedhin and Yohannes Tesfaye.

IROB: A LAND OF EXTREMES

The Irob used to be a pastoral people, moving with their goats and cattle from the mountains on the eastern escarpment of the Ethiopian highlands to the lower plains. It was only in the last two or three generations that they began to pay more attention to cropping. According to the Irob's own reports, they could no longer obtain enough cereals by selling animals, so they began to seek ways to grow their own cereals. The home of the Irob is at the northernmost tip of Tigray's Eastern zone. The landscape is very rugged and stony, with steep slopes and deep narrow valleys carved out by flash floods from the plateau. There is little land suitable for cropping. The altitude varies from 900m (Endeli valley) to 3200m (Mount Asimba) above sea level; most people live in the range of 1500m to 2700m.

Rainfall in the main inhabited area is low (200mm–600mm per year) and highly variable in space and time. A wet season is expected from mid-June to

---

* Asfaha Zigta is an iron craftsman and businessman in Adigrat, Tigray, who records the history of his people in Irobland; Ann Waters-Bayer is an agricultural sociologist with ETC Ecoculture in The Netherlands and adviser in participatory research and extension to ISWC 2

Map 14.1 *Ethiopia (showing Irobland)*

mid-August. However, water and soil eroded from the highlands, which receive more rain over a longer season, sporadically pour from the Adigrat plateau down through the seasonal streams that flow into two perennial rivers. These flow eastwards towards the Red Sea, but sink beforehand into the Danakil depression, 100m below sea level.

As is common in tropical highlands, the daily variation in temperature is greater than the seasonal variation over the year. The annual mean temperature in Alitena (at an elevation of 1850m), the heart of Irobland, is just under 20°C; maximum temperatures can rise above 30°C; minimum temperatures can fall to 5°C. Frost occurs occasionally above 2500m.

Irob is a land of extremes of depths and heights, of droughts and floods, of frost and scorching sun. And it is a land in which the people, trying to survive and even to cultivate in this harsh environment, have proved to be extremely inventive. Even in the early 1970s, when a Swiss geographer studied land use by the Irob, he marvelled at the '*grosse Spielbreite an Techniken und Nutzungsformen*' (the broad gamut of techniques and forms of land use) (Strebel, 1979). He drew attention to the innovativeness of the Irob, who had, without outside assistance, developed site-appropriate SWC methods for crop production within an amazingly short period of time.

When ISWC 2 commenced in Tigray in early 1997 and was seeking indigenous innovators in land husbandry, an obvious place to look was Irobland. Here we introduce two outstanding Irob innovators, Zigta Gebremedhin and Yohannes Tesfaye. The information comes from several semi-structured interviews and informal discussions with these two men, their relatives and neighbours. The interviews were conducted at irregular intervals from January 1996, when the authors first worked together to review a small development project that deliberately built on the innovations of the Irob (Hagos and Asfaha, 1997).

ZIGTA GEBREMEDHIN ROBS THE WATER OF ITS SILT

Zigta GebreMedhin, a man about 80 years old from Awo village, told us the story of Ghebray Hawku from Daya village near Awo who dreamt up a new idea about 50 years ago. In an attempt to catch the soil and water that rushed down the slopes, he piled stones and earth across the stream's path in order to make a field for sowing cereal. His neighbours saw his hard work and pitied him as they thought he was slightly demented. But Ghebray told them: 'Tomorrow you will all be as crazy as I am.' The others laughed but, as Zigta noted, a seed had already been planted in their minds. That seed began to grow when another Irob man, Kahsay Waldu, returned home as an ex-soldier. He had seen traditional soil and water harvesting by farmers near Tripoli. In a valley beside his home, he imitated the North African farmers by constructing a small dam, much like the one that Ghebray had built. Zigta observed this with interest and decided to experiment with the idea himself.

The innovation: *daldal* – a series of silt traps

Zigta started in 1957 by fixing a large stone at the bottom of a seasonal water-course beside his house. Silt collected behind this barrier. He sowed a few seeds on the newly created patch of land and harvested an armful of maize cobs. The next year, he placed more stones to make a somewhat higher barrier, collected more silt and harvested more maize. Over four decades he built a series of checkdams going further up the watercourse and raised and length-ened the walls each year. In this way, he created step-like terraces that are now about 8m wide, with a horizontal distance of about 20m between dams. Some of his checkdams are filled up to 10m deep with silt that had been flowing down from the eroding Agridat plateau. Not only has new farmland been created where there had been only rock before, but also well-filtered water can now be collected from the foot of the lowest dam during most of the year. This innovation for trapping silt and water is known as *daldal* in the Irob language.

Over the years, Zigta watched how the soil and water flowed within the terraces and over the dams, and learned from these observations. He changed the shape of the dams by curving the walls outwards as he widened them so that more soil collected and the force of the water was spread. He scraped down other patches of soil to help fill in the area behind the newly raised terrace walls so that a larger cropping area would form more quickly. He improved the arrangement of the stones on the top of the dam walls and placed some very heavy stone slabs slanting slightly upwards at the outer edge so as to prevent possible damage to his structures from the overflow. He dug

Credit: Ann Waters-Bayer

Plate 14.1 *Zigta GebreMedhin explaining the functioning of his checkdams (Tigray, Ethiopia)*

trenches both to increase run-on at some points and to divert excess water at others. He transferred sods of a tough local grass, known as *tahagu*, on to the deposited silt immediately behind the dam walls. The grass grew down and through the stones, holding them together like gabion wire. An additional benefit is that he can feed the grass to his animals. He planted trees in front of the walls to reinforce them. He did not develop techniques to fertilize the terraced land; he explained that the continuous addition of soil and litter, including tree leaves, with each flood maintains soil fertility.

The innovator: characteristics and motivation

At the time when Zigta began to build checkdams, wealth was measured not in land and crops, but rather in livestock. He was a skilled livestock keeper and had several cattle, goats and beehives. In the mid-1950s, his family consisted of four persons. He used to sell or trade animals to obtain cereals. However, after some years, as the terraces behind the checkdams became bigger, he could produce his own cereals and did not need to sell so many animals. He thus became richer.

His motivation for innovation came partly from necessity and partly from curiosity. When asked why he started the strenuous work of building check-dams, Zigta replied:

The geographical conditions of Irobland were not suitable for cropping, and are still not very suitable today. So our main source of livelihood in the past was livestock. We used to travel to far-away towns such as Zalambessa and Adigrat in order to buy cereals which were carried on the backs of donkeys for days, up and down the mountain paths. Some people who did not have donkeys carried the cereals home on their shoulders. While experiencing these hardships, we were not idle in our minds. We were forced by nature to think for ourselves with a long-term view.

Because of the different seasons, wet and dry, we moved from place to place with our animals to seek grazing land. During these movements, I observed some things again and again: when it rains, dried leaves and fallen trees are washed down the valleys with all the soil. Seeing what Ato Ghebray and Ato Kahsay [the first two Irob men who tried to place stones to catch the soil] were doing motivated me to start trying it myself, but if I had not noticed what I mentioned now, I would not have tried checkdams based only on these two men's examples. The main question I asked myself when I started was: 'Would it be possible to catch soil and water to create land and grow crops where this floodwater passes?'

Zigta described the years of developing his silt-harvesting system like a series of experiments: doing something with a vision of the benefits it could bring (in other words, a hypothesis, although he did not use this term himself), observing the effects, analysing the reasons for them, thinking of new ways to improve the technology, trying it out, observing, analysing and so on, in a process that continued until the Eritreans invaded Irobland in May 1998 and he had to flee to Adigrat. Sometimes in his experimentation, he recalled, he tried something and it worked well. Sometimes it did not, so then he reflected and tried something else. This entire development process went on without the aid of extension services. He depended on his powers of observation, his analytical capacity and his own creativity.

When asked why other people in the community did not make checkdams initially, he suggested two points:

1   *Handtools were not available then. Others saw it as hard work and it was not clear how it would bring benefits.*
2   *Not everybody has a clear vision of what they want now and in the future in their lives. God arranges everyone to be who he is and to have certain qualifications in life.*

Thus, Zigta sees himself as a hard-working and forward-looking man, and as someone who has been blessed with certain aptitudes. Other members of his family and community confirmed this. He is known as a man for whom laziness is next to sin. If he is convinced of an idea, no amount of difficulties and hard work will prevent him from trying to realize it. His perseverance and conviction that he is right (some relatives also refer to this as his 'stubbornness') drive him on, even when others think it is impossible. He is also known as a man who is concerned about the future of the community and its resources. He initiated community action to manage the use of Sangade, a common grazing area in Irob. Upon his suggestion, the boundary was marked, rules were drawn up to regulate land use and the priests say mass there once a year to ensure respect of the rules. When the community has made such decisions, he is strong in seeing that they are put into practice.

Traditionally, a man who has many ideas, is talented with words and can organize activities well becomes a community leader. Zigta came to be recognized as such a leader. In his opinion, 'working closely with the other members of the community is the best way to teach them what one knows'.

Spread of the innovation

Zigta's conviction that he was on the right track and that others should also benefit from his knowledge motivated him to spread his idea. He suggested directly to neighbours who had similar seasonal watercourses near their homes to try building dams themselves and he gave advice to them. He tried to encourage individuals, rather than whole groups of people. Zigta explained:

> *At the time when I started building checkdams, community meetings and discussions were not very common. When we visited each other for other purposes, we advised each other for better continuity of our farming.*

Some of these farmers, motivated by seeing the results that Zigta achieved with his checkdams, copied his techniques in order to create new land for themselves. However, the spread of dams was initially quite slow. Even though many farmers recognized the potential of checkdams, it was not easy to build them because no handtools were available and it took some years before enough land could be created to grow enough cereals to make a substantial contribution to family diet. Zigta recalled:

> *We were using big trees instead of crowbars to push down big stones. Where there is no soil at all, there is no possibility to grow something right away; instead, you have to wait for the soil to collect.*

Over time, communication improved between the farmers in Awo and other villages about how to capture more soil and water – for instance, by changing the shape of the dams and reinforcing them. For structural work on the dams, including major repairs, neighbours started to organize themselves into groups. As pastoralists, the Irob had been relatively individualistic, except for some agreements on the use of common pasture, but their growing emphasis on building and repairing dams to allow cultivation led to an increase in mutual aid. Almost all Irob farmers who live near seasonal waterflows now use the *daldal* technique.

A rapid increase in the building of such dams began after a small project financed by the Catholic Church made metal tools available for quarrying stone and encouraged the Irob to apply their visions for development and their masonry skills to activities that were planned and implemented by the community, such as making cliffside paths and larger communal dams. This project started in 1975 and is known today as ADDA (Adigrat Diocese Development Action). In the case of the *daldal*, as in the following case, there were important interactions between indigenous innovation and an externally supported project that helped to give recognition to and spread the new ideas, and to make the tools and funds (in the form of food-for-work) available to implement them.

YOHANNES TESFAYE EXPLOITS THE RIVER'S STRENGTH TO HOLD WALLS

Yohannes Tesfaye is in his mid-50s, a generation younger than Zigta. His home overlooks the Zararut River (1800m altitude) that passes below the town of Alitena. Annual rainfall is about 300mm. Yohannes was frustrated to see so much water and silt pouring through the riverbeds of their arid, rocky land and disappear into the depths of the Danakil. He had a family of eight persons and wanted to cover at least part of their food needs through their own production. He thought of using the silt carried by the river to build up land beside it and using the river water to irrigate the land. At a place beside the river where silt collected but was often washed away by subsequent floods, he built a wall parallel to the bank. He placed rocks in such a way that the silt-laden water was diverted into the area behind the wall. At first he made the wall like a house wall, with large flat stones laid on top of each other, but when the river flooded, the powerful water lifted up the stones and washed them away. He tried a second time, and the same thing happened.

The innovation: *seytan madewa* – the devil's tie

Yohannes then had the following idea: if the water made the stones stand upright, he would try to set the stones upright already before the water met them. So he found a hard rocky outcrop in the steep wall of the riverbank and used this as the starting point for placing a line of heavy flat stones upright, one standing beside the next, with longer and shorter stones alternately. He wedged further upright stones in a second storey into gaps between the first line of stones, making a small wall. He did this as an experiment, to see what the floodwater would do with the wall. He observed that the water roared over the top of the stones but did not dislodge them; Yohannes had outwitted the river by using the force of its own water to push one stone against the other and, in effect, tie them together through this pressure. This type of riverside wall became known locally as *seytan madewa* (devil's tie), named after the complicated tie which is very difficult to open used to fasten the bag made of a complete goatskin to hold precious gifts for an Irob bride.

It is possible that this innovation was helped by Yohannes' observations, although he did not mention this himself. In the natural sedimentation and accumulation processes in the riverbeds lower down the escarpment, such as around Gunda Gunde and Endeli, small-scale 'devil's ties' also occur: a large boulder acts as an anchor and a sequence of 5–10 flat stones accumulate upstream, pressed against the boulder. Observation of such phenomena in nature may explain why similar spontaneous innovations with upright stones have been reported from rocky areas elsewhere in the world, eg in Ireland (Bruno Strebel, pers comm, 1999).

Yohannes continued to experiment with slightly tilting the stones to go with the direction of water flow and pivoting the stones sideways a bit so that they stood at a slight angle to the flow. He found this to be even more secure and also effective in leading excess water away from the riverside plot. This water went further down the river instead of jumping over the wall into the plot. He managed to build up a riverside plot of about 800m² in which he grew maize, fruit trees (mainly orange) and vegetables (mainly cabbage). He sold his products locally in Alitena; the Catholic priests and nuns were his best customers.

The innovator: characteristics and motivation

Yohannes is often called *kubrare* in the Irob language, which means the 'master builder' or 'engineer'. He is widely admired among the Irob as a naturally gifted person who is creative and has a good sense of design. He learned by working. His father died while he was a youth. He received little formal education – only up to Grade 4 at the mission school. He started working on construction sites in the town of Adigrat; he carried stones and received very little payment, but he was observant and keen to learn. He gradually acquired skills in hewing and laying stones and deliberately sought ways to improve his knowledge of construction.

When Yohannes heard that engineers had come to Adigrat to build a big road, he went to work for them and to observe, eg how they designed bridges, where and how they made the curves in the road. When architects and engineers came to design and supervise the building of the Catholic Church in Adigrat, he worked there as a daily labourer and spent much time observing how the professionals did their work. He also travelled to other towns to see how other churches were designed.

When the ADDA project started in 1975, Yohannes was given the responsibility to supervise the work around Alitena. This involved construction of footpaths, wells and checkdams by community members. Here he applied the knowledge he had acquired on road construction sites to the alignment of community footpaths and he drew up the original plan for a road from Alitena to Zalambessa (to join the main road to Adigrat), built purely with manual labour. Later, when engineers came to improve the road, they changed very little in his design. Yohannes also became active in designing and building rural churches.

His fellow Irob admired him and never regarded him as 'crazy' when he came up with a new idea. By the time he developed the devil's tie, masonry work was already well known in the area. Other inhabitants immediately recognized the usefulness of his idea. Yohannes himself felt that:

> As is always the case in any society, when someone gains a certain advantage, even if it is by his own hard work, there is always someone who opposes. However, this was not a general reaction. Most people were interested in what I was doing. Nobody challenged my right to try to create land; only the water challenged me.

Yohannes has never been rich. At the time he started to claim land from the river, he had no land of his own, only three head of cattle and no other livestock, not even bees. When the people of Alitena discussed the potential uses of this new land protected by the devil's tie, they decided to continue the work jointly and to use the new land as a community nursery for fruit trees. Yohannes accepted a higher plot of land near Aiga, some distance from the river, in compensation for his efforts in creating land beside the river. However, by this time, he was more interested in construction than in cultivation, so he continued to devote most of his time to the community activities.

Spread of the innovation

The riverside field protected by the devil's tie is situated near a major footpath leading through Irobland to Eritrea. Many farmers passed, observed what Yohannes had done, thought it was a good idea and tried it themselves. If Yohannes was in the field when passers-by stopped to look, they sometimes asked him to explain. But the principles were obvious to people in the area who were already used to working with stone to catch soil and control water, so few explanations were needed.

Yohannes created the devil's tie in the early 1970s. Shortly thereafter, the project now known as ADDA commenced. During the difficult years of the struggle against the Derg regime, little external technical advice could be given to the project. The communities in Irobland continued to organize themselves to carry out the activities they had planned and depended largely on their own informal experimentation. The project work included the building of large communal dams. Yohannes noticed that floodwater pouring over the top of a dam pounded heavily on the land below and began to undercut the dam. He suggested trying something like the devil's tie to prevent this. Along the line where the water hit the land below the dam, community teams pounded large flat stones upright into the soil, slanting towards the flow of water, ie pointing from the flat terrace upward to the top edge of the dam above it. The stones broke the force of the falling water and dispersed it more widely within the terrace so that some water remained in the field while the rest poured over the next checkdam lower down the valley.

When Yohannes was supervising the teams working on the large dams, he led the workers to build the devil's tie wherever it was appropriate. Many of these people then used the same technology in their own series of smaller checkdams in the narrow valleys beside their homes and in the walls protecting their own riverside plots. Yohannes also applied the devil's tie to protect the lower part of footpaths built up the cliffsides of canyons through which water flowed seasonally so that the foundations of the paths would not be swept away. Other Irob people observed this technique and copied it when doing their own construction work. In this way, the practice of placing stones in the devil's tie spread throughout Irobland.

SITE-SPECIFIC INNOVATIONS

Zigta recalled seeing Yohannes' first experiments with the devil's tie near Alitena in the bed of the Zararut River. 'I was not very quick to adopt the idea,' he said, and explained that the original devil's tie was suitable for an area to which he did not have access. It was designed to save existing riverside land from erosion during heavy floods and to create new land by capturing soil and water beside the river, but Zigta lived at a higher altitude (about 2500m) and far from the river. He had already established checkdams by then and initially did not see how the devil's tie could benefit them. However, when Yohannes started to adapt the idea of the devil's tie to protect the cliffside footpaths and the larger communal dams, Zigta also began to experiment with inserting lines of upright stones like the devil's tie to break the water falling from the top of each of his own dams down to the next terrace so that the dams would not be washed out from below. He built such a structure only at those spots where he observed a danger of undercutting in order to keep the extra work to a minimum.

It thus becomes obvious that the structural innovations developed by Zigta and Yohannes are suitable only for farmers living in particular sites: near a seasonal watercourse with heavy run-off of silt and water in the case of the *daldal*, or beside a river with steep cliffs in the case of the *seytan madewa* in its original form. Both innovations are suitable for people living in steep rocky areas with excessive run-off. Farmers living at other sites may not be able to copy these feats of indigenous engineering exactly, but they could gain some ideas from them and could adapt them. For example, in areas that are less mountainous than Irobland, it may be difficult to find a rocky outcrop to support the downstream end of the wall of upright stones of the devil's tie; other means, such as a single cement block, may have to be used instead. If creative farmers who were trying to reclaim land from rivers in other areas were brought together with Irob experts in building the devil's tie, ideas for appropriate adaptation would probably emerge. Here, extension agents could help by bringing such farmers together.

The development process could be even quicker if formally educated engineers with an understanding of local practices and conditions would work together with Irob farmers to improve the innovations still further and to help other farmers to adapt the innovations to other conditions. It is this interaction of formal scientific knowledge and indigenous knowledge and creativity that ISWC 2 is trying to stimulate.

ACKNOWLEDGEMENTS

We would like to thank Bruno Strebel, geographer and technical adviser to the ADDA project, for the additional information he has provided, based on over 25 years of studying and supporting technology development and land management by the Irob.

REFERENCES

Hagos W and Asfaha Z (1997) 'How to stop erosion: catch the soil', *ILEIA Newsletter*, vol 13, no 2, pp21–23
Strebel, B (1979) *Kakteenbauern und Ziegenhirten in der Buknaiti Are (Nordäthiopien): ein Planungsbeitrag zur Agrarentwicklung in semiariden Berggebieten*, ('Cactus farmers and goat herders in the Buknaiti Are (North Ethiopia): a contribution to planning agricultural development in semi-arid mountainous regions'), Philosophical Faculty II, University of Zurich

15

A challenge and an opportunity: innovation by women farmers in Tigray

*Fetien Abay, Mamusha Lemma, Pauline O'Flynn
and Ann Waters-Bayer**

ISWC-Ethiopia started its work in 1997 in Tigray region in the far north of the country. An initial coordinator of the programme and the external adviser – both women – gave particular attention to identifying and working with women farmers who innovate in land husbandry. This paper describes the approach taken by ISWC-Ethiopia to recognizing female innovators and the types and processes of innovation that were discovered. It makes an initial analysis of the nature of women's innovations and brings some evidence of ISWC-Ethiopia's impact on encouraging innovation by women.

APPROACH TO RECOGNIZING WOMEN INNOVATORS

In Ethiopia as a whole, research and development efforts related to land husbandry have usually ignored the development potential of women's knowledge and innovation. Women's domestic work has a low status in Ethiopian society and their productive work in agriculture is seldom acknowledged. During the struggle to liberate Ethiopia from the Derg regime, the Tigray People's Liberation Front (TPLF) tried to change gender perceptions, but encountered considerable cultural resistance (see Chapter 29).

---

* Fetien Abay and Mamusha Lemma are past and present joint coordinator, respectively, of ISWC-Ethiopia; the former is a crop scientist and the latter a sociologist and both are lecturers at Mekelle University in Tigray, Ethiopia; Pauline O'Flynn is an MSc student of Rural Development at University College Cork, Ireland, and did her fieldwork in Tigray; Ann Waters-Bayer is an agricultural sociologist with ETC Ecoculture in The Netherlands and adviser to ISWC-Ethiopia

Map 15.1 *Ethiopia (Tigray action area)*

As a rule, women in rural families do not regard themselves as farmers and would not present themselves as innovators in land husbandry. This situation is not unique to Ethiopia. Also among farm families in Kenya, Uganda and Tanzania, the PFI programme found that women did not come forward to show and explain their own innovations; instead, male household members assumed this task, even though they did not understand the innovations as well as the women did (Critchley et al, 1999).

There are many factors that could explain the lack of self-esteem of women with respect to their farming activities. These include the traditional beliefs and attitudes regarding women's role in rural society; the low levels of formal education of women; the limited mobility of women compared with that of men who often migrate to towns or other countries to seek labour, and the poor access of women to external information.

ISWC-Ethiopia has endeavoured to give recognition to women's innovation in land husbandry in order to change gender perceptions of society, including those of the women themselves, and to increase the women's self-confidence and capacity to contribute to development. The first steps were to gather evidence of innovation by women farmers and to make these accomplishments more widely known.

Raising awareness

The search for evidence was part of a wider thrust to discover both men and women innovators in land husbandry (see Chapter 6). Development agents with the government extension service and NGOs, students and staff from Mekelle University and staff from Mekelle Research Centre were involved in a series of workshops designed to raise awareness about local innovation and gender roles. Already the initial workshop included exposure visits to women innovators who had been identified by the female coordinator of ISWC-Ethiopia. Between workshops, the DAs, both male and female, were given assignments to observe and document the innovations and informal experimentation by men and women farmers in their working area. In addition to the workshops for DAs organized by ISWC-Ethiopia in different zones of Tigray region, this awareness raising was incorporated into the established in-service training courses given to DAs on campus.

Gender issues are also addressed in the agricultural extension courses given to the regular students at MU. Senior students, male and female, identify and document innovation by women farmers within the Practical Attachment Programme (PAP) in which all students of agriculture take part (see Chapter 31).

Also, researchers from MRC and MU were encouraged to investigate gender issues in agricultural innovation. The ISWC-Ethiopia coordinators met informally with scientists who appeared to be open to these ideas, discussed with them the objectives and preliminary findings of the programme, and tried to interest them in investigating women's innovations more deeply. Seminars were also organized at MU where the scientists could present and discuss their findings.

In addition, in the Negash area of Eastern Tigray, a detailed study was made of the role of women in agriculture and natural resource conservation. This also looked into the degree of support that was available from various organizations, such as the Women's Association and the BoANR, and the extent to which this situation promoted or constrained innovation by women (O'Flynn, 2000). The purpose was to gain better insight into the roles of women in resource management so that it would be easier to know where to look for innovations. It is often difficult initially to identify women innovators because women tend to be involved primarily in the more 'invisible' activities in the home, farm and community.

Increasing women's self-confidence

Another major challenge was to build up the women's self-confidence. Women often collude with many men's perceptions of their not being productive in economic terms. The ISWC-Ethiopia coordinators contacted women innovators individually and, by discussing their achievements with them, tried to raise their self-esteem and give them the courage to withstand social pressures in their communities. The coordinators suggested to the women that they link with each other in informal networks to give each other mutual support.

ISWC-Ethiopia has undertaken various activities that are designed to give recognition to women innovators and to increase the social esteem of their innovations, for example:

- *Discussing women's innovations at village meetings.* Women innovators are encouraged to present their innovations at village meetings. This helps to increase their confidence and skills to explain what they were doing, also at subsequent district, zonal and regional meetings. It gives the farming communities an opportunity to learn about women's innovations and to change their perceptions of women.
- *Including women innovators in travelling seminars.* Women innovators were invited to join travelling seminars that were organized to visit and discuss local innovations in land husbandry. This gives them access to information and knowledge from localities other than their home villages and inspires them to experiment with some of the new ideas.
- *Giving awards to women innovators.* Some women innovators identified at village level were given awards at ceremonies arranged by the Tigray BoANR during farmer field days and community meetings. Outstanding women farmers were also honoured at regional level in the presence of government policy-makers in an attempt to open the latter's eyes for women's achievements. The awards consist of money and a certificate, and serve as an important incentive for the women.
- *Encouragement to women innovators by local leaders.* ISWC-Ethiopia approached opinion leaders in the rural communities, including priests and members of the local council (*baito*). These people agreed to give moral support to women innovators at community meetings by mentioning them as examples of successful farmers.
- *Encouragement to women innovators by DAs.* The programme devotes much time to organizing and facilitating the awareness-raising workshops and in-service training for DAs as these agents should play a key role in promoting innovation by women. Discussions among the DAs about innovation and gender, and the joint planning of ways to give due recognition to women innovators constitute an important part of this training.

The programme also promotes women's innovation by printing and distributing t-shirts depicting a woman ploughing, displaying photographs and posters on women's innovation, publishing descriptions of women's innovation in newsletters and newspapers in both English and Tigrigna, featuring women innovators on radio broadcasts in Tigrigna, and linking women innovators with organizations that can provide them with the support they seek.

EXAMPLES OF WOMEN'S INNOVATION IN LAND HUSBANDRY

Already during the first year of the project, the DAs in government service and NGOs, the scientists, and the students and staff of MU, including the ISWC-Ethiopia coordinators, discovered a wide diversity of innovations by women, such as:

- digging infiltration pits in backyard gardens and fields;
- making infiltration furrows in cropland;
- using floodways to distribute manure over fields;
- backyard compost-making;
- ploughing by donkey and/or oxen (normally a male task);
- ploughing other (also male) farmers' fields upon request;
- covering farms with crop residues to protect against erosion;
- collecting and planting/sowing grasses, sods and seeds;
- enclosing slopes to reduce flooding and gully formation;
- protecting spontaneous growth of tree seedlings;
- improved sowing techniques, such as soaking seeds to accelerate germination and mixing seeds with sand to regulate the seeding rate;
- growing early maturing crop varieties;
- constructing stone bunds on farmland (normally a male task); and
- conserving local plant biodiversity by growing in their home gardens plants which are not considered by scientists to be of economic importance but are used by women in their homes for cosmetic, medicinal and other purposes.

In order to illustrate the types and processes of innovation by women and the constraints that they face, we present three case studies.

Case I: Leteyesus Gobena

Leteyesus Gobena lives in Central Tigray, where annual rainfall is about 600mm. She is 30 years old and a widow with two dependent children. She started ploughing five years ago, after her husband died. In the local culture where men traditionally plough, this is an innovation in itself. During her first attempt to plough, she created some unintended furrows on each side of the uncultivated grass strips that traditionally divide the small plots. She observed that the teff (an indigenous cereal, *Eragrostis tef*) growing in plots with these furrows brought a higher yield than the teff in her neighbour's adjacent, uniformly ploughed plots. The next season, she made the furrows on purpose. In order to gain more benefits from the uncultivated strips, she dug up sods of grasses preferred by livestock and moved them to the grass strips. She explained that the shade from the grass prevents the water in the furrows from disappearing too quickly in the sun. She also collected seeds of wild grasses that are useful for different purposes, such as making brooms, and sowed these on the strips. The following year, she cultivated where the enriched grass strips had been and made grass strips lined with furrows where she had previously grown teff. She thus seems to have developed a kind of simultaneous enriched fallow on her tiny landholding (0.2ha).

Leteyesus was obliged to innovate by the pressing problems she faced after her husband's death. She had oxen and was sharecropping her land with a man who gave most attention to his own land and little to hers and who did not take good care of her oxen. Because of his delays in ploughing her land, she could not sow on time and she obtained low yields. She therefore decided

Credit: Fetien Abay

Plate 15.1 *By ploughing herself, a woman in Tigray, Ethiopia challenges local cultural norms*

to plough herself. She could find no one to teach her, but, through practice, she learned to adjust the ploughshare properly and to drive the oxen to pull it in the desired direction.

When she started ploughing, neighbours and relatives openly expressed their fears that she would bring God's curse on the community. They ridiculed her with remarks like: 'Let alone feeding your family, you will not produce even enough for a coffee meal.' It is customary in Ethiopia to provide a small amount of food, such as a bowl of roasted cereals, during a coffee ceremony. The insinuation was that the produce from Leteyesus' ploughing would not even be enough to provide this small token. Some gave her the nickname '*Girazmach GebreEyesus*'. Girazmach was a title of a male warrior during feudal times in Ethiopia and GebreEyesus is a man's name. The people were making fun of the fact that she was trying to take charge of a difficult task, like a war commander, and, moreover, a man's task. Some relatives argued that she would injure herself while ploughing and advised her to go back to sharecropping. However, she ignored these efforts to discourage her and continued to plough her land.

Over time, after recognition and encouragement by ISWC-Ethiopia, the local DA and some community leaders, Leteyesus noted a change in community attitude to her practice. She is now often mentioned as a good example in community meetings. Her initiative to plough is being praised in the local elementary schools. She is proud to say that the District Administrator and several women farmers have visited her.

As her innovation challenges the local cultural norms, Leteyesus has not been so forthcoming as some male innovators in making known what she does. It is much more difficult for women to be accepted as farmers and innovators in their own right. Thus far, four women have tried out her innovations of ploughing, making furrows for better water infiltration, and enriching and alternating the grass strips from season to season. First, they visited her farm to learn through observation and by practising the techniques with her. One of them even invited Leteyesus to her own farm to supervise her attempts to plough. Leteyesus has noted that an increasing number of women farmers are coming to her, saying: 'You made our village known to the government. Please make us known as well.' This suggests that women have developed more confidence to innovate and experiment, and they now feel that the community will give them recognition.

Since being in contact with ISWC-Ethiopia, receiving a prize sponsored by ISWC for her innovation and being involved in various workshops, including a travelling seminar to other innovators (see Chapter 18), Leteyesus has developed self-confidence and commitment to spread her innovations. She was featured in an exhibition about farmer innovation organized by the District Council and, in recognition of the development services she is giving to the community, she was awarded 500 *Birr* (circa US$60) and some farm implements. She has also started several new initiatives to improve the life of her family, such as hiring people to build a new house and to dig a well.

Case 2: Askwal Barochy

Askwal Barochy is 50 years old and lives near Wukro in Eastern Tigray. The average annual rainfall in her area is 580mm which falls mainly during the long wet season; the rains in the short wet season are very unreliable for cultivation. She assumed control of the farm (roughly 0.75ha) when her husband died eight years ago. Since then, she has steadily improved the farm, using the resources available to her in an innovative way. Until recently, only a small part of her land was irrigated. In 1999, she decided to divert water from the main irrigation channel to small plots where she had previously been able to grow crops only during the long wet season. Now she is working 0.44ha of irrigated land where she can grow crops throughout the year. She sells vegetables at the market twice a week.

Much of her farmland is on a slope. She allowed trees to regenerate naturally on the hillside behind her house. As a result, she has no problems with soil erosion and she can supply her family with fuelwood. She still uses both dung and wood as fuel, but all the wood is from her own land. She also protects tree seedlings that grow spontaneously on her farmland and, in one area of her farm, she has planted a euphorbia hedge to do this. She claims that she has never received any advice from extension workers about these SWC techniques.

Askwal is an extremely hard-working woman who loves farming; this is her main motivation to innovate. She is fortunate that her father is also a good farmer and she has learned much from him. During a wealth-ranking exercise made by community members in 2000, Askwal was the only female head of household who was considered to be wealthy according to local standards (two or more oxen). All other female heads of household were grouped in the poorest wealth class, except one of medium wealth. When she was asked directly why she was regarded as the only wealthy woman heading a household in the community, Askwal laughed and said that the people were mistaking hard work for wealth.

Case 3: Abrahat Tsagay

Abrahat Tsagay is 45 years old and lives with her husband and four children in Eastern Tigray where they farm an area of about 0.75ha. She was very much influenced by her father who was always trying out different grasses on his farm to use as forage and for various other purposes. After moving to another village when she married, Abrahat likewise experimented with grasses. As she learned this from her father, it might not be considered an innovation. However, in the area where she now lives, she is using the resources in a way that no one else has done before. Over the years, she built up considerable knowledge about how best to grow different grasses and about different ways to use them.

Abrahat set aside a plot of land uniquely to grow grasses. Each year, she harvests the seeds and keeps them to sow later in another part of the farm. She has sown grasses on the borders of all the plots on her family's farm. The family uses the grasses to feed livestock and also recognizes their usefulness in conserving the soil. Last year, through cash-for-work, the community built stone bunds on the family's farm. In the following wet season, Abrahat sowed all the bunds with various grass seeds that she had saved.

The farm is far from any source of water. Therefore, most women in the area grow vegetables and spices in their home gardens only during the wet season. Abrahat, however, generates cash year-round from her garden because she carries extra water (in addition to that needed for the household) on her back from the spring several times a week. She also generates income out of the grasses which she uses to make *miha* (a kind of sieve needed in all farm households during the harvesting season) and other wicker products which she then sells.

The other women in her community are not keen to go to the trouble of carrying extra water to grow vegetables in the off-season, but both women and men in the community have become interested in growing grasses. When Abrahat first started doing this, neighbours laughed at her for using precious land for grass. Now, many of them are beginning to do the same.

INITIAL ANALYSIS OF THE NATURE OF WOMEN'S INNOVATIONS

In Tigray, culture and religion exert very strong influences on people's lives. The women's innovations are basically of two different types: taking on non-traditional roles, and making use of available resources in new ways without challenging social norms.

Taking on new roles

These are cases in which women step outside of their traditional roles and challenge social norms. This type of innovation is more likely to be found among women who are household heads or whose husband cannot do farm work for some reason, such as absence (temporary labour migration), handicap or long-term illness. Such innovations arise primarily out of necessity rather than a deliberate decision to challenge norms. Innovations of this type include a woman carrying out tasks normally done by a man, such as ploughing, threshing or building stone walls. These innovations by female heads of household are more visible (at least to those who are familiar with women's traditional roles) than those by women living with their husbands. Leteyesus is an example of this type of innovator.

Using available resources in new ways

In situations where a woman is not forced through a shortage of male labour to take on traditionally male tasks, innovations are more likely to take the form of doing in a different way something that women have always done or doing something that is new to the community. An example of the latter is Abrahat who collects grasses, grows them on a special plot, saves the seeds and strategically sows them in different parts of the farm. Also her initiative in growing garden products in the dry season, using extra water she carries to the plot, is an example of making more productive use of existing resources without challenging traditional roles. Likewise, Askwal optimizes the use of available irrigation water and the spontaneous regeneration of trees.

The village study in Eastern Tigray also revealed that women in different wealth classes pursue different types of innovations. Poorer women can afford to take fewer risks, but this does not mean that they are less innovative – indeed, the opposite may be the case. This became clear in discussions with men and women farmers about their experience with credit. Most farmers felt obliged to take credit for fertilizers and sometimes improved varieties of seed, reportedly because of pressure from the BoANR. Outside of taking credit for fertilizers, only one interviewee from the poorest category had taken credit for anything else and that was to buy a sheep. Poorer women felt that it was too risky to take credit and preferred to look for ways to use what they have more efficiently. If the women were not being asked to take credit for fertilizers, they might be more inclined to take it for other reasons, such as to buy a goat or some chickens, but they do not feel that they can afford to take credit for more than one thing at a time.

Evidence of their search for ways of using local resources more efficiently without having to take credit can be found in how the women manage soil fertility. Wherever possible, women try to use animal manure. However, the poorer households have few or no livestock and therefore have little access to this source of natural fertilizer. Some of the women who head poorer households found alternative means of fertilizing their land, including allowing animals of relatives or friends to graze on their land in return for access to the dung. Many of the women use cooking ash as fertilizer and one woman who has a particularly large amount of ash from kilning her pottery finds this to be an excellent soil enhancer. While these practices are not new to the area, they do indicate the efforts being made by women to maintain soil fertility with the resources available to them, and it is likely that a deeper study of fertility management would reveal the innovative means that have been developed especially by poorer women. Many of these women are reluctant to respond to the 'encouragement' (through credit) of using artificial fertilizers as they fear they will not be able to pay their debts if the rains fail.

Some female-headed households own no livestock at all, but many do have at least a few chickens. Chicken droppings are an excellent source of nitrogen and phosphorus, which is far better than the manure of cattle, sheep, goats or donkeys. Only a small quantity of droppings is produced per animal, but it is a valuable addition to a compost heap that would otherwise contain no animal manure at all. One woman interviewed who could not afford to buy any chickens of her own had come to an arrangement with a neighbour to take care of some chickens for her in return for half of the eggs. This arrangement allowed her to add the chicken droppings to her compost which otherwise consisted only of plant-based materials.

Many of the innovations that women, whether 'rich' or 'poor', are practising in their home gardens tend to be overlooked or ignored because of the seemingly insignificant size of the plots (on average, about $5m^2$). Here, the women grow mostly spices and vegetables, usually for home consumption and only during the rainy season. However, as seen in Cases 2 and 3, some women have gone beyond this. By diverting water or putting extra labour into carrying water, these women are managing to produce fresh vegetables year-round for home consumption and for sale.

In another case, a widow and her daughter irrigate about 0.125ha of their land (bigger than the average home garden) from a well that was sunk ten years ago by the woman's husband. Each day, they draw water from the well with a bucket. They try to grow as many different species as they can; at the time of the study, they were experimenting with some cotton plants. Home gardens such as these are important for maintaining a diversity of spices and other plants that are used regularly in the home but are not likely to be grown in the fields. A more detailed investigation of women's knowledge about growing and using such plants would be needed to appreciate fully the role of women in preserving these species and experimenting with them. Such innovations are far less obvious than a woman ploughing and are therefore more easily overlooked, even by DAs and students deliberately seeking local innovators.

IMPACT OF ISWC-ETHIOPIA ON INNOVATION BY WOMEN

As a result of the in-service training and workshops, combined with assignments to identify and document local innovations, both male and female DAs have become more aware of innovation by women. The DAs report that they are now spending more time advising women farmers and making women's innovations more widely known. Thus far, however, less than 13 per cent of the innovators documented by DAs are women. This does not include cases of husband–wife innovator teams. Some additional women innovators have been identified by partner organizations and by students of agriculture who were trained in PTD approaches, but records of these cases have not yet been entered into the ISWC-Ethiopia database. Three MU students on practical attachment have investigated specific issues related to gender and innovation, and several other students have given ISWC-Ethiopia oral reports on their field observations of women innovators. One foreign MSc student (O'Flynn) has studied innovation by women and held a seminar about the results at MU.

During the struggle for liberation, both men and women participated in making decisions, and women were trained by the TPLF in various agricultural skills, including ploughing and making farm implements. This history has prepared the ground for recognizing women's contribution to the economy and for encouraging more equitable participation in development, although some sociocultural constraints obviously still need to be overcome. The ideological atmosphere for change in Tigray has favoured the efforts of ISWC-Ethiopia to encourage women farmers to innovate, experiment and share their ideas. Political leaders at regional and zonal levels express support for the approach. The women taking part in the programme have developed confidence to withstand the social pressures that still remain at community level. They have gained access to useful information and knowledge about agriculture and natural resource management through visits to other farmers, informal networking meetings and participation in workshops and conferences.

The women said that the awards they received motivated them to continue innovating: to experiment and to learn from their experiences and the experiences of others. Those who received awards have used them for the benefit of their families. For example, one woman used the money to build a new house and to send her son to school. Women who formerly sharecropped their land with male farmers who did the ploughing now have their own oxen and do their own ploughing. They are getting a better yield and can keep it entirely for their family.

The women farmers are building up their own informal networks. These are not networks of innovators but rather local networks of neighbours with similar concerns. In some cases, the group is named after the woman who was the driving force behind its formation. These groups meet in churches or market-places and come together to practise innovations like infiltration pits in the backyard of a member who provides food for the group. Occasionally men are also involved in this work but, because of the recent war between Ethiopia and Eritrea, usually only women are there to do it. Besides encouraging other women to plough their land and teaching them how, one woman innovator, Mawcha GebreMedhin has formed a network of women who are particularly concerned with preventing the encroachment of others on their enclosed land. They convinced the local council to establish and enforce a bylaw to fine people who illegally cut the trees in private enclosures and homestead plantations. This has encouraged the women to enrich the vegetation in the areas they have enclosed.

Community members are showing more appreciation for the achievements of the women innovators who are mentioned as examples and models in community meetings and in schools. The photographs and posters of women innovators raise great interest and admiration by other farmers, including those outside the villages where they live. The recognition being given to women farmers is also creating social pressure for men farmers to make more efforts to put local resources to their best use, as women have been demonstrating.

CONCLUSIONS AND FURTHER CHALLENGES

Most of the women who were identified as innovators are household heads with little access to farming resources such as oxen for traction. They have little family labour, often because of the death of the husband, divorce or separation in the case of resettlement. They were obliged to innovate in order to gain a livelihood and to ensure the survival of their children. Thus, they seem to have been driven to innovation at least initially by their poverty.

The most striking innovations by women are those that go against traditions, such as ploughing and using donkeys instead of oxen for traction. However, the vast majority of innovations by women involve small-scale, less obvious changes in the farming system, such as digging infiltration pits, planting grasses for multiple purposes or trying out a variety of plant species in their backyards. These innovations show how local resources can be used more intensively in subsistence farming. They require less external inputs than many of the innovations developed by richer male farmers and are therefore of potential interest to many small-scale farmers, not only women.

Most of the women were quite open in sharing their new ideas, at least with other women, but it seems to be more difficult for men to accept women's innovations. These do not spread so easily as men's innovations, often because of cultural reservations in the community. Moreover, female heads of household find it particularly difficult to spare the time from their work to advise others.

The major difficulties faced by women innovators are the social pressures of the community. The women who challenge the cultural norms, in particular, are criticized by other community members. Raising the social esteem of such women and supporting them to withstand these pressures has been a major gender concern of ISWC-Ethiopia. The programme has tried to make women's innovations more widely known. The self-confidence of the women has been increased as a result of, among other things, giving them awards as innovators, encouraging them to present their innovations at village workshops, facilitating their participation in travelling seminars, and stimulating support for them by community leaders and DAs. Other farmers are beginning to appreciate women's innovations and to try some of them out on their own farms. Community members' observation of women's success as innovators has led to greater recognition and acceptance of their contributions to improved land husbandry.

Relatively few women innovators in Tigray have been identified thus far. However, discovering their innovations and encouraging more women to innovate constitute only the beginning of the challenge.

REFERENCES

Critchley, W R S, Cooke, R, Jallow, T, Lafleur, S, Laman, M, Njoroge, J, Nyagah, V and Saint-Firmin, E (eds) (1999) *Promoting Farmer Innovation: harnessing local environmental knowledge in East Africa*, RELMA/UNDP, Nairobi

O'Flynn, P (2000) 'The implications of the gendered division of IK and divisions of labour in land husbandry on innovations by women farmers: case study of Me'ago village in Eastern Tigray', MSc thesis, University College Cork, Ireland

Part 4

Farmers' evaluation and extension of local innovations

Community assessment of local innovators in northern Ethiopia

*Yohannes GebreMichael**

A study of indigenous SWC practices in Northern Shewa and Southern Wello in northern Ethiopia revealed considerable dynamics in the indigenous knowledge system. It also revealed how the values of farming communities lead them to assess the innovators in their midst in different ways than research scientists might assess them.

The study was carried out from 1995 to 1998 (Yohannes, 2000) and covered six Peasant Associations (PAs) located in different agroclimatic zones and receiving either high or low levels of government extension input in SWC. Resource persons from among the farmers in each locality classified all landowning households into different wealth ranks (rich, medium and poor); the criteria they used were mainly livestock and landholdings. Using random and stratified sampling, 371 out of a total of about 1800 households were selected for inclusion in the study.

During the household survey, the 371 household heads were asked to give the names of at least three innovative farmers in their community (from the total number of households, not just from the sample for the study) who conserve their farms best by their own initiative. They were then asked to state why they selected these farmers.

---

* Yohannes GebreMichael, who completed his doctoral studies with the Soil Conservation Research Project under the federal Ministry of Agriculture in Ethiopia and Berne University in Switzerland, is a SWC specialist based in Addis Ababa and chairs the ISWC-Ethiopia Steering Committee

Map 16.1 *Ethiopia (Southern Wello and Northen Shewa research areas)*

CHALLENGES IN IDENTIFYING INNOVATORS

An innovation was defined as something new that has been started within the lifetime of the farmer, not something inherited from parents or grandparents. The farmer innovator is not necessarily a 'model' or 'contact' farmer; rather, s/he creates or tries out new ideas without their having been recommended by extension workers.

'Innovation' is a broad terminology that can refer to discovery of a completely different way of doing something or to modification of an existing technology. This perspective creates a big challenge in distinguishing an innovative farmer.

Every farmer has to be an innovator to some degree. Among the smallholders in the study area, there is a great diversity with respect to characteristics of the household (eg family composition according to gender and age) and plots (eg altitude, slope, soil type, plot size and shape, physical structures). Two plots are not treated identically by the same farmer, let alone by different farmers. This means that a technology cannot be applied in exactly the same way in different plots; some site-specific modifications will be necessary. Only the basic principles or functions of the technology will remain the same.

In addition to the spatial diversity, the time dimension demands innovation. Whether inherited or introduced, technologies undergo some modification with changes in land use that may be caused by natural factors, eg changes in rainfall pattern, or by human factors, eg changes in human population and land tenure policy.

The technologies of SWC are very diverse, involving physical, biological and agronomic measures, and a single technology may have different functions within a plot. Some technologies are fixed, such as bench terraces; some are mobile and may be meant primarily to improve fertility. Some are designed to harvest water, others to drain excess water, perhaps both in the same field at different times of the year. It becomes even more complex when agronomic practices such as frequency of ploughing, seeding rate and crop mixtures are considered, or management practices such as time of sowing. It is therefore often difficult for local people, let alone outsiders, to observe when something new is being done in a plot or farm.

CHARACTERISTICS OF INNOVATIVE FARMERS

Nevertheless, more than 70 per cent of the farmers interviewed thought that nowadays every farmer is becoming an innovator, in the sense of trying out something new in SWC, often picking up an idea from neighbours. By the time a given farmer innovates, the basic idea may no longer be new to the community but it is new to that farmer and s/he experiments with it to adapt it to the specific conditions of his or her farm. Very few of the interviewees (less than 5 per cent) could not point to any farmer innovators in their community. There is a tendency that a large number of farmers in a given area adopt a similar technology that has proved to be useful, rather than isolated individuals continuing to operate with a technology that others in the area do not take up.

In each PA, the interviewees identified 5–10 outstanding innovators. Most of these were mentioned by several farmers in the same locality. A total of 47 farmer innovators were identified. During the interviews, the criteria that emerged for choosing these farmers were:

- low number of *borobere* or *gurebya* (big and small gullies) in their plots;
- good spatial arrangement and integration of physical and biological SWC techniques;
- well-designed and well-implemented SWC works that are not easily destroyed;
- safe drainage of water to natural waterways so that the water has no negative effect on neighbouring plots;
- very good land quality and healthy crop stand or growth; and
- high frequency of maintenance and amount of time devoted to SWC work in the farm.

None of the identified innovators were women, although almost 10 per cent of the interviewees were women. This was because the focus of the discussion was on new physical structures of SWC or changes in traditional structures, most of which are done with the aid of draught oxen; women in the study area do not work with plough and oxen. Moreover, most of the female household heads are engaged in off-farm activities such as petty trade, crafts and beermaking, and do not devote themselves to farming. They usually lend out their plots for sharecropping.

More farmer innovators were identified in the areas with low rather than high levels of extension input. This was probably because the SWC, promoted through Food-for-Work or large-scale campaigns, introduced standardized technologies which led to homogeneity instead of encouraging adaptation to diverse conditions.

By far the majority of innovators were elderly (50 years and above). Some of the middle-aged innovators had come into the area under a resettlement programme and were former soldiers. Their exposure to other parts of Ethiopia possibly gave them ideas to try out in their new surroundings. The level of formal education did not have a significant influence on their propensity to innovate.

Large family size was also not a decisive factor on its own; many innovators were single or had only small families. They did their SWC in a way that did not demand a great deal of labour at one time, but rather spread the work over several months or years of day-to-day work on the land.

The farmer innovators were in different wealth ranks: 46 per cent were 'rich', 33 per cent were 'medium' and 21 per cent were 'poor'. Some farmers explained that the rich, especially, can innovate because they:

- have their own draught oxen and can release family labour for SWC work;
- can use manure from their livestock which adds to the positive effect of the SWC works;
- are usually elders who have experience in experimentation and are better able to assess the potentials and limitations of SWC techniques; and
- have a large number of plots in different agroecological locations which demand different innovations.

According to all farmers interviewed, a fundamental feature of farmer innovators is that they are hard working. They were not usually involved in off-farm activities. The other farmers generally felt that they could learn many things from the innovators, such as their ethic of devotion to the land, their way of 'uprooting' stones for bund construction, their way of designing the structures and their use of different crop varieties.

Another fundamental feature was that many of the innovators' plots were located at critical sites such as on steep slopes, at run-on sites, in depressions and close to big gullies. As mentioned above, the focus in the discussions was on physical structures and it is at such critical sites that physical structures would be a major strategy in SWC.

Land security had little influence on the propensity to innovate in SWC. This is because, at such critical sites, short-term survival would be almost impossible without inputs into land care; the seed would be easily washed away and the farmer would have to use more seed per unit area. Thus, it was in the farmers' immediate interest to find ways of reducing erosion to a minimum in the current year, no matter whether the land would be theirs in future years.

INNOVATORS AND COMMUNITY VALUES

The criteria applied in the identification of innovations and innovators by the community members inherently addresses the interests of the community at large. In other words, both the innovation and the innovators are evaluated as an integral part of the whole system. This was demonstrated during group discussions with different categories of farmers, when the following points emerged.

Incorporation into traditional sharing of ideas and skills

Most of the rich innovators were engaged in sharecropping. For this, they are in high demand by the poor in general and by the female household heads in particular because these farmers take good care of their shareholders' plots. The innovators in the middle and poor wealth ranks were sharing their techniques (innovations) and skills with community members during the traditional exchange of labour for various farming activities, including SWC works.

Mutual support for survival

During crises such as drought, many farmers have borrowed different local seed varieties from the farmer innovators. Many of the innovators have experience in preserving a diversity of local varieties through experimentation, often in their backyards.

Rehabilitation of degraded land

With the prevailing population pressure and scarcity of land, rural communities are forced to divide up the common grazing areas and woodlands that are located on steeper slopes and to use these for cultivation. Usually, the farmer innovators who were allocated land on the slopes experimented with various SWC techniques in their attempts to make the land productive. Other community members learned of such techniques primarily through observation and, where appropriate, adapted them to their own situations. This is a cost-effective way of learning as the less innovative farmers do not consume so much time in trial and error.

Basis for community statement to outsiders

When external agents try to introduce unpopular technologies to the community, members like to be able to use local innovators and their innovations as a point of reference for making community statements about these unwanted interventions. For example, when the technique of graded *fanya juu* terraces with artificial waterways was being promoted by extension in Northern Shewa, the community pointed to the innovation of one local farmer (who had developed a combination of traditional ditches and level bunds) to demonstrate the better quality of local techniques in conserving the soil and producing a better crop yield. At the same time, they were proud of and more confident in the innovation by their community member.

Compliance with community rules, regulations and cultural values

Any innovation that does not fit into community values is not easily accepted or integrated as a common practice and the innovator also has acceptance problems. The following examples reflect evidence along this line:

- *Conflicts with traditional farming practice.* A young farmer (an ex-soldier) in one of the villages sowed earlier than the other farmers and managed to obtain a relatively good harvest. However, the villagers said that this would bring hailstorms and delay the useful rains. Because early seeding could bring a higher risk of bird and wildlife attack, all neighbouring farmers usually sow similar crops at the same time. The young farmer was criticized through the *Edir* (a traditional institution) for his dangerous practice.
- *Undermining the historical land-use changes.* In another village, the community criticized some young farmers who planted marginal hillside plots with eucalyptus trees. From past experience, the farmers feared that the reafforested land would be claimed by the government and would be lost to the community.
- *Misuse of innovators by outsiders.* In the lowlands, a middle-aged farmer was gaining a good yield by using fertilizer and improved seed. This initiative was recognized by the extension agent who praised the farmer's achievement at a community meeting. Through the *Edir* this farmer was criticized because the other community members did not want his success story to be used as a reason to force them to buy inputs at high interest rate – a current thrust of the extension service.
- *Conflict with common resource use.* One farmer built stone bench terraces on his farmstead which also served as a fence against free-ranging livestock. Many other farmers condemned this practice because it prevented the traditional sharing of grazing land after harvest. Many poorer farmers who have only very little land but a few animals depend on free grazing on other farmers' land. Those who have some livestock and no arable land, such as some poorer women, are particularly disadvantaged by such exclusion. Similarly, the rich farmers with livestock also benefit from the free grazing and therefore also criticized this form of land privatization.

It was also interesting to note that the community members did not regard an innovation as a distinct and neutral technology, but rather saw it in relation to the person who developed it. Moreover, the innovation was not regarded as something that arose out of a vacuum, but rather as something that was a further development growing out of common community knowledge. The relationship of the innovator to the community and its values influenced the local perception of the innovation, and the community assumed its right to accept or reject it.

It can thus be seen that a farmer is given recognition as an innovator not only if his or her creativity improves production and protection of the land, but also if the innovation is valuable to the community. An innovation that does not agree with community values will be difficult to disseminate. It is not sufficient therefore for outsiders to assess innovations only according to the increase in productivity they bring to individuals. The view of the wider community must be sought.

REFERENCES

Yohannes G M (2000) *The use, maintenance and development of soil and water conservation measures by small-scale farming households in different agro-climatic zones of Northern Shewa and Southern Wello, Ethiopia*, doctoral thesis, University of Berne/SCRP Research Report 44, Soil Conservation Research Project, Addis Ababa

17

Stimulating creativity among East African farmers: from isolated individuals to interactive groups

*Will Critchley, Patrick Lameck, Alex Lwakuba, Charles Mburu and Dan Miiro**

The project Promoting Farmer Innovation has articulated a methodology that was derived originally from various participatory approaches and modified through experience. This chapter focuses on one of the ten steps in the methodology: forming clustered networks of farmer innovators living close to each other so that they can stimulate each other's creativity.

The methodology developed by PFI has its roots in the family of participatory approaches to rural development, including PRA, PTD and participatory extension. The methodological framework was tested and modified through experience gained in south-west Uganda under the project Conserve Water to Save Soil and Environment (CWSSE). This latter project has now been integrated into the ISWC 2 programme (see Chapter 19). The field-based activities of PFI, operating through the Ministries of Agriculture in Uganda, Kenya and Tanzania, have been organized into a ten-step ladder as shown in Figure 17.1 (see also Critchley et al, 1999). We touch here only briefly on the initial

---

* Will Critchley is an agriculturalist with CDCS, Vrije Universiteit, Amsterdam, the PFI programme and ISWC-Uganda; Patrick Lameck is a rural development specialist with INADES-Tanzania in Dodoma and coordinates PFI-Tanzania; Alex Lwakuba is senior soil conservation/agroforestry officer with the Uganda Ministry of Agriculture, Animal Industry and Fisheries (MAAIF) in Entebbe and coordinates PFI-Uganda; Charles Mburu is a soil conservationist/livestock expert seconded from the Kenya Ministry of Agriculture and coordinates PFI-Kenya from a base in Mwingi, and Dan Miiro is soil scientist with MAAIF and coordinates ISWC-Uganda from a base in Entebbe

Map 17.1 *Kenya, Tanzania and Uganda*

steps of identifying and verifying local innovations, and focus on the fourth step: the formation of clusters of farmer innovators.

In the context of this project, farmer innovators (FIs) are defined as those farmers who have developed or are testing new (in local terms) systems of land husbandry that combine production with conservation. Excluded from this are 'project pets': farmers who have been intensively coached or (to continue the metaphor) groomed by development projects, ie who are practising project-driven innovations.

IDENTIFYING FARMER INNOVATORS

In each country (Mwingi district in Kenya; Dodoma region in Tanzania, and Soroti, Katakwi and Kumi districts in Uganda), the project has identified FIs and their innovations sometimes through a process of PRA and sometimes through the existing knowledge of government extension workers, project and NGO staff, and other local contacts that are familiar with farmers in the area. A common initial problem was that 'master farmers' were put forward, while genuine innovators with a lower profile (because they were less well known to the extension staff) were overlooked. When the grapevine of information about PFI began to grow, some innovators came forward of their own volition. The most productive way to identify local innovators proved to be informal, through discussions with people in the communities. For various reasons (see

	10 FIs as outside trainers
	9 Farmers visit FIs
	8 FIs develop new techniques and experiments
	7 Study tours for FIs
	6 FI to FI network visits
	5 Set-up monitoring and evaluation (M&E) systems
	4 Formation of clustered networks of FIs
	3 Characterization and analysis of FIs and innovations
	2 Verification of innovations and 'recruitment' of FIs
	1 Identification of FIs and innovations

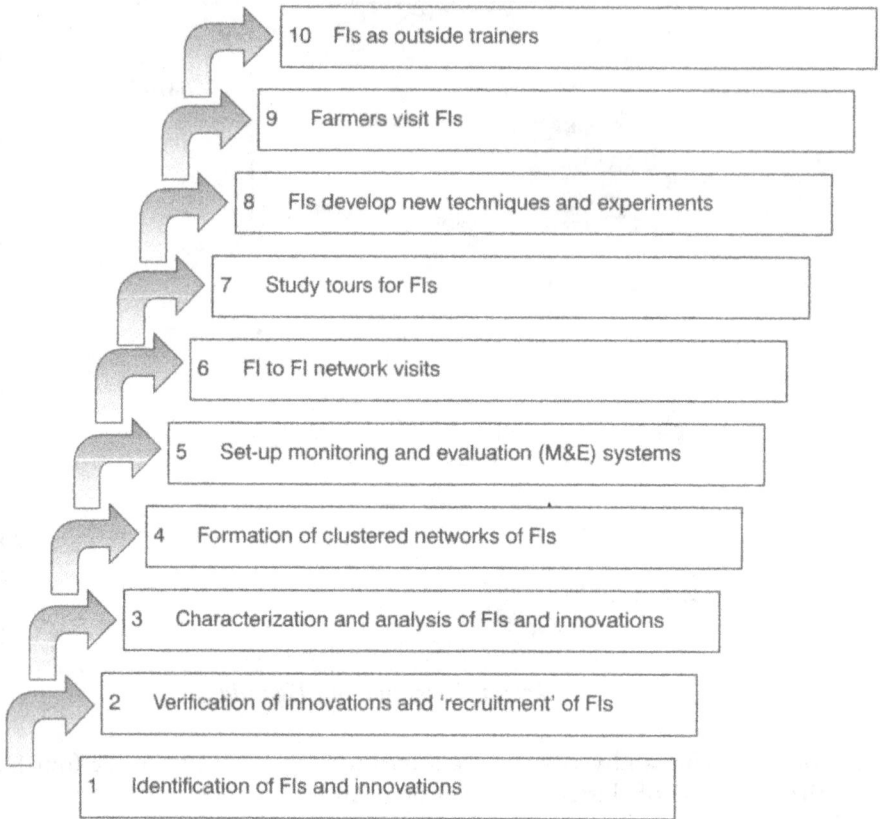

FIs = farmer innovators
M&E = monitoring and evaluation

Figure 17.1 *Ten steps in PFI's field activities*

Chapter 10), it was easier to identify male than female innovators, whatever the true gender balance of innovators in the population may have been.

The next step was to verify whether the innovation was really new and could be useful for other farmers. This was done by the project coordinator and extension staff, though the project acknowledges that an improvement would be to involve researchers and fellow farmers in the process. As the project was designed to promote local innovation by bringing FIs together to stimulate each other's creativity, it was also important to make sure that the innovator really wanted to join a network and to take part in the activities that this would entail. It soon became obvious that the 'recruitment' of FIs would not be a problem: almost without exception, farmers – once they were identified – were eager to join a network of innovators and to become involved in project activities.

Plate 17.1 *A meeting of a cluster of farmer innovators (Uganda)*

NETWORKING BETWEEN INNOVATORS

The concept of the project was that, by bringing together innovators from the same locality, they could share ideas and stimulate each other to experiment with additional new ideas. They could visit each other's site of innovation and travel as a group to other locations that might inspire them to innovate further. In considering the question of group size, the project felt that each cluster should not be so large that reticent members would be intimidated, yet it should be big enough to allow a good exchange of ideas. The size also had to be practical and manageable in terms of logistics. The members of the group needed to be able to reach each other easily. The group had to be small enough that all could gather in a member's front room and could travel together in a Land Rover or a vehicle of similar size when making study tours.

The experience from the ISWC-Uganda project was again a guiding factor. That project had begun work with a group of eight farmers – three women and five men – who met regularly and visited each other's farms. It then added a further three clusters of eight farmers each to its programme. These groups were formed from identified and verified innovators who lived close to each other and had chosen to interact. While there was no predetermined periodicity of meeting, the general concept of the project was that, within a year, the group would visit the farm of each of its members. The process of planning and organizing these meetings was initiated by project staff, in consultation with group members.

LESSONS LEARNED THUS FAR

This model was the basis for designing the clusters of FIs under the PFI programme. The number within a group has differed little from country to country (Tanzania has the largest groups, of ten) and the principle of reciprocal visits between members has been maintained. Thus, these appear to be workable components of the programme. Some further lessons that have been learned from the experiences of both ISWC-Uganda and PFI can be summarized as follows:

- Clusters have provided an important nucleus for networking between innovators. Networking between clusters has followed as the next logical step. This interaction between clusters has stimulated discussion, contributed to participatory evaluation of innovations and prompted new activities. A recent impact assessment of PFI Uganda revealed that, on average, FIs have taken up or begun experimenting with five new technologies each since the project started. According to the farmers, the primary sources for the ideas were study tours and fellow innovators.
- One problem in forming FI clusters has been to maintain a gender balance that pleases both the local community and project designers. Under ISWC-Uganda, women said (informally) that, while they enjoyed being part of a group, there was domestic pressure for them to limit the time they spent away from home. This points to the need for widespread gender sensitization (see Chapter 10) and the importance of not imposing outsiders' gender agenda too aggressively.
- Another problem has been that a small number of cluster members seek some form of compensation for 'time lost'. However, the large majority of FIs have viewed the travel and meetings as worth their while (in either economic or social terms, or both). The general policy of PFI has been no handouts other than food and basic out-of-pocket expenses for tours that involve a night away from home.
- Networking between clusters has proved much easier where population density is high, as is the case in ISWC-Uganda (Kabale district) with around 200 people/km^2. However, in semi-arid areas where density is less than 20 people/km^2 and distances between farmers are correspondingly greater (most of the PFI action areas), the programme of meetings and visits has had to be reduced and/or more transport facilities provided.
- A danger has emerged of building up a select and privileged minority group that provokes envy among other community members because the FIs receive considerable attention in terms of study tours, visits to their fields by outsiders, etc. One possible way to avoid this may be to 'rotate' membership within the groups: after a certain period (eg one year) group members could be awarded a certificate and then make way for newly identified innovators. Another would be to withdraw project support systematically from the older groups, at the same time testing whether

they continue to meet with each other and engage in collaborative research or other related activities on their own. While PFI has not yet started to test alternative approaches in this regard, it is interesting to note that the membership of the oldest group under ISWC-Uganda has evolved naturally: some members dropped out for various personal reasons and others joined. These new volunteers were normally suggested by the other group members and then 'verified' by the project.

- The question of long-term (post-project) sustainability of the innovator groups still remains open. Perhaps here lies an investment possibility for the future: could/should these groups be supported by governments as formal associations of farmer researchers? Could/should such groups be carried by the farming communities themselves and, if so, how could this be encouraged? The experiences made in Latin America with community-operated Local Agricultural Research Committees (CIALS) might stimulate some ideas in this direction (Braun et al, 2000).
- One immediate challenge now is to decide how PFI's relatively small clusters of FIs (8–10 members) can best be meshed with FAO's Farmer Field School (FFS) groups (around 25 members) in a forthcoming joint programme in Kenya. Under this programme, the PFI element will uncover innovative technologies to feed into the FFS. The most likely scenario sees PFI continuing its own research-focused clusters that will serve as resource points for learning-oriented FFS groups.

CONCLUDING REMARKS

In the six years since CWSSE began and during the ensuing and parallel projects of ISWC-Uganda and PFI in East Africa, the project teams have been proactive in organizing farmer groups and in developing networks. Spontaneous formation of groups (ie allowing them to develop on their own initiative, at their own speed and with whatever size) may be preferable according to the theory of participatory development, but our experience has been that intervention in organizing groups of newly identified FIs living close to each other has proved practical and popular among the FIs, and there have been no cases of groups dissolving because of their inability to work together. While no group has yet taken up a life of its own, this is exactly the intention of the four clusters in Mwingi district, Kenya. Having heard that project support will come to an end after 2000, they have scheduled a meeting to discuss how they can form self-standing groups.

The methodology is still developing, and the same approach will not work equally well everywhere. There is room for differences, but that diversity needs to be underpinned by certain common denominators, such as a willingness to work together and logistics of size and proximity, as we seek to continue promoting farmer innovation.

REFERENCES

Braun, A R, Thiele, G and Fernandez, M (2000) 'Complementary platforms for farmer innovation', *ILEIA Newsletter*, vol 16, no 2, pp33–34

Critchley, W R S, Cooke, R, Jallow, T, Lafleur, S, Laman, M, Njoroge, J, Nyagah, V and Saint-Firmin, E (1999) *Promoting Farmer Innovation: harnessing local environmental knowledge in East Africa*, RELMA/UNDP, Nairobi

18

Facilitating farmer-to-farmer communication about innovation in Tigray

*Fetien Abay, Belay Teshome, Mengistu Hailu and Mamusha Lemma**

The search for local innovations in Tigray described in Chapter 6 gave all concerned, but especially the development agents in government agencies and NGOs, a better understanding of the dynamics of indigenous knowledge in land husbandry. ISWC-Ethiopia then developed two main mechanisms for farmer-to-farmer communication to disseminate both the new techniques and the spirit of innovation. During travelling seminars, farmer innovators travel together to each other's farms to see and discuss their innovations. During Farmers' Fora, rural communities meet to discuss and compare local innovations and introduced technologies. In this chapter, the process and outcome of the first travelling seminar in 1998 and the Farmers' Fora held in Central Tigray in 1999 are described. These stimulated farmer innovation and experimentation, and influenced the policy environment for local initiatives to improve land husbandry.

INNOVATORS' TRAVELLING SEMINAR

One of the aims of ISWC is to link up farmer innovators from different areas who have been experimenting with similar ideas and types of technology so

* Fetien Abay and Mamusha Lemma are past and present joint coordinators, respectively, of ISWC-Ethiopia; the former is a crop scientist, the latter a rural sociologist and both are lecturers at Mekelle University in Tigray, Ethiopia; Belay Teshome, former graduate assistant and now lecturer in horticulture, and Mengistu Hailu, language specialist, likewise work at the university

Credit: Fetien Abay

Plate 18.1 *The travelling seminar was filmed by the Bureau of Agriculture and Natural Resources (Ethiopia)*

that they can inspire each other. To this end, a travelling seminar was organized for innovators from all four zones of Tigray (Southern, Eastern, Central and Western). It was designed as a field-based workshop for sharing practical experiences and for mutual learning. It gave the farmers an opportunity to discuss the background of the local innovations, the benefits derived from them and the problems faced in the process of innovation.

The objectives of the travelling seminar were:

- to enhance innovator-to-innovator diffusion of ideas and experiences in the processes of improving land husbandry;
- to analyse together with the farmers the constraints to, and the opportunities for, innovation in land husbandry in different parts of Tigray; and
- to allow farmers to assess the replicability of each other's innovations at different sites.

Methodology of the travelling seminar

ISWC-Ethiopia organized the travelling seminar in consultation with staff from BoANR and two of the NGO-supported projects involved in the programme: Irish Aid's Eastern Tigray Development Project and FARM-Africa's Community-Oriented Rural Development Project. ISWC provided the vehicle to transport the seminar participants, while the partner organizations provided other logistical support. The ISWC coordinators invited nine

Figure 18.1 *Sites visited during innovators' travelling seminar in Tigray*

outstanding farmer innovators who had been identified through the programme – seven men and two women – to join the seminar. A coordinator, a graduate assistant and a driver from Mekelle University, and a cameraman from BoANR accompanied the farmers. The seminar covered the eight sites from which the participants came (see Figure 18.1). Development agents working in these areas joined the travelling group on site to witness the farmers' demonstrations and listen to the discussions.

The seminar was held for eight days in August 1998. The farmers regarded this as a slack period for crop farming (after weeding and before harvest) when it was easier for them to leave their farms for a few days. Moreover, because the crops were standing in the fields, the timing was ideal for observing and discussing crop-related practices.

First, the farmers met in Mekelle, the capital of Tigray. Their travel expenses and food costs were covered by ISWC. At the university, the facilitators explained once again (as had been explained to each innovator during earlier visits to their homes) that the seminar was meant to give them a chance to see and learn from the innovations of their peers. The farmers explained what they wanted and expected to see and to learn, and discussed the draft plan for the seminar proposed by ISWC. They expressed an interest not only in local innovations but also in introduced techniques that might be relevant for their situations. For example, they requested the inclusion of a visit to a microdam at Wukro in Eastern Tigray, not far from one of the participants. On the basis of these suggestions, a new joint agenda was developed.

At each site, the farmers saw a variety of innovations and interesting practices. Meetings with other farmers in each locality took place not only in the fields but also in churches, marketplaces and food distribution centres. In addition to the innovations of individuals, community-based activities such as gully-stabilizing structures were also visited and experiences shared between the local community and the visiting farmers.

At the end of the seminar, the BoANR man who had been seconded to record the event in photographs and on video film showed the participants the unedited film of the entire seminar. Although long, this fascinated them all.

Here, only a few of the innovations at only two of the eight sites visited are briefly described, along with some of the comments and suggestions of the visiting farmers. This is meant to give a flavour of what the farmers saw and heard, and what additional ideas were stimulated by this travelling seminar.

Innovation site in Enticho, Central Tigray

One of the women participants was Leteyesus Gobena from Enticho in Central Tigray. She had decided to plough on her own despite criticism and ridicule from other community members (see Case 1 in Chapter 15). She explained to her fellow farmers that her main problem related to ploughing was acquiring implements which she cannot make herself and therefore has to buy.

Besides the cultural innovation of taking on the male task of ploughing, another innovation that Leteyesus developed is the annual alternation of soil and grass bunds that cross her gently sloping land. She explained that her practice increases yield because it is like fallowing a field but applying the principle only to strips in the cultivated field. All seminar participants agreed that it was a good idea to alternate bunds and cropped land, if not every year then at least every two or three years.

The DA in the area had initially been frustrated because Leteyesus had refused to accept the conventional Global 2000 extension package. However, the good results she has achieved with her own innovations eventually gained appreciation from the local farmers and even from the DA. According to the chairman of the village council (*baito*), Leteyesus' experiments have become an example for other people in the area. He told the gathered farmers that her new techniques were important especially for female-headed families.

On the way to Enticho, the seminar participants visited community work on stabilizing a large gully in Sero, also in Eastern Tigray, done in collaboration with FARM-Africa and the BoANR. The community had received some advice from farmer colleagues in Irob, an isolated area in Eastern Tigray where the people have developed their own techniques and considerable expertise in SWC (Hagos and Asfaha, 1997; see also Chapter 14). When the seminar participants stopped in Sero, the villagers were in the midst of maintaining the conservation structures. The visiting farmers explained some of their own innovations to the people in Sero and commented on the latter's work. They observed that the community had planted grasses on the graded sides of the gully. One male innovator suggested planting grasses also at the checkdams, and two others stressed the importance of treating the entire catchment in order to stop the gully from growing. The *baito* chairman said that Tigrayans have developed numerous innovations in every corner of the region, but the problem is that the experiences are not shared. He said that travelling seminars give an opportunity for such sharing and are therefore important for sustainable agricultural development and food security.

Innovation site in Selekleka, Western Tigray

One of the male participants was Hailu Gebrehiwot, whose home in the Selekleka district is situated on sloping land. The catchment above his farmland is steep. High rates of run-off had led to severe soil erosion and gully formation, cutting up his farmland and decreasing the productivity to such an extent that he could no longer feed his family. To solve this problem, Hailu had proposed to the *baito* that land be enclosed in the upper catchment, but his neighbours rejected the idea. He then decided, on his own, to protect at least the part of the catchment directly above his farm. This steep slope is now totally covered with vegetation, the rate of soil erosion has decreased and his farm has become more productive.

In addition, Hailu used locally available materials to construct some terraces across the slope of the farmland in order to harvest the soil and litter still coming from above. Along the terraces, he planted *gesho* (*Rahaminus perinoides*), a lucrative income-generating tree, the leaves of which are used to make alcoholic beverages. He built checkdams across the largest gully going through his farm and thus managed to stabilize the gully where he has now planted various fruit trees.

Although he himself had more than one ox for ploughing, this innovator developed a ploughing harness that can be used with a single ox. Hailu explained that it could solve the problem of farmers who can afford to feed only one ox. Because landholdings are becoming smaller and feed is more difficult to obtain, he felt that single-ox ploughing would eventually have to be adopted by many farmers in Ethiopia.

Hailu also developed the idea of converting the bare rock on one side of the river into arable land by transporting soil to it from other areas. He did this in order to be able to make use of the water that flows year-round in the river. He cultivated the new land with onions, from which he earned 1500 *Birr* (circa US$190) per year. After two years, however, the land became waterlogged through seepage from the canal for the nursery he had made above it. He then planted eucalyptus trees on the waterlogged land.

To transport the soil and stones to cover the bare rock beside the river and also to construct other structures to protect his land from erosion, Hailu made a kind of wheelbarrow (*menkorkor*) out of local wood and a spent bombshell. It is drawn by a pair of oxen.

One of the visiting farmers asked about the side effects of planting eucalyptus. Hailu responded that he chose this tree species since it tolerates waterlogging and makes good use of the seepage water from the canal. The seminar participants were particularly interested in the one-ox plough because feed shortage is a common problem. Hailu promised to provide a new ploughing harness for Leteyesus and Tensue (depicted in Plate 5.1), the two female participants, saying he appreciated their initiative in doing their own ploughing regardless of the local culture.

In Selekleka, the visiting innovators met with some local farmers who said that their *baito* encouraged innovators to develop their ideas further and that a person who reclaims land is granted the right to use it. Priest Tsige Gebremedhin, a local DA, volunteered the information that a woman in his area, Tsige Tesfay, also ploughs her farmland and does all other farming activities on her own. He said that she and other women like her need encouragement from government agencies and NGOs.

Innovators' evaluations

The travelling seminar included three types of evaluation by the farmers: evaluation of the strengths, weaknesses, opportunities and threats (SWOT) of the innovations at each site, daily evaluations of the seminar each evening and a final evaluation on the last day.

SWOT evaluation of local innovations

Exchange of ideas and experiences was an integral element of the seminar. The discussions were structured around the SWOT evaluation of the local innovations. These assessments were not formally structured; they came from the farmers spontaneously. When the farmers looked at an innovation, they praised and/or criticized it and were quick to make suggestions for improvement. For instance, when they visited Leteyesus, they praised her strength but criticized the fact that she was giving more attention to the plots near the road while neglecting the plots in her backyard which had greater potential. They said that the plots around the compound were more fertile and easier to work because they were closer and had better soil. They advised her to fertilize these plots with organic matter from her compound which would also help to keep it clean.

When discussing in the SWOT mode, the farmers also referred to tenure and policy matters. For instance, Hailu explained that he worked on the hillside land during the night. The seminar facilitators were impressed at his time management, but one of the male farmers surprised the facilitators by raising the tenure issue. He asked Hailu why he did this work by night. Cornered by this question, Hailu had to admit that he did it because he was working a hillside to which he had no legal rights. This shows how the farmers openly discussed even very sensitive issues and challenged each other. These SWOT discussions continued in the vehicles during the journey from one site to the next, and there were heated debates about how the policies of the local *baitos* could be influenced and what role an innovator network could play.

Daily seminar evaluations

Each evening, the farmers reflected together on what they had seen during the day and discussed plans for the next day. They nominated someone to chair the evening session; usually this task went to two of the male farmers whose capacities in this respect were appreciated by all. The farmers expressed their opinions whether they had gained the information they had expected and, if not, why not. They continued the SWOT evaluations of some of the technologies, but also evaluated the process of the seminar and sometimes criticized the behaviour of individuals, eg for splitting away from the group or acting in a superior way. They commented on the organization of the day's activities and suggested improvements. A couple of times they revised the schedule by adding an agricultural development site that one or more of them thought would be interesting for the others.

Final evaluation

At the end of the seminar, a final evaluation was made. Farmers praised the idea of the travelling seminar. They were particularly pleased that, by being invited to join it, they were finally being given what they considered to be due recognition for their work.

They found it valuable that the seminar exposed them to many different farming conditions and systems in Tigray, allowed them to see many new ways to cope with the difficult environment and made them aware that the techniques are specific to place, weather and time. They recognized that some of the techniques could be applied in their own situation, whereas others could not be directly adopted or, in some cases, could not even be adapted to fit the conditions in their own villages because these were too different. Nevertheless, they felt that merely seeing what the other farmers had done gave them an impulse to be innovative in their own way.

During the travelling seminar, the key actors were the innovators themselves who showed their technologies to fellow farmers and outsiders. Through this experience, they gained self-confidence and developed their capacity to explain their technologies to others. This was especially striking in the case of the women innovators.

The seminar also gave the innovators a better understanding of the implications of local land policies. The participants identified policies that encouraged or discouraged them in innovation and experimentation. They expressed their gratitude at having a chance to be exposed to how different *baitos* deal with land-use rights.

As negative aspects, the farmers pointed out that the seminar involved only few people, and was tiring and costly in terms of petrol and time. They suggested that the expenditures could be used more effectively if ISWC and/or BoANR would organize meetings at each innovation site where they could address a large group of farmers.

The evaluation revealed that almost every participant had not expected to see an innovation better than his or her own and was surprised at what others had also achieved. The participants expressed the opinion that such travelling seminars encourage not only themselves but also other farmers to innovate. They were keen to share the experience they had gained with the farming community in their home area. However, they saw a need for audiovisual aids such as mobile film units and documentary photographs showing indigenous practices and innovations. In this respect, they appreciated very much the video film made by the BoANR cameraman and recommended that it should be shown more widely.

Aftermath of the seminar

After the travelling seminar, seven of the nine innovators organized formal meetings in their own village areas. In most cases, these meetings were arranged with the assistance of DAs and in consultation with their *baitos*. The *baito* head called the people together and the innovators told about the seminar and the innovations they had visited, including their own. All nine innovators shared their experiences at more informal meetings on their own initiative, in a few cases at meetings specifically for this purpose, but normally taking the opportunities of church gatherings and *tsebels* (gatherings in the home for cultural events). They talked with the farmers sitting around them or, at the end of a church ceremony, they stood up and addressed all the people in the church, telling them what they had seen. In addition, they responded to the enquiries of curious neighbours who knew that the trip had taken place.

Some of the innovators tried out technologies they had seen during the seminar. For example, Leteyesus dug a well like the one she saw during the visit to Sinkata in Eastern Tigray, one man made a wheelbarrow like the one he saw at Selekleka, and another made one-ox plough implements like those he saw at the same site. One participant, Embaye Kindeya, invited another participant, whose innovation he wanted to test, to advise him directly on his farm. After having seen how the community in Sero planted elephant grass to stabilize gullies, several innovators started spontaneously to think of sites in their home areas where elephant grass should be planted and, upon returning home, convinced the *baito* to begin this activity.

FARMERS' FORA ON WATER MANAGEMENT

Whereas the travelling seminars focus on bringing farmer innovators together with each other, the Farmers' Fora bring together innovators and other farmers in their area. ISWC-Ethiopia collaborated with the BoANR in organizing field-based workshops at village and district level focused on farmer innovation in land husbandry. Because a main concern of farmers in Tigray is water management and a main thrust of BoANR is therefore in this direction, the field workshops were focused on this. In Tigrigna, they are called *Teli Waela*, literally 'days of moisture', which could be described in English as 'Farmers' Fora on Water Management'. These are workshops designed not only for farmers but also for local administrative and extension staff to visit and discuss improvements in water management.

Conventional field days to transfer technologies

The BoANR already has a tradition of holding field days to demonstrate the merits of introduced technologies. These tend to be of a ceremonial nature, involving officials from the BoANR and community leaders. The BoANR experts carefully choose a site where a farmer has applied a single technology successfully. The model farmer explains how he did it. The other farmers observe and discuss his success. The model farmer is praised and given an award such as fertilizer or a farm implement. This is meant to stimulate other farmers to register themselves for credit to be able to apply the technology themselves. The other farmers may make remarks that challenge the demonstrated technology, but these are usually submerged by the remarks of the BoANR experts.

Sometimes, an outstanding local innovation has been mentioned or shown during these field days, but the focus in recent years has been on 'modern' technologies applied by model farmers who received credit to do so. The central figures from the farming community, besides the above-mentioned model farmers, are the 'agricultural cadres' and 'farmer promoters'. The cadres are local people who are selected by the DAs to organize and monitor agricultural development in that area and are given continuous training in a specific field. The promoters are local people selected by the cadres to assist them and the DAs in promoting new agricultural technologies.

The tradition of *gumgum* (evaluation) in Tigray, established during the struggle for liberation, is reflected in the annual meetings held in each hamlet and village area which involve the inhabitants in an intensive evaluation of local performance in development. The meetings include an appraisal of the advantages and disadvantages of the techniques applied by farmers in the previous year. Meetings are then held in each district to bring these critical appraisals up to a higher level. Farmers are therefore accustomed to expressing their opinion, both positive and negative, to officials. Thus, although the conventional field days were designed to demonstrate success, farmers' dissatisfaction with some of the introduced technologies was not unknown to development planners.

Fora for technology assessment by farmers

ISWC-Ethiopia introduced the idea of designing farmers' meetings around the comparison of introduced technologies and local innovations that were trying to address the same problems. Over the past two years, the programme and BoANR have jointly organized annual Farmers' Fora throughout all 35 districts in Tigray. The project covered the cost per day for the farmers (10 *Birr* or about US$1.20 for food and drink), facilitated some of the meetings in the first year, coached BoANR facilitators and provided guidelines for conducting the workshops in the other districts. As reporting on the process and outcome of the workshops in the first year was fairly weak, ISWC-Ethiopia, at the request of the Head of the BoANR, designed a reporting format for better documentation and monitoring of the subsequent meetings.

The BoANR staff, in particular, who had been directly trained by ISWC in workshop facilitation techniques, were keen to conduct these workshops.

At the village-level fora on water management, the participants included male and female innovators who had been identified in the area, model farmers, other farmers, representatives of various local associations (Women's Association, Youth Association, etc), members of the village and hamlet *baitos*, and sometimes BoANR experts from district or zonal level. Usually about 80–100 people participated.

The innovative farmers and the model farmers explained that their technologies related to water management, and BoANR experts were also given a chance to bring in their ideas. The workshops were not designed to 'sell' any particular technology. Various options were discussed and much room was given for critical comments. The facilitators encouraged the participants to discuss freely the strengths and weaknesses of the options, and to suggest improvements. The discussions were not confined to technical aspects; they also covered issues such as resource tenure and policy lobbying. For instance, farmers openly criticized the conventional extension packages, the inflexible transfer-of-technology approach and the quota system for recruiting farmers into the credit scheme for purchasing agricultural inputs.

Example of Farmers' Fora in Central Tigray

As an example, the Farmers' Fora on Water Management held in Central Tigray in 1999 are described here, these having been the meetings best documented by BoANR staff. As it was not possible for the ISWC team to be involved directly in preparing and facilitating all the fora in this or other zones in Tigray, the team knew that the initial workshops would not meet their expectations completely. The analysis of the workshop documentation allowed the team to assess the degree to which the concept of farmers' fora to compare local and introduced innovations had been understood and put into operation. This analysis served as a basis for discussing the workshops within the BoANR and improving subsequent workshops.

In the Central zone, the focus of the Farmers' Fora in 1999 was on conserving moisture in agriculture. The categories of fora participants mentioned in the DAs' reports included:

- BoANR agricultural experts from zonal and district levels;
- DAs from *tabia* (village) level;
- outstanding local innovators;
- model farmers;
- agricultural cadres;
- farmer promoters;
- some other men and women farmers in the district or villages, often as representatives of local farmers', women's or youth associations;
- *tabia* administrators;
- staff from the Bureau of Planning; and
- elders and other prominent persons in the locality.

The composition indicates that these meetings were widely based and included people from all stakeholder groups that can directly or indirectly influence and be influenced by farmer innovators. In each of the ten districts in Central Tigray, the total number of participants that were involved in these fora ranged from 370 to over 2000. However, according to the DAs' reports, innovators identified through the ISWC programme were included deliberately in the district-level fora in only five of the ten cases.

Some of the indigenous innovations and introduced technologies discussed during the fora at both village and district level were digging and maintaining trenches, harvesting water, diverting water for irrigation, preparing compost, leaving land free for grass growth (locally called *a'rmo*), sowing grasses and tied ridging.

Some DAs made considerable efforts to persuade local innovators to take part in the fora held in their village areas, encouraging them to explain their innovations and giving them moral support and coaching in preparing their presentations. The praise and honour given in the presence of their colleagues made the innovators feel 'like we were ten feet tall', as one farmer expressed it. They felt proud of what they had achieved and gained confidence to disseminate their ideas to others.

The presentations made by local farmers about their innovations served as springboards for discussion. For example, Gebremichael Beza's innovation was discussed at the Farmers' Forum in Maiberazio. He explained to the participants how, together with other members of his family, he heaped up stones and soils inside a long gully that was threatening his plot and thus managed to plug the gully. He also sowed grasses and fruit trees along the trenches that he made above the treated gully.

After listening to this, the workshop participants discussed the strong and weak points of the innovation. They found that the work of Gebremichael and his family, specifically the drawing of trenches, had succeeded in protecting the land from flooding and allowed the family to extend its farmland. They praised the biological treatment of the gully, referring to both the agronomic and the economic values of the grasses and other plants he had sown.

As a weakness showing room for improvement, the participants observed that, if the trenches had been made with a water-level instrument, the water would not collect in the lower spots and create a threat of sudden flooding. Secondly, some mentioned that Gebremichael could take the soil deposited in the trenches and spread it over his plots during the dry season. Thirdly, some thought that the flooding could be reduced still further by digging additional trenches at the top of the hill. These are recommendations for consideration by the innovator, but they also stimulated thought among the other participants about what they could do on their own land. During the workshops attended by ISWC-Ethiopia coordinators, the innovators were very open for constructive criticism by their peers. It seemed to motivate rather than frustrate them.

Other locally developed innovations that were discussed during the technology assessment workshops at district level included treating flood-stricken land, sowing different grass species and using them for a variety of purposes, rehabilitating gullies and converting them into arable land, planting trees for animal forage, and digging trenches and reinforcing them with sisal and other plants.

At some of the district-level Farmers' Fora, the BoANR gave awards to outstanding innovators. This reportedly inspired many other participants in the meetings to try out the new ideas on their own farms.

Signs of change

The BoANR has taken over the responsibility of organizing the Farmers' Fora which are basically technology assessment meetings focused on specific themes. At the district level, these meetings still have a bias towards the technologies being promoted by the formal extension system, and the general thrust is more towards transferring specific technologies (whether introduced or local) rather than stimulating farmers to compare technologies and then choose what would be the most suitable to try out under their specific conditions. This may be because the district-level employees are more conscious that non-compliance with the official programme of technology transfer may affect their careers negatively. They feel more responsible to higher levels in the system and do not yet perceive a change in official policy.

At the village and hamlet level, the DAs were more open to recognizing and encouraging local innovation. This may be because of the closer day-to-day contacts between the DAs and the farmers, and the DAs' greater feeling of responsibility to meet the site-specific needs of the farmers among whom they live.

At the meetings organized as regular extension meetings (ie not Farmers' Fora partly supported by ISWC), the DAs are inviting local innovators as resource persons. The innovators tend to be used as an entry point for discussing how these people deal with local problems, leading into consideration of the various alternatives. At these meetings, it is usually the innovators who are most outspoken in posing challenging questions about the technologies proposed by the BoANR experts and in urging farmers to consider what is really best for their own circumstances.

COMPARISON AND CONCLUSION

The travelling seminars allow only a small number of farmers to visit other areas but, by inviting local farmers and DAs to join the seminar participants at the sites of innovation, a larger number of people can be involved in the sharing of information. The exchange visits and farmer-to-farmer communication appear to be very inspiring experiences. Government agencies, NGOs and farmer associations should combine forces to create more opportunities for such events. Farmers and BoANR staff still have to give much consideration to how travelling seminars can be organized in a more cost-effective way and how funds for their implementation can be generated.

The Farmers' Fora show how farmer-to-farmer exchange about local innovations can be promoted. They enrich an established mechanism in the extension system for presenting technologies. The BoANR is starting to incorporate site-specific comparison of the merits of local versus introduced innovations into its regular extension activities. The fora have given farmers an opportunity to be particularly outspoken about their views on conventional extension packages, including Global 2000. It is partly as a result of this that this scheme is gradually becoming more responsive to agroecological and socioeconomic diversity within Tigray. Matters such as the type and amount of external inputs to be applied under different conditions, the conditions of credit and the role of the DAs are being publicly debated. The BoANR now recognizes that the system of expecting DAs to meet annual targets in convincing farmers to take up credit for fertilizers and other inputs had actually been constraining the farmers from exploring (ie experimenting with) possible improvements to the proposed technologies.

The fact that local innovation is now part of the agenda in the large gatherings of farmers organized regularly by the BoANR indicates that many of its staff members have changed their attitude toward indigenous knowledge and creativity. In district-level meetings, some attention is now being given to the achievements of local innovators: they are invited to explain their techniques and are given awards. During *gumgum* at district level, the DAs are publicly asked whether they encourage local innovators to participate in all meetings concerned with agricultural development held at village and hamlet levels. This suggests that the concept of indigenous innovation in land husbandry is indeed infiltrating into the formal procedures of the BoANR.

The presence of district and zonal experts and administrators in the technology assessment meetings and at some innovation sites during the travelling seminar is of remarkable importance. These people are very close to policy-makers or are policy-makers themselves, and it is policy that has a particularly great effect on the inclination of farmers in Tigray to innovate. For instance, in the past, a farmer who treated a gully in no man's land and started to grow field crops or trees was blamed for using land that s/he did not deserve. This discouraged local initiative. Now, however, attitudes of policy-makers are changing. It is being realized that the considerable efforts invested in reclaiming a gully should entitle the farmer to use the land thus created. Discussing the benefits of innovation from both economic and agronomic perspectives in the presence of BoANR experts and administrators is a very effective way to lobby for policy change, in addition to the fertile ground it creates for encouraging further innovation by farmers.

REFERENCES

Hagos W and Asfaha Z (1997) 'How to gain from erosion: catch the soil', *ILEIA Newsletter*, vol 13, no 2, pp16–18

19

Innovation and impact: a preliminary assessment in Kabale, Uganda

Dan Miiro, Will Critchley, Alie van der Wal and Alex Lwakuba[*]

The project Conserve Water to Save Soil and the Environment was started in south-west Uganda in 1994 with the aim of building on local traditions of SWC and developing a methodological approach for wider application. In 1997 it was incorporated into the ISWC programme. Two years later, the impact of the more than five years of project activities was assessed. This chapter presents the main findings and lessons for programmes focused on promoting local innovation in land husbandry.

BACKGROUND AND RATIONALE

The project

In early 1994, an innovative project called Conserve Water to Save Soil and the Environment[1] started in Kabale district in south-west Uganda. Its objective was to build on local traditions of SWC. First, the project set out to validate specific traditional practices through joint monitoring and evaluation by scientists and farmers. The next stage was to add value to these practices through participatory technology development. Then, improved practices were to be spread through participatory extension. A further objective was to develop and document a methodology to achieve these aims (Willcocks et al, 1992; Critchley, 1999).

---

[*] Dan Miiro is a soil scientist with the MAAIF in Entebbe and coordinates ISWC-Uganda; Will Critchley is an agriculturalist with CDCS, Vrije Universiteit, Amsterdam, and provides support for PFI and ISWC-Uganda, Alie van der Wal is a geographer who was with CDCS until early 2000 when she joined SNV in Bhutan; and Alex Lwakuba is a senior soil conservation and agroforestry officer with MAAIF and coordinates PFI-Uganda

Map 19.1 *Uganda (showing Kabale district)*

A practice was regarded as traditional if it had been used by the community in the area since time immemorial or had been introduced but had been practised by the community or at least by some members for so long that they considered it part of their normal farming.

By the end of 1996, the CWSSE project had studied and tested improvements to two traditional practices that had been developed jointly by farmers and scientists: banana mulching and trash lines (Briggs et al, 1998; Ellis-Jones and Tengberg, 1998). These had been identified as farmer priorities through PRA methods. At the same time, a programme of network visits was taking place among eight farmers whom the project had selected as representative of those practising the two traditions. The network visits were supplemented by study tours, often into other districts. As a result, the improved traditional practices were diffusing rapidly to others, beyond the eight focal farmers. The methodology was defined and recorded (Critchley, 1999; Critchley et al, 1999b).

However, as so often happens with projects of this nature, the funding came to an end before the objectives could be fully achieved. Fortunately, this coincided with the start of the second phase of the ISWC programme[2] which had a similar philosophy and approach. The Kabale project was taken on board and continued as ISWC-Uganda. Under its new name, the project continued to thrive, with two significant changes resulting from being under a new umbrella programme. Firstly, a lower budget reduced the input from scientists and tilted the balance towards farmer-to-farmer extension of 'best bet' practices. Secondly, the project expanded spatially and threw the net wider to include farmers who were actually innovating in land husbandry. A further 24 farmers were included, making 32 in total, in four clusters of eight farmers each.

Thus, in line with ISWC 2, the focus changed from tradition to the dynamic process that creates tradition: local innovation. By the beginning of 1999, it was clear that farmer-to-farmer visits were having considerable local influence. The impression was that many farmers were trying out new technologies, but hard evidence was lacking. The time was ripe to assess the project's impact systematically, not only to support 'warm stories' with 'cold fact', but also to test assessment methods that could be useful for other projects. One of the project's original objectives was methodology development and a preliminary impact assessment was duly carried out in July 1999.

Why assess impact?

Before reporting details of the exercise, we look at the rationale for impact assessment (IA). The case is set out clearly by Guijt (1998):

> *The pressure is on to prove the effectiveness of efforts that claim to lead to more sustainable development... Now that the honeymoon period of participatory NRM [natural resource management] is coming to an end, funding agencies are asking advocates of such approaches to prove their many claims.*

In all rural development projects, it is necessary to know what impact has been achieved. In some interventions, such as building access roads, it is easy to monitor basic impact parameters (eg growth in commercial activity). In other projects, such as adult education, it is not so easy to put a finger on the effect of the project (eg increased confidence and empowerment). However, simple or not, without some form of quantitative and qualitative assessment, we cannot know just what has been achieved. It is not good enough to continue saying: 'The project is a great success' or 'A lot has been done with a little money'. Such comments are often made by people with a vested interest in a project, but can they be believed? We need to have a more objective analysis of what has really been done, and how the benefits compare with the costs. That is the main purpose of IA.

An associated benefit is that IA helps in planning for the future. Projects focused on farmer innovation appear to have great promise and popularity, but we must be careful that unsubstantiated claims do not fabricate a new 'development narrative' (see Chapter 26). Such projects need to be subject to analytical appraisal, otherwise there is no solid justification for scaling up the farmer innovation approach.

In development circles, there is a growing interest in and focus on (participatory) monitoring and evaluation (M&E) and IA. Numerous recent publications set out the rationale, stake out the concepts, define terms and provide some initial guidelines (eg Estrella and Gaventa, 1997; Abbot and Guijt, 1998; Arevalo et al, 1998; Guijt, 1998; Harnmeijer et al, 1999; Herweg et al, 1998; IDS, 1998). This is not the place to review the literature in detail, but we draw attention to three points arising from these writings:

1   The debate is still ongoing and is based as much on theory as on hard evidence from the field; in other words, this is still an 'open book'.
2   There is an energetic movement to promote wider participation in M&E and IA in order to increase empowerment of the beneficiaries and to give them a greater sense of ownership of project activities. However, some authors point out that participatory monitoring and evaluation (PM&E) is actually a subset of a broader M&E and adds value to, but cannot replace, the latter (Abbott and Guijt, 1998; Critchley et al, 1999a).
3   Terms are often used in different ways. Guijt (1998) defines the main concepts, but she suggests pragmatically that what is most important is to make terms clear *for a given situation*. So this is what we shall do now.

In the context of the exercise reported here, our working definitions were as follows. *Monitoring* is a continuous exercise, usually involving measurements of one sort or another. *Evaluation* is putting judgemental value on actions, systems or developments. It is partially continuous and partially discrete. *Impact assessment* describes a discrete exercise which attempts to bring together information, evidence and opinion to weigh up the effect of an intervention on people, production, the environment, policy, and so on. In this case, all the farmer innovators and field agents participating in the ISWC 2 project were involved in the record-keeping and discussions for M&E, whereas the computing and analysis of results for the impact assessment involved primarily the researchers (authors).

OBJECTIVES, FOCUS AND METHODOLOGY OF IMPACT ASSESSMENT

Objectives

The overall goals of the IA were to track changes since the start of the initial (CWSSE) project that would show the project's effect on local farmers with respect to their ability to produce food and cash crops in a more sustainable way.

The specific objectives were reflected in the following questions:

- Have the farmers' innovations/initiatives spread? If so, through what channels?
- Does the 'building on tradition'/'farmer innovation' methodology work?
- Is the methodology replicable? And has it indeed spread?
- What are the technical and economic impacts of the innovations?

From the beginning of the IA process it was realized that not enough information was available to answer these questions fully, especially not the last one. Only two traditional practices (banana mulching and trash-line management) had been studied in technical and economic detail. These were not the techniques that had actually spread the most widely as a result of the project (these were compost-making and water harvesting). Moreover, not all farmers had followed exactly the scientists' technical recommendations as they adapted the techniques to suit their local conditions. This rendered an economic assessment of total impact impossible at this stage.

Focus

In order to ensure that the IA exercise was manageable (and in the knowledge that it could only be partial and preliminary), initial foci were identified. These were:

- the adoption[3] and spread of promising ('best bet') localized traditions and improved land husbandry practices derived from both farmer innovation and scientific research;
- comparative characteristics of innovators and adopters;
- perceived benefits by farmers who had taken up the above-mentioned practices;
- the impact of local network visits and more distant study tours;
- indications of the institutionalization of the overall methodology; and
- testing of the IA methodology.

Methodology

The main part of the exercise took place during a single week in July 1999. The authors planned the main methods in detail before going to the field. These included field visits and interviews with:

- six farmer innovators (the local team purposively selected them from the 32 collaborating farmers so that those members of the IA team who were not familiar with the outstanding local innovators could gain a good appreciation of what they were doing);
- six adopters (a random sample from those recorded by the innovators);
- field agents (the three closely associated with the project's activities);
- district level MAAIF officials (District Agricultural Officer and three colleagues); and
- the Commissioner of Farm Development (responsible for the project at Ministry Headquarters).

There was also a participatory assessment meeting with innovators (25) and adopters (15), using the following tools:

- relative scale/ladder for participatory assessment of cross-visits (both network visits and study tours);
- SWOT analysis of cross-visits; and

- group discussion on change and stimulus of change.

Data was collected using forms for characterizing innovators and innovations (see Critchley et al, 1999a). This was followed by analysis using EXCEL/SPSS (Statistical Package for the Social Sciences) after codebook development and data entry of:

- farmer innovators (full cover of 31, excluding one with an incomplete set of data);
- related innovations; and
- adopters (sample of 37).

RESULTS AND DISCUSSION

Characterization of innovators, innovations and adopters

Before the concentrated week of IA, the local ISWC-Uganda team had characterized the farmer innovators and their innovations. Farmers were designated as 'innovators' if they had developed or were still testing technologies that were new to the locality. It remained to characterize a sample of 'adopters'. The IA team took a random sample of adopters from the records kept by the innovators themselves and carried out interviews using a form for characterization of adopters (Form N; Critchley et al, 1999a).

The IA team assumed that every farmer who was known to have tested an idea had adopted it. That may prove wrong. There will be cases in which a farmer tried something, perhaps even for two or more seasons, and ultimately rejected it. The assessment reported here was a snapshot exercise, not a longitudinal study, and therefore could not differentiate between short-term and long-term adoption. Additionally, several adopters told the IA team that they had modified the techniques to suit their situation. They had effectively become adapters. Again, these are lumped together as adopters.

Table 19.1 shows that the socioeconomic characteristics of the innovators[4] differed from those of the adopters. The innovators identified by the project were older, better educated and richer in resources than the farmers who adopted their innovations. Moreover, judging by the higher proportion of women than men in the sample of adopters, women appeared to be more responsive than men to new ideas. Why? The assessment did not delve into this question which is evidently worthy of further study.

Compared with the characterization exercise carried out by the Promoting Farmer Innovation project (PFI[5] in Kenya, Tanzania and Uganda; see Critchley et al, 1999a), innovators were somewhat older in Kabale (average age 44 years under PFI) and the gender balance is better (81 per cent men and 19 per cent women under PFI, although this had improved by mid-2000 to 67 per cent men and 33 per cent women).

Table 19.1 *Characterization of farmer innovators and adopters*

Characteristic	Farmer innovators (n=31)	Adopters (n=37)
Sex	58% men	38% men
	42% women	62% women
Average age	50 years	40 years
Average family size	7.6	6.4
Level of education:		
secondary or above	39%	11%
Average annual income	US$500	US$380
Average farm size	3.9ha	1.6ha
Percentage of land cultivated	52%	68%

As for the innovations themselves, Table 19.2 shows that various forms of organic matter management comprised the most popular types of land husbandry innovation in Kabale. This is hardly surprising, as soil fertility had long been recognized as a serious constraint to long-term production in the area. Water harvesting came a close second. Three of the four clusters of eight farmers working with ISWC-Uganda were operating in the driest part of Kabale (800mm annual average), where bananas, the major crop, can be sustained only by making use of every drop of water available.[6] With respect to the technical categories of innovations, there was little difference between innovators and adopters.

A comparison of results from Kabale with those from the PFI project revealed two major differences. Firstly, organic matter management was more important in Kabale, whereas in PFI (which operates in semi-arid areas) water harvesting was the most common category of innovation and organic matter management came second. Secondly, gully control measures were of relatively little importance in Kabale, which can be explained by the relative lack of gullies in Kabale compared with the PFI areas of operation. Although some innovations are recorded in Table 19.2 under 'Agronomy' and 'Forestry', these had not been the focal areas of interest for the ISWC-Uganda project which restricted identification principally to innovations related to SWC.

The IA team asked the original 32 innovators to number and name those farmers who they knew had followed their example. The team reviewed these names with the local extension agents and came to a figure of just over 500 'first-generation' adopters. The adopters interviewed by the team were also able to name people who had copied from them (ie second-generation adopters). Although this information was not verified, the average number taking up the practice from each adopter was 2.6. This implies a spread to another 1250–1500 farmers, in addition to the 500 first-generation adopters.

Questions about the perceived benefits of the innovations were included in the characterization forms for both innovators and adopters. As the innovators, by definition, had more experience with the innovation, the benefits that they perceived were more explicit: nearly two-thirds claimed to have benefited

Table 19.2 *Characterization of the innovations*

	Farmer innovators (n=31)	*Adopters (n=37)*
Technical categories	Organic matter management 42%	Organic matter management 43%
	Water harvesting 39%	Water harvesting 27%
	Agronomy 10%	Agronomy 14%
	Forestry 6%	Forestry 5%
	Gully control 3%	Other 11%
Average no of known adopters from each innovator/adopter	17	2.6
Main perceived benefits from the innovation*	Increased production 65%	Increased production 59%
	Moisture improvement 58%	Moisture improvement 27%
	Control of erosion 23%	Control of erosion 27%
Main problems experienced/ that may deter others*	Labour shortage 74%	Lack of awareness/skills 41%
	Shortage of tools/equipment 32%	Labour shortage 22%
	Shortage of capital 19%	Shortage of land 14%
		Shortage of livestock 14%

* More than one answer was possible

from increased production. It is not surprising that farmers perceive and appreciate production advantages more clearly than SWC benefits. This is a common finding across Africa and elsewhere. The main problem experienced by the innovators was shortage of labour. As most adopters had just started with the new practices and evidently had experienced few immediate problems, the team decided to ask the adopters: 'What are the main problems that might deter others from taking up the practice?'. It was striking that, in their opinion, issues like ignorance, lack of awareness and lack of skills were more important than shortage of labour (these answers are in italics in Table 19.2).

Of the total of 24 female adopters, 13 were involved in organic matter management activities, such as building trash lines and applying compost or manure, whereas only 3 of the 13 male adopters were doing these. New agroforestry techniques were adopted by women only. The technologies under 'Agronomy' were mainly adopted by men and referred to banana management which is a male task in Kabale. There was little gender difference in the 'Water harvesting' and 'Other' categories, which included very labour-intensive work such as digging ditches and constructing stone bunds.

Impact of network visits and study tours

Farmer-to-farmer visits constituted a key element of the CWSSE/ISWC-Uganda project. These took two main forms:

1   network visits (NVs)[7] during which farmer innovators hosted field days for fellow innovators and other interested farmers; and
2   study tours (STs), during which innovators were taken to sites of interest outside their villages indeed, mainly outside Kabale district.

Both these types of visit were evidently popular and productive. However, in order to find out how the farmers thought they had benefited from the visits, questions to this effect were put to the 25 innovators and 15 adopters who attended the participatory assessment meeting. Three participatory assessment tools were used: a relative scale or ladder (Guijt, 1998), a SWOT analysis and asking network groups to reflect on tangible initiatives or technology adoption by group members.

Relative scale/ladder

The participants divided into two groups: one focused on the impact of network visits, the other on study tours. To each group, ten similar questions were posed on which they were asked to reach a consensus on a scale between +3 (very strongly agree) and –3 (very strongly disagree). Seven possible answers were displayed in a 'ladder' and the appropriate level was marked by an facilitator chosen by the group:

+3 = very strongly agree
+2 = strongly agree
+1 = agree a little
 0 = neither agree nor disagree
–1 = disagree a little
–2 = strongly disagree
–3 = very strongly disagree

The results were as follows:

Group One: Network visits (ie visits from innovators-to-innovators and from ordinary farmers-to-farmer innovators)

Have the NVs opened your eyes to new ideas in farming?	+3
Have the NVs improved your farming skills?	+3
Have the NVs increased the number of technologies you test?	+3
Have the new technologies required additional labour?	+3
Have the new technologies required additional external inputs?	+3
Have the NVs taken away your valuable time?	–3
Has your participation in NVs led to jealousy?	+1
Have the new technologies led to increased production?	+2
Have the NVs opened your eyes to ideas other than farming?	+3
Have neighbours copied what you have learned on NVs?	+2

Group Two: Study tours (visits by innovators to sites mainly outside Kabale)

Have the STs opened your eyes to new ideas in farming?	+3
Have the STs improved your farming skills?	+2
Have the STs increased the number of technologies you test?	+2
Have the new technologies required additional labour?	+3
Have the new technologies required additional external inputs?	+3
Have the STs taken away your valuable time?	−3
Has your participation in STs led to jealousy?	+1
Have the new technologies led to increased production?	+2
Have the STs opened your eyes to ideas other than farming?	+2
Have neighbours copied what you have learned on STs?	+3

Not only did the farmers clearly appreciate both the network visits and the study tours, but these events were also a positive source of inspiration for new ideas and the ideas appeared to be passed on to neighbours. However, the assessment revealed two main points of concern. Firstly, the farmers (on average) perceived the new technologies as being labour intensive and, secondly, the visits and tours led to some feelings of envy (see Chapter 17).

SWOT analysis
The outcome of the SWOT analysis that was made during the participatory assessment meeting is presented in Table 19.3. These results support the findings from the above-mentioned ladder exercise.

Group discussions on new initiatives
The farmer innovators grouped into their four respective networks (clusters) and discussed *who* had taken up *what* new initiatives since they joined the programme and what had been the source or stimulus. Table 19.4 gives the results. Once again, it shows that women were more responsive to new ideas than men: the uptake by women was an average of 4.5 initiatives each, but only three by men.

The project had organized more network visits than study tours and the farmers judged the overall impact of the former to be greater, but the average proportional uptake of new ideas after network visits was less. The four study

Table 19.3 *Results of SWOT analysis of network visits and study tours*

Strengths	Weaknesses
Acquire more friends	
Visit new places	
Learn new things	
Acquire new varieties of crops	
Innovators become focal persons	Tiring: people become weak
in the community on their return	Meal times irregular
Opportunities	Threats
Extend length of the study tours	Jealousy/envy

tours[8] (with, on average, six innovators involved in each) led to an average uptake of 1.25 initiatives per person per study tour. The 16 network visits (with, on average, six innovators involved in each) led to an average uptake of 0.4 initiatives per person per network visit.[9] This suggests that the study tours to the outside were more effective (on a one-to-one basis) than the local network visits and was probably due to the fact that a typical study tour covered about ten different sites.

A detailed analysis of one study tour carried out by PFI in Uganda (which brought farmers from Eastern Uganda to Kabale as well as to other locations in the country) showed that each farmer involved subsequently tried out four new ideas each. (Some of these were not directly related to land husbandry. Farmers keep their eyes open for any interesting ideas, including the subjects of cooking, sanitation and microenterprises.) However, study tours were more costly[10] and not available to everyone. The local network visits were easier to arrange (in Kabale, where population density is around 200 people/km^2, the farmers live fairly close to each other) and are much less exclusive as other farmers who did not belong to the innovator cluster were invited to join.

It should be noted that the visits organized by the project were not the only sources of inspiration for new initiatives. The category 'Other' rates quite highly and refers mainly to advice from the local extension agent and from other development projects in the area. 'Private visits' refers to innovators seeing and being impressed by what they saw while travelling afar or nearer home. The project cannot take direct credit for these two categories but, of course, welcomes them. The project *may* have played a role in stimulating a heightened interest in innovation among farmers.

The most popular new technologies adopted by the innovators were:

* improved composting/manuring (22 innovators);
* infiltration ditches/water harvesting (20);

Table 19.4 *New initiatives of 32 farmer innovators since joining the ISWC-Uganda activities*

Cluster (8 farmers in each)*	Numbers			Source			
	Total uptake of new ideas	By men (n=18)	By women (n=14)	Network visits	Study tours	Private visits	Other
A	39	15	24	14	11	6	8
B	27	17	10	1	19	3	4
C	14	8	6	7	0	4	2
D	36	13	23	19	0	3	14
Total	116	53	63	41	30	16	28

*Cluster A is 3 years old, Cluster B is 2 years old, and Clusters C and D are 1 year old

- improved mulching of bananas (16);
- agroforestry (15);
- improved banana management (12); and
- improved trash lines (5).

Impact on methodology development and institutionalization

The IA exercise did not go into detail to determine to what extent the project methodology had worked and become institutionalized. This had already been studied by Critchley (1999), who found that the methodology had proved broadly workable within the extension service. With respect to research, the greatest value of the on-farm work to validate and add value to local innovations had been indirect by providing a focal point for the project, rather than direct in the sense of generating useful technical recommendations. Indeed, the recommendations forthcoming from the on-farm research, although proven in scientific and socioeconomic terms, were only dimly recollected by farmers and extension staff alike. Furthermore, those technologies that had spread most rapidly were not closely related to the on-farm research (Briggs et al, 1998; Critchley et al, 1999b).

The key impact of the methodology development component of the ISWC-Uganda project was probably its decisive influence on formulating the methodology that underpins the PFI programme. In this way, the methods have spread and are being integrated into extension and research policy in Kenya, Tanzania and Uganda itself.

CONCLUSIONS AND RECOMMENDATIONS

We acknowledge that the IA exercise reported here is only partial and preliminary. Only some aspects of the project were investigated in detail and only certain assessment tools were used. It could be argued that, because a central element of IA – namely, cost–benefit analysis (CBA) at both farmer and project level – was lacking, the exercise was flawed. However, a full CBA was not possible because inadequate technical data were available about the technologies. It might also be said that the exercise was not participatory enough, relying too much on questionnaires (characterization forms) and generating too many 'dry' numbers of limited value. Equally, though, it could be said that too much participation in planning and implementing such an exercise could lead to the collection of subjective information biased towards the views of dominant members of the participating groups. There is surely room for both participatory and non-participatory elements in IA. Even with its evident limitations and constraints, the approach adopted for this IA proved to be at least functional and of some value, and can serve as a point of departure for other similar projects. In particular, the approach will be useful as a guide for similar exercises that PFI will undertake before the end of its first phase.

Much interesting and immediately usable information was generated and recommendations could be formulated to help steer the next phase of the project. Insights were gained not only through the data generated, but also through reflection during the IA process. Among the most important findings and corresponding recommendations were the following:

- Building on tradition/innovation through participatory methods can have considerable, tangible impact in terms of farmers testing, adapting and adopting new ideas.
- Farmer-to-farmer dissemination of ideas works well. Outside tours are particularly potent, but visits between innovators also produce measurable results and are simple to organize.
- The involvement of scientists is needed to validate innovations in technical and economic terms. Without this information, quantitative benefits cannot be assessed objectively. However, farmers' perceptions of benefits can be used as a proxy and these are, for a given locality, the acid test of the acceptability of technologies.
- Regular M&E needs to be stepped up for many reasons, including the provision of information to assist in IA. For example, farmers should be encouraged to keep simple records of inputs and outputs related to their innovation compared with a control plot.
- Too close a focus on a limited number of farmers (in this case, innovators) can raise problems of envy in the wider community; to remedy this, members of farmer clusters could be replaced after a certain period, and 'ordinary' farmers should be invited more frequently to network visits and field days.

The next logical step would be to plan a follow-up IA after two years. That exercise, while updating results about innovation, adaptation and adoption, should examine aspects of costs and benefits, building on data from the scientific validation of innovations. It could also incorporate specific assessments such as gender aspects of innovation and adoption, decision-making in innovation and adoption, and environmental impact based on indigenous indicators.

ACKNOWLEDGEMENTS

We are most grateful to Shem Turyamureeba, John Turyagenda and Pasco Bizimana for their assistance in carrying out fieldwork for this study and for their dedication during the day-to-day operation of the project. The results of the study have already been presented in a back-to-office report by Alie van der Wal and Will Critchley, dated August 1999; an article by Alie van der Wal in the newsletter *Farmer Innovators in Land Husbandry* (November 1999); and in the proceedings of the ISWC 2 anglophone regional workshop in February 2000 (Mitiku et al, 2000).

NOTES

1 Funded by the Overseas Development Administration and coordinated by Silsoe Research Institute with technical assistance from CDCS; local management MAAIF

2 Funded by The Netherlands Government (DGIS) and coordinated by CDCS with partners; local management by MAAIF

3 In this chapter, the term 'adoption' has been used very broadly: see definition in Results and Discussion

4 These results reflect only those innovators and innovations identified by the project, but not necessarily the characteristics of the population of innovators in the community at large

5 PFI is funded by The Netherlands Government, managed by UNDP's Office to Combat Drought and Desertification, executed by national governments and technically backstopped by CDCS. Its approach and methodology is basically the same and indeed was derived from experience under CWSSE

6 The fourth cluster is in an area with 1000mm annual rainfall; moisture conservation is less of a priority here, but soil fertility issues are at least as important as in the drier zone

7 In the typology of visits given in Critchley et al (1999b), these network visits are a combination of two types: a) farmer-innovator-to-farmer-innovator and b) farmer-to-farmer-innovator. The other two types of visits in the typology are c) farmer-innovator-to-farmers and d) study tours

8 Data on study tours and network visits pertain to 1998 and 1999 only. Accurate records before this date do not exist. Only the results of Network A would be affected by this omission

9 These calculations differ from those originally presented in reports, because new and more accurate information subsequently became available about the number of NVs and STs and the number of farmers attending them

10 Study tours vary considerably in cost, depending on the number of people travelling, whether a vehicle must be hired and the period of travel. An indicative figure of US$1000 can be used to compare with US$100 or less for a local network visit

REFERENCES

Abbot, J and Guijt, I (1998) *Changing views on change: participatory approaches to monitoring the environment*, SARL Discussion Paper 2, IIED, London

Arevalo, M, Guijt, I and Saladores, K (1998) *Participatory monitoring and evaluation*, PLA Notes 31, IIED, London

Briggs, S R, Critchley, W R S, Ellis-Jones, J, Miiro, H D, Tumuhairwe, J and Twomlow, S (1998) *Livelihoods in Kamwezi, Kabale District, Uganda, a final technical report from Environmental Research Project R4913, Conserve Water to Save Soil and the Environment (1995–1998)*, unpublished report IDG/98/11, Silsoe Research Institute, Silsoe

Critchley, W R S (1999) 'Harnessing traditional knowledge for better land husbandry in Kabale District, Uganda', *Mountain Research and Development*, vol 19, no 3, pp261–272

Critchley, W R S, Cooke, R, Jallow, T, Lafleur, S, Laman, M, Njoroge, J, Nyagah, V and Saint-Firmin, E (1999a) *Promoting Farmer Innovation: harnessing local environmental knowledge in East Africa*, RELMA/UNDP, Nairobi

Critchley, W R S, Miiro, H D, Ellis-Jones, J, Briggs, S and Tumuhairwe, J (1999b) *Traditions and innovation in land husbandry: building on local knowledge in Kabale, Uganda*, RELMA, Nairobi

Ellis-Jones, J and Tengberg, A (1998) 'The impact of conservation on soil productivity: examples from Kenya, Tanzania and Uganda', in S R Briggs, J Ellis-Jones and S J Twomlow (eds) 'Modern methods from traditional soil and water conservation technologies', proceedings of a land management workshop, White Horse Inn, Kabale, Uganda, January 1998, unpublished report IDG/98/10, Silsoe Research Institute, Silsoe, pp164–181

Estrella, M, and Gaventa, J (1997) *Who counts reality? Participatory monitoring and evaluation: a literature review*, IDS Working Paper, University of Sussex, Brighton

Guijt, I (1998) *Participatory monitoring and impact assessment of sustainable agriculture initiatives*, SARL Discussion Paper 1, IIED, London

Harnmeijer, J, Waters-Bayer, A and Bayer, W (1999) *Dimensions of participation in evaluation: experiences from Zimbabwe and The Sudan*, Gatekeeper Series 83, IIED, London

Herweg, K, Steiner, K and Slaats, J (1998) *Sustainable land management: guidelines for impact monitoring*, Centre for Development and Environment, Berne

IDS (1998) *Participatory monitoring and evaluation: learning from change*, Policy Briefing 12, University of Sussex, Brighton

Mitiku H, Waters-Bayer, A, Mamusha L, Mengistu H, Berhan G A, Fetien A and Yohannes G M (eds) (2000) 'Farmer innovation in land husbandry', proceedings of anglophone regional workshop, 6–11 February 2000, ISWC-Ethiopia, Mekelle

Willcocks, T J, Twomlow, S J, Ellis-Jones, J and Critchley, W R S (1992) 'Conserve Water to Save Soil and the Environment', project proposal, Overseas Development Division, Agriculture and Food Research Council, Silsoe Research Institute, Silsoe

20

Three models of extension by farmer innovators in Burkina Faso

*Aly Ouedraogo and Hamado Sawadogo**

In the early 1980s, farmers in the Yatenga region of the densely populated Central Plateau in Burkina Faso developed a method of rehabilitating degraded land by improving the traditional planting pits known as zaï. Some of the farmers who have contributed to this development are also making great efforts to promote the spread of this technology. This chapter describes three extension approaches of farmer innovators – approaches that are, in themselves, innovations.

The *zaï* are pits that farmers dig in rock-hard barren land, into which water otherwise could not penetrate. The pits are about 20–30cm in diameter and 15–25cm deep. The farmers put organic matter into them. This attracts termites which dig channels and thus improve the structure of the soil so that more water can infiltrate and can be held in the soil. By digesting the organic matter, the termites make nutrients more easily available to the plant roots. Thus, the termites are major allies of the farmers in their battle to rehabilitate degraded land.

In most cases farmers grow millet or sorghum or both in the *zaï*. Sometimes they sow seeds of trees directly together with cereals in the same *zaï*. In this way, the young trees also benefit from the concentration of manure and water in these pits. When harvesting grain, the farmers cut the stalks off at a height of about 50–75cm. The parts of the stalks that remain standing protect the tree seedlings from grazing animals. Over the years, thousands of farmers in Yatenga have used this locally improved traditional technique to reclaim

* Aly Ouedraogo is an agronomist with PRA Network (Réseau MARP), based in Ouagadougou, Burkina Faso, and Hamado Sawadogo is a conservation agronomist, and a PhD student at the Faculty of Agricultural Science in Gembloux, Belgium

Map 20.1 *Burkina Faso (Yatenga and Zondoma action areas)*

strongly degraded land. Any experienced observer can see that thousands of hectares of land have been rehabilitated with this technique in the last 20 years or so, but 'hard' data are not available. The major objective of most farmers is to increase food production in order to secure food for the family, but some farmers are now using *zaï* as a technique to (re-)establish woodland, with a view to selling timber and other products.

Some of the farmers who have contributed greatly to developing the improved *zaï* technology have also made major efforts to promote its spread and further improvement. Three 'extension models' developed by innovators are described here.

THE 'MARKET DAY' MODEL

In the village of Gourga, 4km west of Ouahigouya, the capital of Yatenga region, Yacouba Sawadogo uses a market day model to promote the spread of *zaï*. He started improving the traditional planting pits around 1980. The *zaï* have since become recognized, also by scientists, as the most efficient technique in the Sahel for rehabilitating severely degraded land. Since he started improving the *zaï*, Yacouba has experimented with a range of other related innovations. He uses the *zaï* for growing trees (see Chapter 4).

Since 1984 Yacouba has been organizing market days in Gourga to give farmers an opportunity to share their experiences with *zaï*. These started as small events, but now each market day involves people from more than 100 villages. The events are held twice a year. The first market day is shortly after the harvest and farmers bring a sample of the crop varieties (millet, sorghum, maize, cowpea) they have cultivated in their *zaï*. Yacouba stores these seeds on his farm. The second market day is held just before the wet season. Farmers can then select the species and varieties they would like to plant in their *zaï*, taking into account the improved growing conditions. Yacouba's main problem is that his facilities for storing the varieties are poor.

Each market day has a specific theme. For instance, during the last market day the focus was on growing sesame. An earlier theme was the use of *zaï* for growing trees through the system of direct seeding. At each market day, there is also a display of the local tools used to dig the *zaï*. This allows farmers from outside the region to see for themselves which tools can be used and to find out where they can buy them.

Yacouba has created an Association for the Spreading of *Zaï*, of which he is the chairman. The General Assembly of this association takes place during the market days. He created this association to mobilize external financial or material support for the spreading of the *zaï* technology. The external support has always been very modest. In 1997 the Association received three motorcycles, fuel and some cement from an NGO. Before 1997 Yacouba used his own motorcycle and paid for his own fuel to visit villages to spread his message. The national television of Burkina Faso made a programme about the market day and the radio made two broadcasts about Yacouba's experience with managing natural resources.

Many visitors come to Yacouba's farm and receiving them costs him a substantial amount of time. The solution he has found for this problem is to request an input from each visitor. Those who come from abroad are asked to plant a tree seedling, which Yacouba raised in his own small nursery and groups of farmers from elsewhere in Burkina Faso or West Africa are requested to dig some *zaï* on his land. This also functions as a kind of on-the-job training.

What motivates Yacouba Sawadogo to innovate and to spread his innovations so actively? He says that he wants to prove that environmental degradation is not irreversible and that it is possible to make a living in Yatenga. At the same time, he wants to be recognized as an innovator and this public recognition is a major incentive to him.

THE 'ZAÏ SCHOOL' MODEL

In the village of Somyanga, also in the Yatenga region, Ousséni Zoromé initiated the 'zaï school' model. In 1992 he started training some local farmers in how to make good *zaï*. He chose the poorest possible site, immediately next to the tarmac road linking Ouahigouya and Ouagadougou, the capital city. The soils on the site had been completely destroyed by bulldozers constructing the road. The farmers managed to achieve a millet harvest of 400kg of millet/ha on this very poor land. All people travelling along the main road saw this immediately because it was a year of extreme drought and many crops had failed. Also, the Minister of Agriculture saw the plot and called in a team from the national television to film it.

Ousséni immediately started to create new groups which he calls '*zaï* schools'. Each group has to rehabilitate collectively a piece of degraded land. In this way, all participants are trained on the job. The yields obtained on the field rehabilitated by the members of the *zaï* school are shared between the members. Whenever possible, Ousséni organizes demonstrations for groups either on his own fields or on the fields of members of the *zaï* schools. He has received no external material support except occasionally some fuel for his old motorcycle from the Regional Department of Agriculture. He continues to use his own motorcycle for this extension work and usually pays for his own fuel.

Currently, there are 21 *zaï* schools with a total of more than 1000 members and their numbers are increasing rapidly. Ousséni is now seeking external support to expand the number of *zaï* schools and to improve them. Each group pays a contribution of 5000CFA (circa US$8) to become a member of a regional union of *zaï* schools which Ousséni created in 1997.

An interesting entry point for collaboration with the ISWC programme is that the farmers in some of the *zaï* schools are carrying out their own experiments on their fields. These experiments include comparing the impact of compost and non-decomposed manure and testing an early maturing variety that is rare in this region.

THE 'TEACHER–STUDENT' MODEL

In the village of Gourcy, Ali Ouédraogo, a very experienced farmer innovator, has invested heavily in improved *zaï* in combination with applying compost, planting trees and protecting the naturally regenerating trees and shrubs. He is training individual farmers in five villages around Gourcy and visits them regularly to work with them directly in their fields to show how he manages *zaï*, to give the farmers advice and to exchange ideas with them. His 'students' in turn train other farmers in improved *zaï* techniques. Some of the students do not simply adopt what Ali suggests; they go on to experiment with his original ideas and to develop adaptations of it. For example, one farmer (Hamadé Bissiri) felt that the *zaï* made by Ali are extremely large and require a great deal of working time and physical strength to dig. Not everyone is able to do this. Hamadé therefore modified the layout and dimensions of the *zaï* to suit his capacities.

Since 1993 Ali has trained 12 farmers, each of whom has started training other farmers at their request. They are not paid for these services. Their major reward is social esteem, but this is sometimes sweetened by gifts of appreciation (chickens, colanuts or a meal). Ali keeps close contacts with these 12 farmers who live in three villages within 6km from Gourcy.

These three models of farmer-led extension were all developed on the initiative of the farmers themselves who have become, in fact, public service providers who receive no remuneration for their time. At the most, they receive some limited external support for travel from local NGOs or individuals. These farmers have few links with the government extension service, except through the above-mentioned regional union of *zaï* schools, which received some support in developing project proposals to acquire tools and receives information about relevant regional or national meetings. The Regional Department of Agriculture provides travel and a subsistence allowance to the farmers who attend these meetings.

TOWARDS FOOD SECURITY AND WEALTH

The farmers in Yatenga and in other parts of the densely populated Central Plateau of Burkina Faso are becoming increasingly interested in *zaï*. Under such dry conditions this is not surprising. The pits collect and concentrate run-off water, allowing farmers to use small quantities of manure and compost very efficiently.

The use of *zaï* allows farmers to expand their resource base and to increase household food security. In the early 1980s, all three of the above-mentioned farmers faced structural food deficits. Since then, through their innovative energy and investment in rehabilitating degraded land, they have created larger areas that are suitable for crop and tree production (see Chapter 4) and have attained food security for their families – and in years when rainfall is below average. Ali has even built up a stock of food from which he could feed his family for three years. All three farmers now have many more trees on their fields than they had 20 years ago. These help to maintain the productive environment, increase biodiversity, reduce the risks of production and, thus, strengthen food security.

These innovative farmers do not want to monopolize their knowledge. They are generous in sharing their experience with others and their benefits are primarily in the form of personal satisfaction and higher social recognition. These appear to have been their main motivations to develop their own extension models for giving practical training and advice to other farmers who, in turn, are keen to learn from them.

Part 5

Stimulating and supporting joint experimentation

Participatory Technology Development on soil fertility improvement in Cameroon

Paul Tchawa, Pierre Kamga, Christopher Ndi, Christopher Vitsuh,
*Samuel Toh and Antoine Mvondo Zé**

The development of the system of night paddock manuring as a farmer
response to declining soil fertility, and the chain of innovations that this initial
innovation triggered, have been described in Chapter 9. This chapter describes
the process of joint experimentation by farmers, development agents and scien-
tists to gain a better understanding of this system and to develop it further,
and presents the initial results.

BACKGROUND

In recent years, production of black nightshade (*Solanum nigrum*) in North-west Province of Cameroon has increased tremendously. It has become the major income-generating vegetable crop of market gardeners in Babanki, a village in Tubah subdivision of the province. Babanki is situated on the Nkambe road about 30km from Bamenda, the provincial capital. This city has about 135,000 inhabitants. Babanki lies at an altitude of 1970m in the south-eastern part of the 'grassfields', so named after the grassland savannah vegetation that

* Paul Tchawa is a geographer and senior lecturer at the University of Yaoundé I and coordinator of ISWC-Cameroon based at SNV in Yaoundé; Pierre Kamga and Christopher Ndi are animal scientists with the Institut de Recherche Agricole pour le Développement (IRAD) Bambui in Bamenda; Christopher Vitsuh and Samuel Toh are farmer innovators with Kedjom Ketingoh Union Farmers Group (KEKUFAG) in Bamenda, and Antoine Mvondo Zé is head of the Soils Laboratory at the University of Dschang, Cameroon

Map 21.1 *Cameroon*

dominates there. The average annual rainfall is about 2000mm and the wet season lasts from mid-March to mid-November. The soils were originally fertile, coming mainly from volcanic rocks. These natural conditions explain the relatively high rural population density (around 150 persons/km^2). However, intensive use led to a decline in soil fertility, a problem increasingly felt by the farmers. Nomadic cattle pastoralists (eg Mbororo Fulani) from further north were attracted by the grass savannahs and, since about 1940, have been settling in the area (Vabi, 1993). As described in Chapter 9, farmers started to see advantages in the presence of the Mbororo cattle to solve the problem of declining soil fertility and developed the system of night paddock manuring.

Farmers in Babanki were keen to improve their technology, especially in view of the high demand for nightshade in Bamenda. They were pleased therefore at the interest that scientists showed in the local innovations and experimentation in nightshade production and were eager to collaborate with the scientists in further experimentation.

PREPARING FOR JOINT EXPERIMENTATION

The farmer innovators who had been identified by the ISWC-Cameroon programme and the coordinator of the programme organized a one-day workshop in June 1999 to identify priorities for joint experimentation. About 50 men and women farmers from Babanki, together with five scientists and four development agents, met at the palace of the traditional chief (*Fon*). The Fon himself attended part of the workshop. Initially, both farmers and scientists felt somewhat apprehensive as it was the first time that the villagers were meeting with so many scientists at once. Three staff members from CIPCRE (Cercle International pour la Promotion de la Création) – Patricia Nkouamo, Mary Mbafor and Jules Talom – facilitated the workshop and immediately started with an 'icebreaker'. Each participant was invited to do an imitation of the most frequent attitude they take at work. This indeed broke the ice. All participants felt more at ease and no longer appeared to be inhibited in expressing themselves. At an early point in the workshop, one of the farmer innovators, Samuel Toh, clearly expressed the views of the farmers (quoted in Tchawa, 1999):

> *We are aware that the researchers have their priorities. We also have our priorities, and they are probably different from those of the researchers. Let's first address our priorities and then we will also work with the researchers on what they want to study.*

He then went on to summarize their priorities:

> *We sometimes have the impression that we enclose too many cattle for too long in our night paddocks, so our fields become too well fertilized. We would like to know how many cattle should be enclosed for how long in order to get the best levels of soil fertility. Secondly, we would like to know which crops we could best cultivate and for how long, so that we get the best results from the good fertility.*

The scientists from the University of Dschang were so impressed by Samuel Toh's analysis and presentation at the workshop that they nicknamed him 'Professor'. After some discussion among scientists in group work and further discussion between scientists and farmers, the head of the University's soils laboratory agreed to follow the priorities set by the farmers. The other outsiders (scientists and development agents) had become very excited about exchanging ideas with farmers about experimental design, parameters to observe and measurement tools.

In early July 1999, a second workshop was organized with almost the same participants to define how the joint experiment would be set up. The various tasks were listed and the roles to be played by each of the research partners were defined. A research team was formed that included four farmer innovators, two scientists from the University of Dschang (a soil scientist and an agroforestry specialist), a geographer from Yaoundé I University, three development agents from CIPCRE and four graduate students from the University of Dschang.

Two animal scientists and one agricultural economist from Bambui Station of IRAD (Institut de Recherche Agricole pour le Développement) joined the team three weeks later. After the other scientists briefed them in detail about the options chosen in the workshop, they met the farmer innovators who showed their paddocks and explained their research priorities. It took some time for one of the scientists to understand the aim of the collaboration. He claimed initially: 'You want to open doors which are already open. The data on soil fertility enhancement using cattle dung are available in many scientific books. I can give you some references.' Later, after becoming very active in the process of joint experimentation, this very scientist confessed that the data available in those many books come from research done in a quite different environment for quite different purposes from those of the farmers of Babanki.

The four farmers involved in the experimentation – Samuel Toh, Philip Ndong, Christopher Vitsuh and Peter Mabong – were selected by the community on the basis of the following criteria: they had long experience with the night paddock manuring system, they owned the land they were cultivating, they were eager to experiment and, within their farmers organization KEKUFAG, they had the reputation of being open and willing to share their experience.

During the July workshop, the research partners agreed to divide their tasks as follows:

Farmers:
- make sure that the cattle stay on each plot for the duration of the period agreed;
- note the time that the cattle enter the plots each night and leave them each morning;
- construct the fencing;
- cultivate and manage the plots;
- observe and measure plant growth once a week;
- observe weed growth on each plot;
- register data on labour inputs for each plot (time and number of persons);
- observe plant diseases and treat, if need be, with a solution made from tobacco leaves;
- weigh the number of bags of nightshade harvested on each plot;
- record all the above-mentioned quantitative and qualitative data,
- solve any problems related to the experimental plots (livestock disease, fencing, etc).

Scientists:
- assist in the selection of sites for the test plots;
- analyse the cultivation history of the plots;
- determine the initial soil fertility status of each plot (pH, C, N, P, K, C/N ratio);
- determine the soil fertility levels after the fields have been manured;
- determine soil density in the plots before the cattle are enclosed in them;
- determine soil density on each plot after manuring ends but before tillage starts;
- determine the rate of mineralization of the organic matter;
- analyse all other data forwarded by the farmers.

NGO staff:

- support the farmers in their various tasks associated with the experiment;
- organize visits by other farmers to the experimental plots.

After the workshop, ISWC-Cameroon recruited a field agent to monitor this experiment, as well as others being set up elsewhere in West and North-West Cameroon. He helped the farmer experimenters to prepare a format for keeping records, based on the agreement reached by all research partners about indicators, means and methods of measurement, and means of recording the data.

SETTING UP AND CARRYING OUT THE EXPERIMENT

The four farmers selected the experimental plots to represent different topographical conditions. Peter Mabong's plot had the steepest slope, followed by the plot of Samuel Toh. The two other plots were on relatively gentle slopes. The farmers and scientists agreed that each replication would be 500m^2 in size, consisting of one control plot of 100m^2 and four plots of equal size where the cattle would be kept overnight for one, two, four and six weeks respectively. The layout was as shown below. The plot size was suggested by the scientists as appropriate for five treatments with the number of cattle available. The farmers thought that the experimental area should be no larger than this in view of the scarcity of cropland in the village area. The trial plots were arranged in such a way as to avoid lateral run-off from one plot to another. The multiple-stakeholder research team jointly marked out the plots, as shown in Figure 21.1.

At the beginning of the experiment, the scientists requested the following background information from the farmers:

- When did you start cultivating this field?
- What did you cultivate last year in this field?
- What techniques do you normally use on your fields (eg contour ridging, irrigation, burning, application of chemical fertilizer)?

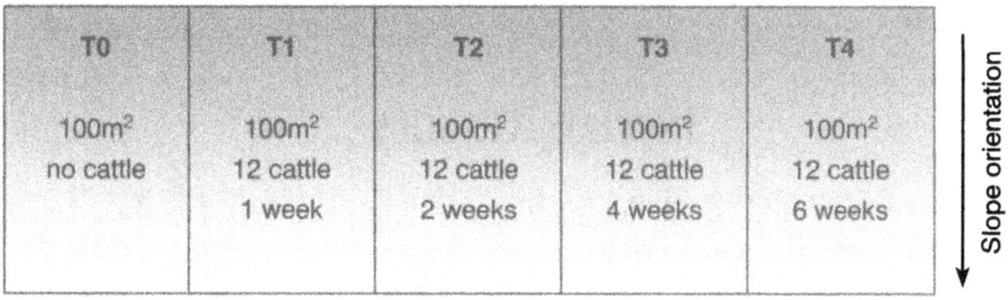

T0	T1	T2	T3	T4	
100m^2	100m^2	100m^2	100m^2	100m^2	Slope orientation
no cattle	12 cattle	12 cattle	12 cattle	12 cattle	
	1 week	2 weeks	4 weeks	6 weeks	

Figure 21.1 *Layout of the trial plots in the farmer-led research*

Plate 21.1 *One of the farmers' trial plots (West Cameroon)*

The scientists also made an inventory of the plants growing on the fields and took several soil samples. The four farmers involved in the trials joined the scientists in the field during sampling. This offered an opportunity for the scientists to ask each farmer about the history of his plot and their local indicators of change in soil fertility levels.

Each farmer experimenter did his own recording, using notebooks as well as charts pasted on the wall of the farmhouse. This does not mean, however, that the experimentation was only the business of those four farmers. It was also supported and observed by the *Fon* and the other villagers.

The manuring on the farmers' trial plots was commenced on 15 July 1999 and lasted, in the plots manured for six weeks, until 26 August. The soil was tilled on 1–3 September and then, in line with the normal practice of farmers applying this technique, the plots were left for almost a month to allow the dung to mineralize before they started to ridge the land and to sow at the end of September.

INITIAL RESULTS

The first harvest of nightshade leaves was possible in early December. Leaves are harvested several times in one growing season. As examples, Tables 21.1 and 21.2 show the number of bags of leaves that two of the farmer experimenters obtained after the first six harvests (a bag of fresh leaves weighs about 40kg).

Table 21.1 *Bags of nightshade leaves harvested by Samuel Toh*

Date of harvest	T0 (control plot)	T1 (12 cattle/ 1 week)	T2 (12 cattle/ 2 weeks)	T3 (12 cattle/ 4 weeks)	T4 (12 cattle/ 6 weeks)
3 December 1999	0.5	0.75	0.75	0.5	0.5
4 January 2000	1	1.5	1.5	1	1
4 February 2000	0.5	2	2	2	1.5
7 March 2000	0	1	1	2	2
7 April 2000	0	0	0	3.25	3.25
5 May 2000	0	0	0.5	1.25	2.75
Total	2	5.25	5.75	10	11

Table 21.2 *Bags of nightshade leaves harvested by Christopher Vitsuh*

Date of harvest	T0 (control plot)	T1 (12 cattle/ 1 week)	T2 (12 cattle/ 2 weeks)	T3 (12 cattle/ 4 weeks)	T4 (12 cattle/ 6 weeks)
3 December 1999	0.25	0.50	0.75	0.75	1
4 January 2000	1	1.25	1.5	1.5	2
4 February 2000	0.75	1.25	1.5	1.5	2
7 March 2000	0.5	1.5	1.75	1.75	2
7 April 2000	0	0.5	1	1.75	1.75
5 May 2000	0	0.25	0.75	1	1
Total	2.5	5.25	7.25	8.25	9.75

During the first round of harvesting in December, the yields were still low, but they doubled the following month and even tripled in the case of one farmer. The total number of bags harvested on 500m^2 (ie the total experimental area) over six harvests was 36 bags by Peter Mabong, 34 bags by Samuel Toh and 33 bags by Christopher Vitsuh and 29 bags by Philip Ndong.

The four farmers explained the differences between the plots by their cultivation history, the availability of water for irrigation and the size of the animals used for manuring the plots. Nevertheless, they felt that the differences could not hide the fact that, without manure, nightshade did not produce well. During the fifth harvest in April, none of the farmers harvested any leaves from the control plot. The plots that were manured for one or two weeks (T1 and T2) showed some effect of fertilization, but this effect disappeared after the fourth harvest. The farmers assumed that nightshade would not grow well on these plots a second year without once again enclosing cattle on the plot. The plots that were manured for four and six weeks (T3 and T4) brought the highest yields. Altogether, the four farmers harvested 35.5 bags of nightshade leaves from T3 and 41.25 bags from T4. This difference is insignificant and the additional yield does not compensate for the extra costs required for feeding and guarding the animals.

SUPPORTIVE ON-STATION RESEARCH

After starting the joint experimentation with farmers, the scientists felt that it would be useful to conduct supportive on-station research on the IRAD station in Bambui. They invited the farmers to assist them in setting up the experiment and in cultivating the soil and sowing the nightshade in order to ensure that this work would be done properly and as similarly as possible to that in the on-farm experiments. The on-station research allowed the scientists to measure certain parameters which the farmers were not interested in measuring and recording themselves as they appeared to be time-consuming for very little additional benefit to the farmers. For example, the scientists measured the effect of manuring on weed growth, whereas the farmers merely observed this and made a qualitative assessment. The on-station experiment started in mid-January 2000 and involved a cycle of manuring in the dry rather than the wet season. As in the farmers' experiments, 12 head of cattle were used per treatment.

Table 21.3 *Additional data collected in the parallel experiments on-station*

	T0 (control plot)	T1 (1 week's manuring)	T2 (2 weeks' manuring)	T3 (4 weeks' manuring)	T4 (6 weeks' manuring)
No of weeds/m^2 12 days after sowing nightshade	82	149	106	278	196
No of nightshade plants/m^2 12 days after sowing	200	223	230	322	423

Quantification of weed growth

Weed growth was particularly strong on the plot manured for four weeks (T3) and slightly less on that manured for six weeks (T4) which had the highest number of nightshade plants after 12 days. Farmers had observed that competition between weeds and nightshade was severe on well-manured plots; this quantification supported their observations.

Quantification of fertilization

Three months after the July 1999 workshop, during the process of monitoring the joint experiments, the scientists requested a meeting of the research partners to work out a system for collecting and measuring the quantities of manure and urine produced by the cattle. The scientists explained to the farmers that this would allow them to calculate how long the animals should be kept in the paddock for optimal manuring of the nightshade crop. The scientists and farmers jointly designed a small shed for stabling one animal.

The shed is closed on all four sides and has a cement floor to facilitate manure collection. The urine is evacuated through a drain and collected in a graded bucket on which the quantity can be read directly. The scientists and development agents designed the measuring system and explained to the farmers how it worked. The scientists then called yet another meeting to examine the expenditures related to this particular project. The costs were shared according to the abilities of each partner to pay (75 per cent by ISWC-Cameroon, 15 per cent by CIPCRE and 10 per cent by the farmers). In addition, the farmers provided labour and assisted the bricklayer. Three such sheds were constructed, one beside Samuel Toh's house, another beside Christopher Vitsuh's house and a third one in the research station. The station provided the poles, labour and the animal, kept in the on-station shed; all other costs were covered by ISWC-Cameroon.

In the first shed, a 4-year-old white Fulani steer was kept overnight, in the second a $3^{1}/_{2}$-year-old red Fulani cow and in the third a 4-year-old red Fulani steer. The scientists proposed these differences in sex and breed of the animals because, knowing the structure of the herd in the village, they wanted to be able to extrapolate the results to apply to an entire herd used for manuring purposes. The animals spent 9–10 hours per day grazing in the fields and then 14–15 hours in the shed each night for 36 nights. The quantities of manure and urine were weighed and read by the two farmers. About once a week, a scientist from IRAD visited them to check the recording and to take samples of dung and urine. During the 14–15 hours in the shed, the steer produced an average of 7.8kg of fresh dung and 6 litres of urine, whereas the cow produced 7.5kg of fresh dung and 3.5 litres of urine (Tchawa and Kamga, 2000).

The scientists developed the following equation to calculate the quantities of urine and fresh dung deposited on the experimental plots:

$$Q = (D \times N)a + (D \times N)b$$

whereby: Q expresses the quantity of fresh dung in kg and urine in litres
D is the number of days the herd spends in the night paddock
N is the number of animals in the herd kept in the paddock
a and b are coefficients representing average quantities of manure and urine produced as a function of sex, age, weight and breed of animal, and season.

The season plays an important role in the nature of pastures and therefore in the amount and quality of manure produced. However, this could not be considered in the calculations derived from the experiment described here as it covered only five weeks.

In the farmers' experiments:

$D = 0, 7, 14, 28$ and 42 days
$N = 12$

a = 7.7 kg/day
b = 5 l/day

Table 21.4 shows the estimated quantities of dung and urine deposited on the farmers' experimental plots in Babanki. These calculations provide an example of the contribution that scientists can make in supporting farmers' experimentation. The scientists and development agents organized a preliminary feedback meeting in Babanki to inform the farmer experimenters and the other villagers about these first results. The research protocols proposed by the scientists helped the farmers to find answers to their question about the length of time the cattle should be kept in the night paddock in order to obtain the optimal levels of soil fertility. Without both the farmers' careful recording of the data and the scientists' elaboration of the formula, the calculations would not have been possible.

Table 21.4 *Estimated quantities of dung and urine deposited on the farmers' experimental plots*

Treatment	Urine (litres)	Fresh dung (kg)	Time of cattle in paddock (h)
T0 (control)	0	0	0
T1 (1 week)	420	647	100
T2 (2 weeks)	840	1294	201
T3 (4 weeks)	1680	2587	402
T4 (6 weeks)	2520	3881	603

These findings have stimulated additional research questions among the partners. The scientists are now interested in assessing the risks to human health resulting from infestation of the nightshade by nematodes and bacteria because they know from the literature that such risks are associated with consumption of vegetables grown with dung. They would also like to measure residual urea in the plants cultivated on plots fertilized with animal faeces.

ADD-ON EXPERIMENT BY CHRISTOPHER VITSUH

Christopher Vitsuh, one of the farmers who collected dung and urine data on a daily basis, developed the idea of setting up his own experiment to find out the respective effect of urine and dung on soil fertility. He also decided to compare cow dung and urine with chicken manure. He grasped the opportunity of the ongoing joint research to graft on his additional experiment. His basic questions were:

- Is the cattle dung more efficient than chicken manure in improving soil fertility?

- Of the dung and the urine that the cattle deposit in the night paddocks, which contributes most to restoring soil fertility?

He felt confident that he could easily carry out this additional experiment because:

- he was already collecting the dung and urine separately in the joint experiment;
- he had easy access to chicken manure which is available in their local markets;
- he had a field close to the shed where he could conduct his experiment;
- his experiment could be designed very simply;
- he could use nightshade yield to measure the effect of the different organic fertilizers;
- he could use the material for the joint experiment also for his own experiment; and
- through the joint experiment, he had contact with a scientist whom he could ask for advice about how best to design his own experiment.

Christopher had intended to make four ridges, each 20m long and 50cm wide, in his additional plot. One of these would be a control and the three others would be treated with urine, cow dung and chicken manure respectively. Instead of this, the scientist advised him to try a simple randomized layout with three groups of four ridges each, keeping the same size (20m long and 50cm wide). This meant that he needed three times as much space for the experiment than he had originally planned. Two weeks after sowing the ridges with nightshade, Christopher applied 70 litres of urine to one ridge in each of the three groups, 20kg of chicken manure to one ridge in each group and 67 litres of liquid cow dung mixed with urine to one ridge in each group.

Table 21.5 shows the number of bags of nightshade that he harvested in each treatment. The yields on the ridges treated with urine were 20 per cent higher than on the ridges fertilized with cow dung only and 30 per cent higher than on the ridges fertilized with chicken manure. Although the yield differences between the three treatments were not statistically significant, they were clearly perceptible and were therefore significant to the farmer. The yields of nightshade on the ridges treated with urine and cow dung started to decline after the fourth harvest, whereas the yields on the ridges treated with chicken dung started to decline after the third harvest. However, the ridges treated with chicken manure had the highest yield in the first harvest (1.5 bags).

In addition to examining the yield data, which he collected and recorded himself, Christopher made the following observations:

- The nightshade leaves were darker on the ridges treated with urine and chicken manure than on those treated with cow dung; the colour was lightest on the control ridges.

Table 21.5 *Bags of nightshade harvested on Vitsuh's add-on experimental plot*

Harvest dates	Urine (3 ridges)	Cow dung (3 ridges)	Chicken manure (3 ridges)	Control plot (3 ridges)	Total (12 ridges)
17 December 1999	0.75	0.75	1.5	0.25	3.25
18 January 2000	1.5	1.25	2	1	5.75
21 February 2000	2	1.5	1.5	0.75	5.25
19 March 2000	2.5	2	1	0.5	6
20 April 2000	2	1.75	0.5	0	4.25
Total	8.75	7.25	6.5	2.5	24.5

Credit: Chris Reij

Plate 21.2 *Christopher Vitsuh records trial data himself*

- The leaves were smaller on the ridges treated with chicken manure and on the control ridges in the last two harvests.
- The leaves became yellow and dried out when some urine was accidentally dropped on them directly.

Christopher first shared the results with his fellow farmer experimenters who then invited the ISWC-Cameroon coordinator and a scientist from IRAD to discuss them. Now the farmer experimenters have suggested that IRAD and Samuel Toh also try this experiment, as they also have a cowshed available. They then plan to discuss the results and share their conclusions within the farmers' union (KEKUFAG).

GENERAL CONCLUSIONS

This experience by scientists, development agents and farmer innovators clearly shows the flexibility and improvization that form part of the PTD process. The changes during the course of collaboration did not harm the research; instead, they were decisive in the strategy to generate improved technologies. If everything is done according to the basic principles of the PTD approach (building on existing techniques, maintaining flexibility, involving farmers at all stages, sharing costs and results), the partners quickly make the research their own. All that was known at the outset was that a process of joint experimentation had started and with what it had started, but it could not be foreseen when and where it would end as the research partners worked together in a creative manner.

The experience in Babanki also showed the importance of the role of outsiders in the process of farmer-led experimentation. The scientists helped make the process more efficient. A next step will be to replicate the experiments in order to confirm the initial findings and to validate them scientifically. At the same time, however, the joint experimentation should continue, turning to the question of the duration of soil fertility improvement after night paddocking, ie after how many years a plot that was fertilized by cow dung should be fertilized again. Again, this will require the participation of scientists.

It will also be important for the farmers, development agents and scientists to reflect together about how visual tools can be developed for feeding back the information from this joint research to farming communities. This will contribute to a more rapid farmer-to-farmer extension of the night paddocking technology beyond its already wide and spontaneous spread within North-west Cameroon. It will be necessary to present the findings from the research in a form that will enable the formal extension system to incorporate the technology into their messages.

Finally, the scientists should take every opportunity to make clear to all concerned – and especially to decision-makers in research and development services – that the PTD process can make an important contribution to the development of strategies for the sustainable management of natural resources.

REFERENCES

Tchawa, P (ed) (1999) 'Rapport annuel CES 2 Cameroun' SNV/CES 2, Yaoundé

Tchawa, P and Kamga, P (2000) 'Essai préliminaire de quantification des déjections bovines produites sur quelques parcs de nuit à Babanki (Nord-Ouest Cameroun)' in P Tchawa and JM Diop (eds) 'Paysans innovateurs en gestion durable des ressources: atelier régional francophone du programme CES 2, Bamenda, Cameroun, 29 novembre–2 décembre 1999', SNV/CES 2, Yaoundé

Vabi, M B (1993) 'Fulani settlement in Northwest Province of Cameroon', *Pastoral Development Network Paper* 35d, Overseas Development Institute, London

22

Farmer-led experimentation in the drylands of Central Tigray

GebreEgziabher Miruts and Fetien Abay[*]

Before the ISWC 2 programme began, a project supported by FARM-Africa in Tigray had gained experience in facilitating research by resource-poor farmers. The lessons learnt by this project made an important contribution to ISWC-Ethiopia. Above all, the farmer researchers, who could show and explain to others their experiments and results, helped to convince scientists and policy-makers in Ethiopia of the potentials of a PTD approach to agricultural improvement.

Resource-poor farmers in the Tigray region of northern Ethiopia face harsh environmental conditions, such as unreliable rainfall, recurrent drought periods, severe soil erosion, attacks on their crops and livestock by pests and diseases and infestation by weeds. These farming households seek mechanisms to cope with stress situations and to adapt to changes they perceive in their environment. However, they have been reluctant to adopt the new technology packages introduced by conventional extension programmes because these packages do not address their priority problems, such as moisture stress, or cannot be adopted easily by families with very limited resources. In the past, the farmers in Tigray were not encouraged to experiment with components of the packages in order to adapt them to their own conditions. Another constraint was that the extension agents were not motivated by the way they were expected to work. Such a situation is not conducive to dynamism, innovation and change.

---

[*] GebreEgziabher Miruts is a field officer with FARM-Africa's Community-Oriented Development Project (CORDEP) in Enticho, and Fetien Abay is a lecturer and researcher in crop science at Mekelle University and one of the original joint coordinators of ISWC-Ethiopia

Map 22.1 *Ethiopia (Axum action area)*

There was clearly a need to find a way to generate technologies that would meet the farmers' preferences, suit their economic circumstances and fit into the local agroecological conditions. The farmers must be at the centre of development efforts and the new technologies must be created together with them. This was the core hypothesis behind the farmer-led research approach introduced by an NGO, FARM-Africa, in the Community-Oriented Development Project (CORDEP) in Central Tigray.

To assist in this effort, CORDEP approached interested staff members of Mekelle University College (MUC, now Mekelle University), Mekelle Research Centre (MRC) and the Tigray Bureau of Agriculture and Natural Resources (BoANR). These scientists and development agents (DAs) began working together with farmers in Ahferom district.

Participatory Situation Analysis

The process of preparing for situation analysis with farmers started with meetings with BoANR staff at subdistrict and then district level.

Subdistrict-level meetings

In the spring of 1996, field staff from CORDEP and district-level staff from BoANR organized meetings in two subdistricts of Ahferom district for the 'agricultural cadres'. These cadres are local farmers who have been trained in specific fields to support the work of the BoANR DAs. The meetings were attended by 144 agricultural cadres in Egela and 132 in Enticho subdistrict, coming from a total of 27 *tabias* (a *tabia* is the area of the lowest level of formal local government and is roughly the equivalent of a parish). Development agents in each subdistrict helped to organize and open the meetings. The agricultural cadres and DAs brainstormed about major problems in agriculture and natural resource management in the district. They listed but were not asked to rank these problems since it was felt that this would have tended to reflect official thinking rather than farmers' priorities.

District-level meeting

After these meetings, the CORDEP field staff, members of the Ahferom District Council and staff from the District Office of Agriculture of the BoANR met for one day to discuss the problems that had been listed. Experts from CORDEP and BoANR facilitated and documented the discussions. The participants identified the most appropriate village areas for commencing the new approach to technology development. On the basis of existing information about agricultural problems in the district and the BoANR's judgement about areas that best represented Central Tigray as a whole, they identified four 'test *tabias*' (see Table 22.1).

A district-level 'research partnership team' was formed with eight members: four DAs from the selected *tabias*, the District Council officer for economic development, a CORDEP field officer, a supervisor from BoANR and a female lecturer–researcher from MUC. A woman who was responsible for monitoring and evaluation in CORDEP helped to document the process. The MUC researcher supported the planning, follow-up and monitoring activities. The District Council assisted in establishing contacts with the local councils (*baitos*) and farmers in the selected *tabias*.

Tabia-level meetings

The research partnership team organized a planning meeting with *tabia*-level DAs and *baito* members and explained the new approach. They agreed on a timetable for meetings with farmers in each *tabia* and these meetings subsequently took place. The number and gender of farmers who attended are shown in Table 22.1.

The CORDEP field officer, the BoANR supervisor and the four DAs from the research partnership team attended all the *tabia*-level meetings. Each

Table 22.1 *Number and gender of farmers attending* tabia-*level meetings for situation analysis in Ahferom district*

Subdistrict	Village area (tabia)	Male	Female	Total
Egela	Erdi Jeganu	94	33	127
	Medeb	102	46	148
Enticho	Tahtay Megaria Tsemri	405	245	650
	Mezbir	160	35	195

meeting was facilitated by a member of the *tabia baito* and the local DA with responsibility for that *tabia*. They explained that the aim of the meeting was to identify the subject areas in which the research partnership team and farmers could do joint research with a view to improving farming in the *tabia*. They invited the community members to identify and rank their major problems related to agriculture and natural resource management. In each of the four *tabias*, the brainstorming exercise led to long lists of problems which the research team clustered into four major categories. These were problems related to:

- drought (a shorthand expression for moisture stress);
- shootfly in teff (*Eragrostis tef*);
- striga infestation in sorghum; and
- shortage of animal feed.

Some of the problems were specific to one or two *tabias*, eg parasitic weeds, while others were common to and priorities in all, such as shootfly damage, drought and shortage of animal feed. The CORDEP field officer and the DAs facilitated the farmers in drawing 'problem trees' to show how the problems were related to each other and recorded the results.

The farmers and the scientists considered various local and external options that were available and might help to solve these problems. They decided to focus first on problems related to shootfly, striga and drought, but still not to lose sight of the problems related to animal feed. They then discussed where joint experiments should be conducted, taking into account the differences in biophysical and socioeconomic conditions within each *tabia*. Some criteria for site selection were suggested by the research partnership team. The farmers took these into consideration when they agreed that the test sites should be:

- hot spots where shootfly infestation or late-season drought (moisture stress) is severe;
- in areas where teff or sorghum, the staple cereals in the area, are major crops; and
- on land with representative soils and topography for farms in the area.

On the basis of these criteria, the participants in each of the four *tabia*-level meetings suggested two hamlets (*kushets*, subdivisions of *tabias*) as sites for the joint research. Consideration was also given to the closeness of the *kushets* so that technical assistance by CORDEP and visits between farmers would be easier.

The CORDEP field officer gave further explanations about what 'farmer-led research' means: the farmers choose the topics, agree on the layout of the trial, decide on the factors that should remain the same across all treatments (eg dates of sowing and weeding), manage the plots themselves and evaluate the results according to their own criteria. He explained that the experiment-ing farmers and the research partnership team would decide jointly which treatments to compare, what to record and how to record it. He made clear that no compensation would be provided if the experiment failed. Then he asked for interested farmers from the selected *kushets* to volunteer to do the trials in their fields. In each *tabia*, more than 50 farmers volunteered. From these, the field officer selected 61 farmer researchers in total from the three *tabias*, including three women, and the field officer registered their names. The farmers were chosen randomly from those who raised their hands. The field officer explained that he could give technical assistance to only a limited number of those who volunteered. In the end, only 47 of the selected farmers (9 in Erdi Jeganu, 5 in Medeb, 19 in Tahtay Megaria Tsemri and 14 in Mezbir) continued throughout the three years of joint research. Others withdrew for various reasons, such as migration away from home or lack of sufficient labour for regular weeding.

Further meeting at district level

CORDEP organized another meeting at Enticho, involving BoANR experts, scientists from MUC and MRC, and the volunteer farmer researchers. The CORDEP field officer and a farmer leader selected by the farmer researchers presented the main problems identified by the communities, their prioritiza-tion of these problems and a problem tree showing the chain of cause and effect.

The farmers divided into three subgroups according to their primary inter-est in dealing with shootfly, striga or drought. After half a day of discussing the ideas that came from the scientists, advisers and farmers, the solutions shown in Table 22.2 emerged as the ones that each subgroup of farmers wanted to explore.

Table 22.2 *Suggested solutions for the problems discussed by subgroups of farmer researchers in the selected* tabias

Tabias	Problems identified	Suggested solutions
Medeb, Tahtay Megaria Tsemri, Erdi Jeganu	Teff shootfly infestation	Different rates of applying fertilizer Different rates of applying pesticides Use shootfly-resistant varieties
Mezbir, Tahtay Megaria Tsemri	Striga infestation	Test different cereal varieties Apply manure Combine cultivation techniques, weeding, variety choice and manuring Use different rates of fertilizer
Mezbir, Medeb, Erdi Jeganu	Moisture stress (late-season drought)	Use early-maturing varieties Use tie ridger

Participatory Planning of Experiments

The research partnership group and the experimenting farmers together adopted the name of Farmer Managed Participatory Research (FMPR) group. A week after the above-mentioned meeting at district level, this group met for a one-day workshop to plan the experiments. In what follows, we focus on the experiment on shootfly in teff.

The FMPR group discussed the trial design and made decisions about:

- the potential solutions to try out;
- the type and number of treatments;
- plot size;
- the manner and timing of ploughing, pegging out the treatments, supplying the external inputs, sowing, weeding, applying fertilizer and pesticide, harvesting and threshing;
- duration of the trial; and
- timing of assessment of results within and across *tabias*.

Choosing the potential solutions to test

The farmers suggested three potential solutions that could be tested: 1) different rates of applying fertilizer; 2) different rates of applying pesticides and 3) using varieties resistant to shootfly. Scientists from MRC suggested a fourth possible solution: a later time of sowing. However, the farmers decided not to include it in their experiment, as they had already done their own informal experiments on this. They explained to the scientists that rainfall distribution influenced sowing time. They had come to the conclusion that early sowing is good for the crop, even though it leads to a higher incidence of pest infestation and the crop has difficulties to compete with the pest. They agreed that late sowing would be good to help the plants escape the pest, but the crop would be seriously affected by the very common phenomenon of late-season drought. Thus, both early and late sowing had disadvantages, but the farmers felt that the risk of not obtaining a good harvest was higher with late sowing.

The issue in relation to fertilizer application is that this produces a burst of fresh soft growth of teff that provides a very attractive environment for shootfly; many farmers are therefore reluctant to apply fertilizer to teff. The purpose of including this treatment was to assess the trade-off between increased shootfly attack and damage stimulated by fertilizer and the normal yield-increasing effect of fertilizer through its influence on soil fertility.

After discussing these various aspects of problems in teff production, the FMPR group agreed that the experiment would involve different application rates of fertilizer on three varieties of teff – two local and one cross-bred (DZ Cr-37) being promoted by Sasakawa Global 2000 through the government extension programme – in order to test the effects on shootfly infestation rates and damage, grain and straw yield, and other parameters suggested by farmers during the trial.

Determination of treatments

For the experiments, the BoANR experts proposed a combination of two commercial fertilizers: diammonium phosphate (DAP) and urea. Two rates of application were selected: the one recommended by the BoANR and a 'farmers' rate of application'. In order to determine the latter, an exercise was carried out by two farmers who were selected by the FMPR group because of their long experience with using fertilizer. This exercise was done to determine the rate of fertilizer application used by farmers trying to combat shootfly infestation in teff. The first farmer was given 1000g of a mixture of DAP and urea in equal proportions; the second one was given 500g of the mixture. Each applied the fertilizer as he saw fit to a 5m² plot on his farm. What was left over from the fertilizer was weighed: this was 650g in the case of the first farmer and 250g in the case of the second. This meant that the first applied 350g and the second 250g. The average of these two rates was taken as the 'farmers' rate of application' to be used in the trial. Thus, the fertilizer treatments were as follows:

Recommended rate	25kg DAP and 25kg urea/*tsimdi* (a local unit of land measurement treated as equivalent to 0.25ha)
Farmers' rate	15kg DAP and 15kg urea/*tsimdi*
Control	No fertilizer application.

Local red and white teff varieties and the cross-bred short-season variety (DZ Cr–37) were tested in combination with these three rates of fertilizer application.

Plot size

During the planning workshop, when discussing the size of the trial plots, the farmers took into account the amount of land that was available for crops, the risk of failure in the experiment, and the work involved in sowing, weeding and monitoring the trial. They decided that the entire plot with all treatments should cover 5m². The scientists suggested keeping a 0.5m border bare on all sides of the plot; this was agreed.

Determination of non-treatment factors

It was difficult to reach consensus on the timing of sowing, weeding, etc, as the farmers wanted to be able to react to local climatic conditions. Eventually, they agreed that the farmer researchers in each *kushet* would sow on the same day and that all of them in all *kushets* would try to come as close as possible to sowing on the same day. As for weeding, each farmer would weed all treatments in one day, but the frequency and timing of weeding at each site was left up to the individual farmer.

Each farmer researcher received the following inputs free of charge through CORDEP: 75g of seed of the cross-bred teff variety for each treatment; 500g of DAP and urea in pure form for the variety and pesticide trials; and, for the fertilizer trial, 500g of DAP and urea in pure form for the extension recommendation treatment and 300g DAP and urea in pure form for the farmers' rate. The farmers were expected to use their own seed of the local teff varieties.

Duration of trials

The CORDEP field officer asked the farmers for how long they wanted to conduct the trial. The farmers decided that two years was the minimum, but that three years would be better in order for them to make a good decision whether to reject or accept what they were testing. They said that this length of time was needed because of the erratic distribution of rainfall and their interest in getting stable results from their farming. This explanation fits well with the scientific definition of stability which is observed across the years, unlike the adaptability of a technique or cultivar which is observed across locations.

Assessment

The measurements and observations to be made by the research partners were divided into three sets of data:

1 Independent assessments by BoANR and CORDEP staff (area of shootfly infestation scored 21 days and 36 days after sowing, number of damaged tillers per plant on sample patches in infested spots scored 25 and 40 days after planting).
2 Joint assessment by the farmer researchers and the above-mentioned staff (grain yield and straw yield per plot after harvest which the farmers measured on 1m^2 in each treatment).
3 Farmers' perceptions and other assessments suggested by the farmers (plant vigour, severity of pests, tillering).

Selection of trial plots

The FMPR agreed that, in selecting trial plots, it would be interesting to include plots with different soil types (*Mekayho, Baekhel* and *Hutsa,* according to the local classification system), at different altitudes (mid-altitude and lowland) in areas with different amounts of rainfall. The group agreed on the following criteria for selecting farmers and plots for the trials:

• Of the original 61 volunteer farmer researchers, those should be included who were particularly interested in combating problems of shootfly and who had enough land for the treatments.
• The plots should represent the type of fields normally used by farmers for teff.

Table 22.3 *Division of tasks between research partners*

Farmer researchers	DAs	CORDEP field officer
Preparing the land	Facilitating group meetings	Facilitating, monitoring and supporting the research process
Supplying local seed	Providing technical support	Providing transport for farmers and DAs
Planting, weeding, applying pesticide, harvesting	Monitoring and evaluation	Supplying and transporting inputs for testing (fertilizer, introduced seed)
Preparing pegs to mark plots		
Preparing activity plan and reporting on implementation	Making introduced seed available	Providing allowance for lunch at meetings
Monitoring and evaluation		
Providing feedback for BoANR experts and MUC/MRC scientists		Preparing meetings and workshops
Participating in exchange visits		Reporting feedback to farmers, BoANR experts and MUC/MRC scientists

Division of tasks

The FMPR group discussed the various tasks involved in the research and agreed to divide them among those partners who were able to give most attention to the detailed monitoring of each farmer's trials, as shown in Table 22.3. The district office of the BoANR provided the introduced seed given by CORDEP to the farmer researchers. The collaborators from CORDEP, MUC and MRC organized the workshops for sharing experiences, and experts from MRC and the regional level of BoANR visited the plots to deliberate together with farmers and CORDEP staff about the trials.

MONITORING AND EVALUATION OF THE TRIAL

On the farm of each farmer researcher, the differences between the treatments were observed in four levels of evaluation: within *tabias* (comparison across *kushets* in each *tabia*), across *tabias*, with other members of the community and within the FMPR group (district level).

Evaluation within *tabias*

Each farmer researcher was responsible for observing what was happening in his or her plot on a daily basis. The farmer, the CORDEP field officer and the DAs monitored the incidence and extent of shootfly damage at different stages of crop growth. The farmer provided oral information about this to the field officer who kept the written records. At the stages of seedling emergence, vegetative growth and crop reproduction, the DA brought the farmer researchers in each *tabia* together to compare their plots and recorded their perceptions. The meetings included peer evaluation by the farmer researchers with regard to the management of the trial plots and the quality and timing of their work on the treatments. Records from the earlier meetings were reconsidered in the later evaluation meetings. For these evaluations by farmers within *tabias*, no transport was provided.

Evaluation across *tabias*

Shortly thereafter, at each of these three stages in crop growth, CORDEP brought together the farmer researchers from all the *kushets* to make comparisons between the *tabias*. Transport and lunch allowances for the farmers and DAs who attended these meetings were provided by CORDEP. The BoANR experts recorded in writing their own measurements of shootfly infestation and the farmers' comments.

Evaluation with other community members

Once a year, when the crop was reaching maturity or during harvest, each *tabia baito* called all interested farmers together for a day for a plotside community evaluation. The people came to the site on foot. In each *tabia*, one farmer selected by his/her peers showed and explained the trial to 15–30 neighbouring farmers. The staff of BoANR and CORDEP assisted in giving further explanations about the experiment. The community members scored the treatment plots as better, the same or worse than the control plot. The BoANR and CORDEP staff explained their quantitative records on pest infestation, damaged tillers and agronomic performance. The questions raised by the community members were answered mainly by the farmer researchers. The CORDEP field officer documented these meetings.

Evaluation within the FMPR group

This evaluation was done annually at district level, facilitated by CORDEP and the Head of the District Office of Agriculture. Representatives of the farmer researchers, selected by their peers, attended these meetings. They described the process of the trial, the results obtained, their general impressions, the strengths and weaknesses of each farmer's trial and their plans for the next year. They also presented the costs and benefits of the trials from their viewpoint. Right from the beginning, the farmers had kept track of the inputs

Plate 22.1 *Discussing group experimentation in Central Tigray (Ethiopia)*

and/or costs incurred on the trial plot. These included mainly labour (own and/or family) and cash outlay for fertilizer and pesticides. At harvest, they estimated the value of the grain and straw yield, and compared this with the costs. The DAs presented their own observations and reported on the records they had kept about the trials. At the end of this meeting, the various members of the FMPR group expressed their impressions of the constraints, opportunities and strengths of the trials.

Results

The farmer researchers were proud to see their findings confirmed by both their fellow villagers and the BoANR and CORDEP staff: the farmers' rate of fertilizer application produced better results in terms of both pest control and crop yield than the rate recommended by the government extension service. They argued that the latter would bring benefits only if pesticides were also used, but these are costly and seldom available. The farmers decided to use the farmers' rate of fertilizer application, but the BoANR – with the exception of individual experts who were familiar with the farmers' trial – still believes that farmers should use the recommendations made by scientists.

By the third year of the trial, the farmers were also able to say that the introduced variety of teff (DZ Cr–37) performed well with external inputs of fertilizer and pesticide. They found that it matured earlier than the local variety and could thus escape the moisture stress caused by the early cessation of the rains. However, the farmers felt that the local red variety of teff was better able to withstand shootfly attack and brought a more stable yield without external inputs than did the introduced variety.

When reviewing the process of the joint research during the first year, some of the male farmer researchers said they had faced strong challenges from their wives who had refused to weed the trial plots, saying: 'That's your plot.' In other cases, the wives had assisted the men in doing much of the crop husbandry work and almost all the post-harvest work. These farmers felt that their wives should be given recognition as part of the research group. After this, CORDEP organized meetings to inform the wives about the joint research and involved them in all aspects of it, including the evaluation sessions. Recognition of the importance of a household-based approach to participatory research grew as part of a learning process by all the collaborators in this and similar experiments. In the case of participatory variety selection (Fetien et al, 2000) in particular, women came to play a central role in evaluating the results with regard to post-harvest traits and pest infestation.

IMPACT OF THE PROCESS

The approach of bringing farmers, development agents and scientists together in farmer-led research brought benefits for all involved. In addition to identifying the most suitable rates of fertilizer for reducing the negative impact of shootfly and becoming acquainted with a new variety of teff, the farmers felt that they benefited from the collaboration in the following ways:

- They had become better able to identify and decide on potential solutions and criteria for assessing them.
- The time for meetings and work had been used efficiently, ie they did not feel they had been wasting their time.
- They were proud to be doing their own experiments and felt that they had been genuinely involved in decision-making at all stages of the joint research.

The staff members who had been involved from CORDEP, BoANR, MRC and MUC felt that they had benefited as follows:

- greater recognition and appreciation of farmers' capacity to experiment;
- stronger partnerships and friendships with farmers;
- recognition of the value of farmers as resource persons and partners in extension, especially in those areas where there is no DA;
- recognition of the benefits of taking a household-based approach to participatory research rather than focusing on male household heads;
- wider range of criteria for selecting technologies of potential interest to farmers (the scientists particularly felt that they had learned from the qualitative indicators mentioned by farmers);

- appreciating the capacity of farmers to challenge scientists by posing research questions and by rejecting their recommendations;
- development of a group spirit among all partners; and
- development of confidence among DAs to inform scientists about the constraints and potentials of their area.

The staff of CORDEP presented this experience in farmer participatory research at a regional research review organized by MRC in 1998. The scientists involved in the review felt that this type of joint research with farmers, facilitated by CORDEP, should continue. For this purpose, MRC agreed to open an office in each of the four zonal capitals in Tigray. As part of the university coursework, the lecturer–researcher from MUC brought students to visit the trial sites where they discussed the experiments with the farmers.

In addition to contributing to a farmer-oriented education for students of agriculture, the experience gained by the farmers working with CORDEP in Central Tigray became a source of inspiration for other farmers, DAs and scientists throughout Tigray when the ISWC 2 programme started in 1997. In the first meetings to introduce the concept of Participatory Technology Development in land husbandry, self-confident farmer researchers could already address the participants. The lessons that had been learnt by CORDEP, BoANR, MRC and MUC could be incorporated easily into the programme as all of these organizations are in the ISWC-Ethiopia Steering Committee.

Moreover, at a policy workshop hosted by ISWC-Ethiopia in 1998, decision-makers at zonal and regional level were highly impressed by the explanation given by one of the farmer researchers, Gebre Egziabher Legese, about the process of planning, carrying out, monitoring and evaluating his trial. This demonstration of a farmer's capacity to conduct and explain his research helped to change their attitudes towards farmers.

ACKNOWLEDGEMENTS

We would like to thank the farmer researchers, the *baitos* of Erdi Jeganu, Medeb, Tahtay Megaria Tsemri and Mezbir *tabias*, the Ahferom District Council and the DAs, Desta Tafere, GebreTsadik Berhe, Welday Berhane and Fethanegest Legesse for their collaboration; the former CORDEP research coordinator Beyene Birru and the present and past CORDEP coordinators Mulu Tesfay and Stephen Sandford for their support to this work; Mengistu Haile for assistance in compiling information for this paper; and Christie Peacock, Executive Director of FARM-Africa in the UK for permission to publish this experience.

REFERENCES

Fetien A, GebreEgziabher M and Amare B (2000) 'Participatory crop improvement approaches in drought-prone areas of Central Tigray', in Mitiku H, Waters-Bayer A, Mamusha L, Mengistu H, Berhan G A, Fetien A and Yohannes G M (eds) 'Farmer innovation in land husbandry: proceedings of anglophone regional workshop, 6–11 February 2000', ISWC-Ethiopia, Mekelle, pp16–27

23

Farmer innovation and plant breeding: the case of maize K525 developed by Emmanuel Kamgouo of Bandjoun, West Cameroon

Paul Tchawa, Noubissié Tchiagam Jean-Baptiste,
Antoine Mvondo Zé and Eric Mujih[*]

Scientists in Cameroon had introduced high-yielding cultivars of maize, but farm families preferred the taste of the local variety. A farmer innovator, Emmanuel Kamgouo, persevered with his own informal experimentation and managed to create a new variety of maize that had not only high yields but also a good taste. Other farmers and scientists are now evaluating his variety.

The hilly region of West Cameroon has a long agricultural history and is well known for the originality of its production system (Fotsing, 1994). The hills are covered by poor ferrallitic soils. Annual rainfall is about 1400mm. Although this region constitutes only 10 per cent of the country's area, 25 per cent of its population is concentrated here. In some parts, population densities can reach 600 persons per km². The demographic pressure on the land, the economic crisis that affects especially the urban centres, the social demands of the 1990s and the decreased incomes from cash crops have all contributed to recent transformations in the agrarian systems of Cameroon. The key elements of these transformations are the splitting up of farms, the expansion of agriculture to marginal areas (valley bottoms and steep slopes), the diversification of

---

[*] Paul Tchawa is a geographer and senior lecturer at the University of Yaoundé I and coordinator of ISWC-Cameroon; Noubissié Tchiagam Jean-Baptiste is a plant geneticist at the University of Ngoundéré; Antoine Mvondo Zé is head of the Soils Laboratory at the University of Dschang, and Eric Mujih is a field agent with ISWC-Cameroon

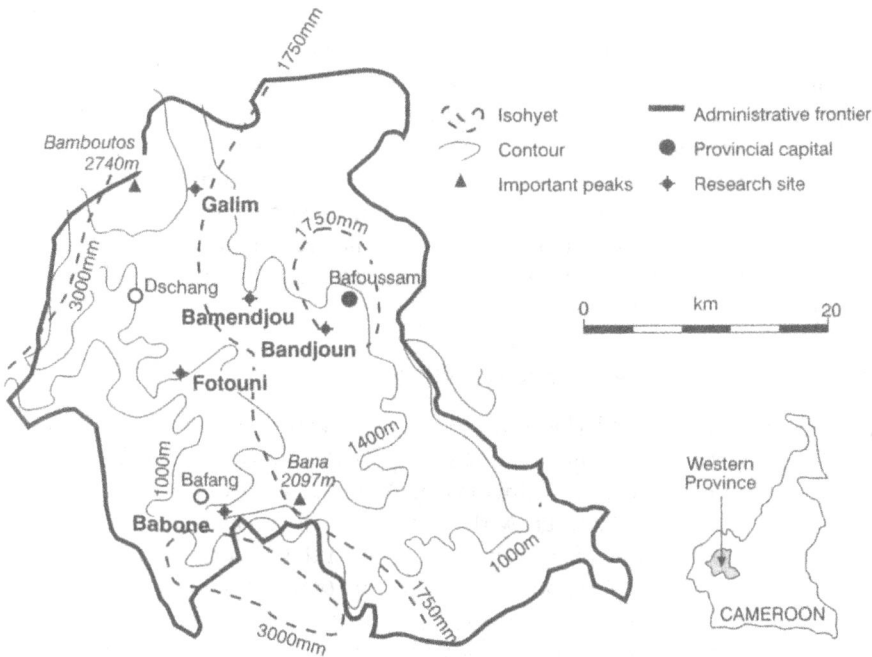

Map 23.1 *Western province of Cameroon*

agriculture (eg into growing vegetables and soybeans) and the introduction of improved varieties of cereals in an effort to increase income.

This region supplies 69 per cent of the national maize production (Dunstan, 1994). Recently, the Institut de Recherche Agricole pour le Développement introduced new and potentially more productive cultivars. The local maize variety is grown mainly by women for home consumption and the taste of its porridge is highly appreciated, whereas farm families often complain about the taste of the new varieties. The seeds of the high-yielding hybrids are expensive and can be used only for one growing season. In the case of the introduced medium-productive varieties, seeds can be saved and sown for three growing seasons. The farmers are interested in early maturing varieties that bring a high yield, have a good taste when prepared as *couscous* (local name for cooked cornflour paste) and are resistant to diseases. This was the context in which Emmanuel Kamgouo, a farmer in Bandjoun village, created his own maize variety.

THE INNOVATOR AND THE HISTORY OF MAIZE K525

Who is Emmanuel Kamgouo?

Emmanuel Kamgouo was born in 1954 and his life history is full of setbacks. He left secondary school without a certificate. His first wife died in 1993, but he remarried and now has a two-year-old child. From 1979 to 1986, he worked for a bailiff in the town of Nkongsamba, about 100km from his native village of Magom. Then he left for Douala and opened a hardware shop, but this was burnt down during the civil unrest in 1990. He then started working for a chemical pesticide firm called Hydrochem, but he earned very little. When his wife died, he decided to leave Douala and become a full-time farmer. This move was facilitated by the fact that he had inherited 6ha of land in his native village. When he started farming in 1993, he grew maize, cabbage, root crops and beans. In 1995, he took the initiative to create a farmers' group in his village; it now has 14 members.

The creation of a new variety: K525

In 1992 Emmanuel received 1kg of maize seed (the variety could not be determined) from an American who was on a support mission to Hydrochem, the firm for which Emmanuel was still working. While he still lived in Douala, he went to Nkongsamba to try to grow the exotic seed. The yield was good, but the taste was not. The following year, he mentioned this to a German agronomist, Alfred Müller, who worked in a farmers' training school in nearby Ndoungué. The German told him how to cross-breed maize. Emmanuel courageously took on this new challenge, determined to combine the potentials of the local maize variety with those of the exotic one.

He started cross-breeding the two varieties on two fields and followed Alfred Müller's instructions. The result was a maize variety with a good yield, but the taste was still poor. Nevertheless, Emmanuel was fascinated by these first results. He persevered with unrelenting enthusiasm to practise his newly acquired skills in cross-breeding, even trying to see if he could cross-breed maize with other plants, such as potatoes and beans, and whether pollination by night brought better results than pollination by day. These attempts testify to his creativity and excitement to explore areas of science hitherto unknown to him. Many things he tried did not work, but this did not discourage him. He continued his attempts at cross-breeding and eventually achieved exceptional results: a maize variety with 3–4 cobs on each stalk, each cob full of sweet kernels. A peculiarity of his maize variety is that the plant has a pivotal root.

In 1995, several scientists from the provincial Ministry of Agriculture office in Bafoussam collaborated with Emmanuel in on-farm trials on his new maize variety. They organized two open days when other farmers came and tasted it. Since then, many women and other farmers of the village of Magom-Bandjoun started to grow this new variety which Emmanuel baptized K525. K stands for his family name Kamgouo and 525 for the number of days he says it took him to create it.

The first impacts of the innovation

From 1996 onwards, Emmanuel became a producer of maize seed. He has an isolated field where he multiplies the K525 seed for sale. He estimated that, in

1999, he sold more than 600kg of seed directly to women, farmers' groups and shops. K525 is also spreading through the networks of farmer innovators in West and North-West Cameroon supported by the ISWC 2 programme. Whenever Emmanuel visits other villages in his region or in other parts of the country, he always takes small bags of seed with him. K525 is appreciated because it has a good taste, matures early and resists strong winds. The only negative aspect mentioned by the growers is that the variety requires the use of mineral fertilizers. In fact, K525 grows best when both mineral and organic fertilizers are combined. All farmers grow maize during the rainy season, but Emmanuel grows his maize variety also out of season. To limit the use of external inputs, he started to dig compost pits. He uses part of the cash income from the sale of seeds to hire labour and to buy mineral fertilizers, chicken manure and pesticides.

JOINT EXPERIMENTATION ON MAIZE K525

In order to gain more information about the potential of this new maize variety, the ISWC-Cameroon programme and its partners initiated joint experimentation involving farmers, scientists and field staff of CIPCRE. All partners contributed to all stages of the research, from developing the protocol to synthesizing the results. As quantitative data about the performance of K525 were lacking, the three main objectives of the experiment were:

- to compare the yield of K525 with that of a local variety of maize and of Kasaï, a variety promoted through the extension system in the region;
- to compare the response of all three varieties to organic fertilizer (chicken manure); and
- to verify the adaptability of K525 in some sites.

The experiment was conducted in four villages from the end of July 1999 onwards, ie out of the main growing season, primarily because ISWC-Cameroon was eager to enter into joint experimentation as quickly as possible and to learn from the initial experience. The farmers involved in the experiment were selected by their peers in each village. They were:

- Emmanuel Kamgouo of the farmers' group in Bandjoun;
- Jean Tagheu of the *Sikati* group in Bamendjou;
- Jacqueline Nguegang of the farmers' union *Madzong La'azizi* in Bangang; and
- Jeannot Yonteu in Galim.

Each farmer cultivated six plots of 100m² each (three varieties and two treatments). Three plots were not fertilized and three were fertilized with chicken manure. The farmers managed their crops in their customary way. Table 23.1 shows the division of tasks between the experimenting farmers, the scientists and the field agents.

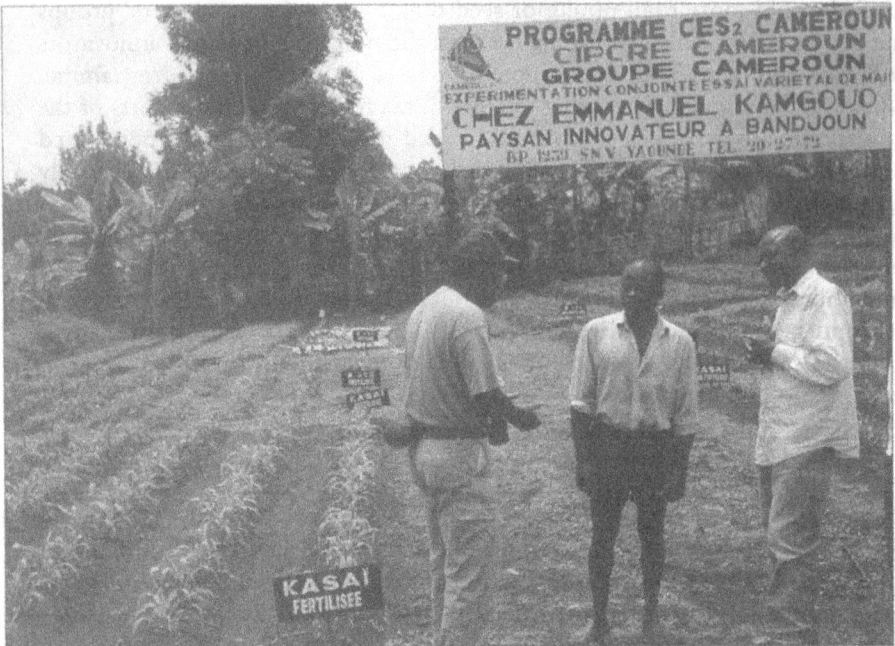

Plate 23.1 *Emmanuel Kamgouo (middle) with researchers on his experimental plot*

Table 23.1 *Roles of the different partners in the joint experimentation*

Activity	Farmers	Scientists	Field agents
Selecting sites	✗	✗	✗
Selecting farmers	✗		
Layout of experiment	✗	✗	✗
Describing plot history	✗		
Measuring soil fertility levels		✗	
Preparing plots	✗		✗
Sowing	✗		
Monitoring	✗	✗	✗
Harvesting	✗	✗	✗
Evaluation	✗	✗	✗
Documentation		✗	✗

Twenty randomly selected maize plants were marked and monitored in each plot. The farmers monitored the following parameters: rate of germination, thickness of the stalk, height of the plant, number of leaves per plant, duration from sowing to flowering and length of the maize cobs. The production was estimated on the basis of the number of grains per plant and the weight of 100 grains.

Table 23.2 *Extrapolated average maize yields (t/ha) on each treatment*

Maize variety	Bangang		Galim		Bamendjou	
K525 fertilized	4.76		5.62		3.58	
non-fertilized	3.77	**4.25***	4.24	**4.9**	2.41	**3.0**
Kasaï fertilized	3.98		5.70		4.35	
non-fertilized	3.30	**3.65**	3.20	**4.4**	3.09	**3.7**
Local fertilized	3.02		3.92		3.89	
non-fertilized	2.98	**3.00**	3.00	**3.4**	2.67	**3.28**

*The figures in bold give the average yield of both fertilized and non-fertilized plots.

First results

Because the maize was grown out of season, the plants suffered from lack of water as well as from plant diseases and parasite attacks. This negatively influenced the results, particularly in Bandjoun where the maize was planted with some delay.

The data in Table 23.2 show that K525 had higher yields than the other two varieties in two of the three sites. The yields on plots fertilized with chicken manure were always significantly higher (at a 5% level) than on non-fertilized plots. Data for the fourth site in Bandjoun are not included; this trial failed because of a problem with the manure.

The experiment was conducted during a second season which started in mid-March 2000 and ended in July 2000. Another two sites were added: the villages of Babone and Fotouni. During this second season, the growth and productivity data at all sites were essentially the same as those obtained in the first season of experimentation. Only Bandjoun was an exception. Whereas the trial failed in the first season, the yield of K525 maize obtained by Emmanuel in the second season was 6.08t/ha. The data confirm that the new variety requires good manuring and that it can be grown out of season if enough water is available to the plant. However, in the second season the new variety was sensitive to attacks from diseases and parasites.

Evaluation by farmers

The farmers involved in the experimentation made the following observations:

- K525 is particularly interesting to farmers who want to sell a large part of their production because its seeds are less expensive than those of the Kasaï variety (600–700CFA/kg compared to 1000CFA/kg).
- The farmers in Bandjoun, Bamendjou and Galim noted a big difference in the yield of K525 between fertilized and non-fertilized plots. The difference was less in the case of Kasai. They noted that K525 often had three cobs per plant on fertilized plots.

- The farmers in Babone and Bamendjou remarked that the yield of K525 was lower than that of Kasaï. They attributed this to the low soil fertility at these two sites. The farmers of Fotouni and Galim stated that they have already adopted K525 because of its better yields without fertilizer compared to the other two varieties.
- According to the farmers of Bangang, the greatest benefit that they drew from this experiment is that they now know that K525 performs well out of season.

In all villages, but particularly in Bandjoun and Bamendjou, the farmers made clear that they do not intend to abandon the local variety, although its yields are lower than those of Kasaï and K525. The main reasons are its taste and the fact that it does not need a substantial quantity of manure to grow.

In the meantime, several other farmer innovators working with ISWC-Cameroon have achieved considerable success with K525 maize. For example, already in the first year that Martin Nkegne in Fotouni village tried this new variety, he sold three-quarters of the harvest for 70,000CFA (circa US$115).

NEXT STEPS

According to Gilbert et al (1993), maize cultivation in Africa is in a deadlock because research does not produce varieties that are appreciated by farmers for taste and adaptability. This case shows that Emmanuel Kamgouo, although he did not follow modern methods of plant breeding, produced a variety which has at least the same potential as varieties produced by the formal research system.

The data for the second campaign of joint experimentation made it clear to all involved that K525 can be grown in both the wet and the dry season. The observations of the farmers and scientists reveal that the variety is quite sensitive to diseases. This is an aspect that would require more involvement of scientists to help the experimenting farmers to improve K525. It will also be important to involve maize breeders of IRAD Bambui so that they can compare K525 with other varieties.

More farmers, both men and women, need to be approached about their opinions on the taste and other characteristics of this new variety. A better understanding of the influence of specific site characteristics on its performance is also needed. Once this information has been obtained, there will be greater justification for a large investment of efforts to spread K525, giving recognition to Emmanuel Kamgouo as its creator.

ACKNOWLEDGEMENTS

This chapter could not have been written without the contribution of the farmer innovators of West Cameroon, particularly the four farmers involved in the joint experimentation: Emmanuel Kamgouo, Jeannot Yonteu, Jacqueline Nguegang and the late Jean-Pierre Tagheu. We also thank the researchers and the staff of CIPCRE, Jules Talom, Jean Daniel Nde, Patricia Kouamo and Mary Mbafor, for their support. A more detailed report on this work can be obtained from Paul Tchawa, SNV, BP 1239, Yaoundé, Cameroon (ptchawa@iccnet.cm).

REFERENCES

Dunstan, S C S (1994) *Fourteen years of farming systems research*, National Cereals Research and Extension Project, IITA/USAID/IRAD/MRST, Yaoundé

Fotsing, J M (1994) 'Evolution du bocage bamiléké: exemple d'adaptation tradition- nelle à une forte démographie', in E Roos (ed) *Introduction à la gestion conservatoire de l'eau, de la biomasse et de la fertilité des sols*, FAO, Rome, pp293–309

Gilbert, E et al (1993) *Maize research impact in Africa: the obscured revolution*, World Bank, Washington, DC

24

Joint analysis of the sustainability of a local SWC technique in Burkina Faso

*Fidèle Hien and Aly Ouedraogo**

In Burkina Faso, a team of scientists, extension agents and farmer innovators analysed the sustainability of zaï, a local technique widely applied in Yatenga to rehabilitate degraded land and to increase agricultural production. The research was designed to identify the limits to the performance of the zaï developed by farmer innovators and the conditions affecting the development of this technique in the context of scarce resources. Tools for PRA were combined with on-farm experiments involving all partners throughout the process. The experience revealed to the scientists and extension agents how participatory research can support agricultural extension.

THE RESEARCH FRAMEWORK

The ISWC 2 programme supported the training of research and extension staff in PRA and PTD and the training of NGO development agents in methods of on-farm experimentation. The research activities were initially focused on a well-known local technique to harvest water and manage soil fertility. This technique, called *zaï*, is used to rehabilitate strongly degraded land. It concentrates water and organic matter (OM) (manure or compost) in shallow pits dug during the dry season; these are commonly dug during the last months of the dry season (March–April/May). Usually millet and sorghum are grown in the pits during the wet season (May/June–September/October). The long-term average annual rainfall for the regional capital Ouahigouya was 560mm over the period 1950–1987.

---

* Fidèle Hien is an ecologist, formerly with INERA and now Minister of Environment and Water in Burkina Faso; Aly Ouedraogo is an agronomist with the PRA Network (Réseau MARP), based in Ouagadougou, Burkina Faso

Map 24.1 *Burkina Faso (Yatenga and Zondoma action areas)*

The yields obtained on fields treated with *zaï* are consistently higher than those obtained on non-treated fields. Over time, the farmers have improved the layout and size of the pits and have started using them for different purposes, such as for growing trees. Trees and shrubs that start to grow spontaneously from the seeds in the manure and compost placed in the pits are protected against livestock. Some farmers even sow in the pits the seeds of tree species they would like to have in their fields, ie they use *zaï* for afforestation.

The *zaï* have been studied quite extensively (eg Roose et al, 1992, 1994). Most of the studies have concentrated on the biophysical and socioeconomic characterization of the technique and, in particular, its impact on yields. They have rarely tackled the question of efficiency of using resources in the context in which these are scarce. The research initiated by the ISWC-Burkina in Yatenga and Zandoma provinces is aimed at answering questions about the potential for expansion of *zaï* in a farming system based on very limited resources – specifically, about the efficiency of using agrobiological resources, the socioeconomic constraints and the sustainability of the practice at field, farm and village level.

METHODOLOGY

The research, which started in 1999, involved about 20 farmers, the extension agents of an NGO based in Ouahigouya, ORFA (Organisation Formation Appui au Développement des Communautés de Base), the PRA Network in Burkina Faso and a team of scientists from INERA. They used an action-research approach that comprised four stages (see Table 24.1).

First, a diagnosis was made of the agropastoral resources of two villages (Kao and Somyaga). Based on their interpretation of aerial photographs, the scientists mapped the geomorphological units, land use and vegetation cover. Farmer innovators and traditional leaders in the two villages participated in validating these maps, identifying the village boundaries and making an inventory of the local agropastoral resources. Then, 12 volunteer farmers, supported by the scientists and extension agents, jointly sketched the flows of OM into the *zaï* on their farms. This diagram served as the basis for the scientists and extension agents to make a quantitative evaluation of the flows.

In a third stage, joint experimentation was undertaken in the fields of six of the 12 farmers in order to measure the impact of the quantity and quality of organic matter being used. This was combined with a change in the spatial layout of the *zaï* proposed by the farmers. They introduced a small earthen ridge between the rows of *zaï* so that more water would be retained in the fields. In contrast to the simple *zaï* system, with pits 20–25cm in diameter and 15–20cm deep, the pits in the *zaï*-ridge system have a diameter of 30–40cm and a depth of about 10cm. This system is often made on fields with a gentle slope (up to 3 per cent), whereas the simple *zaï* system is usually made on flat land. The on-farm experiment was designed, at the same time, for measuring the efficiency of using the organic matter and the mineral elements it contains during the crop cycle.

In the fourth stage, a participatory evaluation was made of the use of human and economic resources and the time management of the 20 volunteer farmers, including the 12 involved in the on-farm experiments. The ISWC-Burkina team selected these farmers on the basis of their knowledge of the local innovators, their innovations and with a view to achieving an adequate representation of socioeconomic differentiation (eg land owners versus those who borrow land, differences in livestock ownership and agricultural equipment). Using the agricultural calendar and labour use data for specific activities supplied by the farmers, the scientists and PRA specialists designed a working tool ('production diary'). If volunteers among the farmers filled this in from beginning to end of the different farming operations, the data would provide the scientists with first-hand data and would allow the farmers to use the data for management and planning purposes.

ASSESSMENT OF AGROBIOLOGICAL RESOURCES AT VILLAGE LEVEL

Based on the land-use maps and the map of the vegetation cover, the scientists made an inventory of the agropastoral resources of both villages, following transects which covered all the morphological, vegetation and land-use units. At the same time, they made an inventory of the plants and structure of the woody biomass and assessed the fodder resources available by the end of the

Table 24.1 *Four stages in the joint research with farmers in Yatenga province*

Stage	Method	Tools	Actors
1 Evaluation of agrobiological resources at the level of village territories	Interpretation of aerial photographs. Identification of village territories. Cartography and measurement of land-use units (soil and vegetation). Inventory of herbaceous and woody plants. Estimation of the standing biomass at the end of the wet season.	Aerial photographs taken. Maps. Transects. Semi-structured interviews (SSI). Yield plots for biomass. Inventories.	Scientists. Farmer experimenters.
2 Evaluation of the flows of OM in the farm	Definition by farmers of flows of OM used in the *zaï*. Measurement of flows (according to origin), quantities and quality of OM used by 12 farmers. Analysis of criteria and factors limiting the quantity of OM produced.	Flow diagram of OM. Diagrams. OM flows for each farmer. SSI. Weighing scales. Data sheets: matrices of *in situ* measurements.	Scientists. NGO staff. Farmer experimenters.
3 On-farm experimentation	Tests and experiments in fields of six farmers: comparison of two application rates of OM (farmer's rate and half this rate) in two types of *zaï* systems (simple and with ridges), eight treatments and four replications per treatment. Monitoring of rainfall data, crop development, absorption of nutrients and measurement of the yields of grain and stover.	Plots treated and managed by farmers. Rain gauges. Measurement tools: crop growth, yields.	Farmer experimenters. NGO staff.
4 Socio-economic evaluation (use of resources and time management)	Selection of 20 farmers. Development of tool for monitoring activities. Validation of tool and training the farmers in its use. Collection and interpretation of data. Feedback of data to farmer experimenters.	Group discussions. SSI. Matrices. Recording booklets of farmer experimenters.	Farmer experimenters. PRA specialists. Scientists. NGO staff.

wet season (Table 24.2). Aware of the limitations related to calculating livestock carrying capacity because of large annual fluctuations in rainfall and biomass, the scientists regarded the data simply as approximations.

Table 24.2 *Theoretical livestock carrying capacity of Kao and Somyaga village areas at the end of the 1999 rainy season*

Unit	Somyaga		Kao	
	Carrying capacity (TLU*/ha/yr)	Theoretical available grazing (TLU/yr)	Carrying capacity (TLU/ha/yr)	Theoretical available grazing (TLU/yr)
Low-lying areas/ alluvial valleys	0.182	27	0.271	42
Lower slopes	0.167	27	0.142	9
Upper slopes	0.132	66	0.192	137
Hills and hillocks	0.186	66	0.276	105

* TLU = equivalent of 250kg cow

The scientists reported these findings back to the farmers and the data allowed them to judge the capacity of each village territory to support the existing livestock numbers kept by the villagers, as well as the livestock that come in from neighbouring villages during the dry season. The farmers drew the conclusion that the livestock cannot survive on grazing alone and depend on other resources, such as crop residues. This was not a new insight for the farmers. Indeed, their innovations were often aimed at producing more fodder, such as by stimulating the growth of forage trees in their fields (see Chapter 4).

ASSESSMENT OF ORGANIC MATTER FLOWS AT FARM LEVEL

Figure 24.1 represents the flow paths of organic matter that eventually enter the *zaï*. All 12 farmers agreed with this diagram, but pointed out that differences between the farms determine the quantity and quality of compost used. These include differences in the availability of sources of organic matter, in livestock numbers and composition, and in the purchasing power of the households. In general, however, the different components of the compost used, in order of importance as ranked by the farmers, was: cattle manure, sheep and goat manure, household wastes, donkey manure and litter, clay, cereal chaff mixed with sand, natural phosphates, wild grasses and ash.

The scientists adapted the basic diagram to fit each of the six farmer experimenters. In the composting period between December 1999 and March 2000, a student on practical training systematically monitored and measured the quantities and sources of OM used by the six farmers to produce compost. At the end of the composting period, the student assessed the total quantities of compost and analysed the nitrogen, phosphate and potassium content in the INERA Soils Laboratory in Kamboinse. The totals could be compiled for only

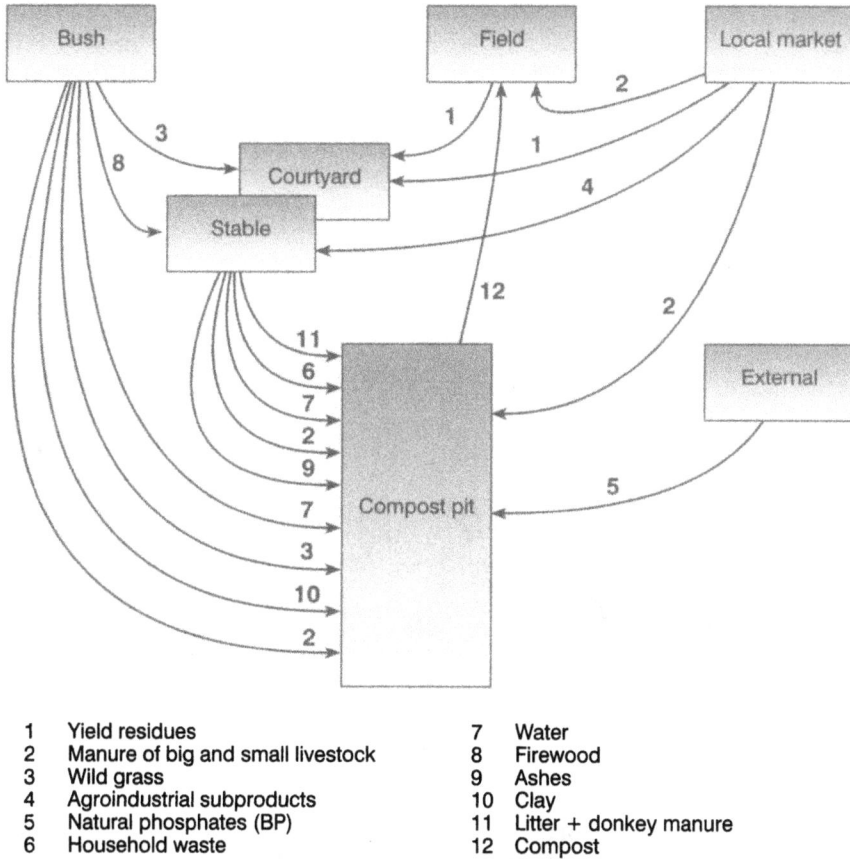

1	Yield residues	7	Water
2	Manure of big and small livestock	8	Firewood
3	Wild grass	9	Ashes
4	Agroindustrial subproducts	10	Clay
5	Natural phosphates (BP)	11	Litter + donkey manure
6	Household waste	12	Compost

Figure 24.1 *Flow of organic matter and other inputs entering the* zaï: *schematic presentation based on the flows identified by 12 farmers*

11 of the 12 farmers. Figure 24.2 shows the quantities of OM produced (*x*-axis), the number of tropical livestock units (250kg liveweight) kept by each farmer and the estimated area of *zaï*-treated fields that could be fertilized with this OM.

Obviously, the possibilities of fertilizing the *zaï* adequately will vary considerably from farmer to farmer. Not surprisingly, a close fit was found between the area actually treated with *zaï* and the quantities of compost available per farm household. In other words, farmers do not rehabilitate degraded land with *zaï* when they do not have enough OM to put into the planting pits. This analysis of OM flows and their relative contribution to compost production has made the farmers more aware of how their land depends on various sources of OM. This stimulated one of the innovators to start experimenting with the composting of weeds, including striga.[1]

When linked to the estimation of the agropastoral resources of each village, the scientists can use these data to assess the constraints to the expansion of *zaï* based on the biophysical resources in general.

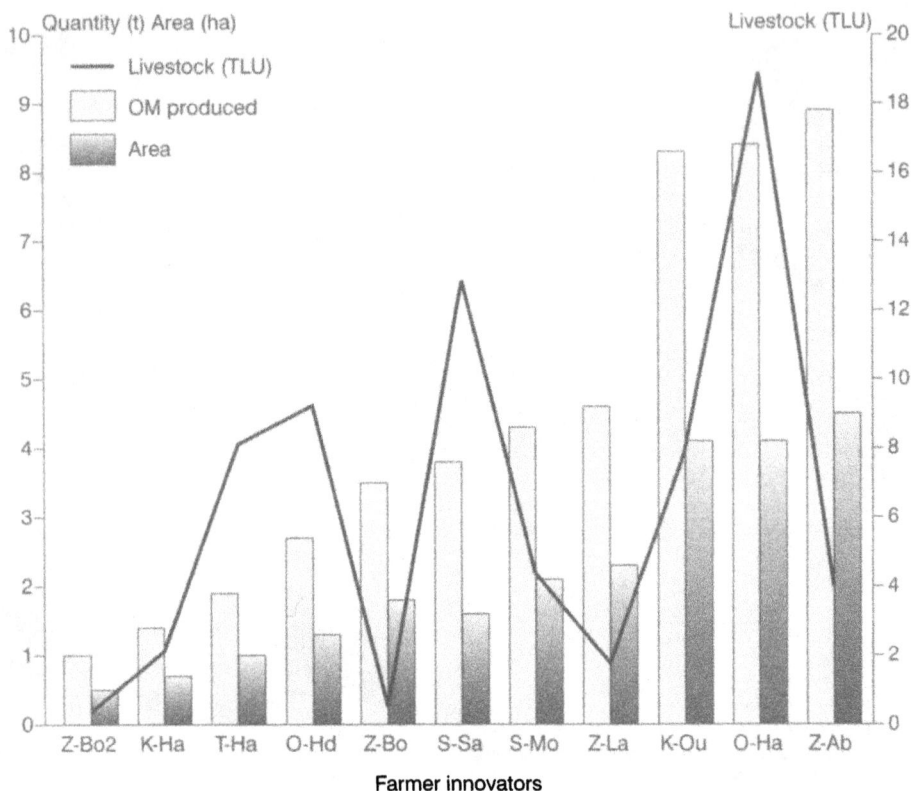

Figure 24.2 *Number of animals, quantity of compost produced and potential area of* zaï *per farm household*

ON-FARM EXPERIMENTATION

Figure 24.3 compares the average yields of stover and grain obtained by four of the farmers using their normal application rates of compost, compared with half rates, in *zaï* that was newly dug in 1999 on entirely barren and denuded fields with a hard crust. The normal application rate is the average quantity of compost that each farmer usually applies to the pits. This varies significantly between farmers, as does the number of pits per hectare. Some farmers dig fewer but bigger pits (eg Ali Ouédraogo), while others dig smaller but more pits per unit area (eg Ousséni Zoromé). Table 24.3 indicates the quantities of compost used and the number of pits per hectare usually made by each of the six farmer experimenters.

The scientists proposed the use of a half dose for the sake of comparison. It is clear from Table 24.3 that this likewise varies from farmer to farmer. Because the experiment was based on the farmers' practices, the quantities of OM were not standardized. One major finding was that the yields of grain and stover for the full dose and the half dose did not differ significantly, at

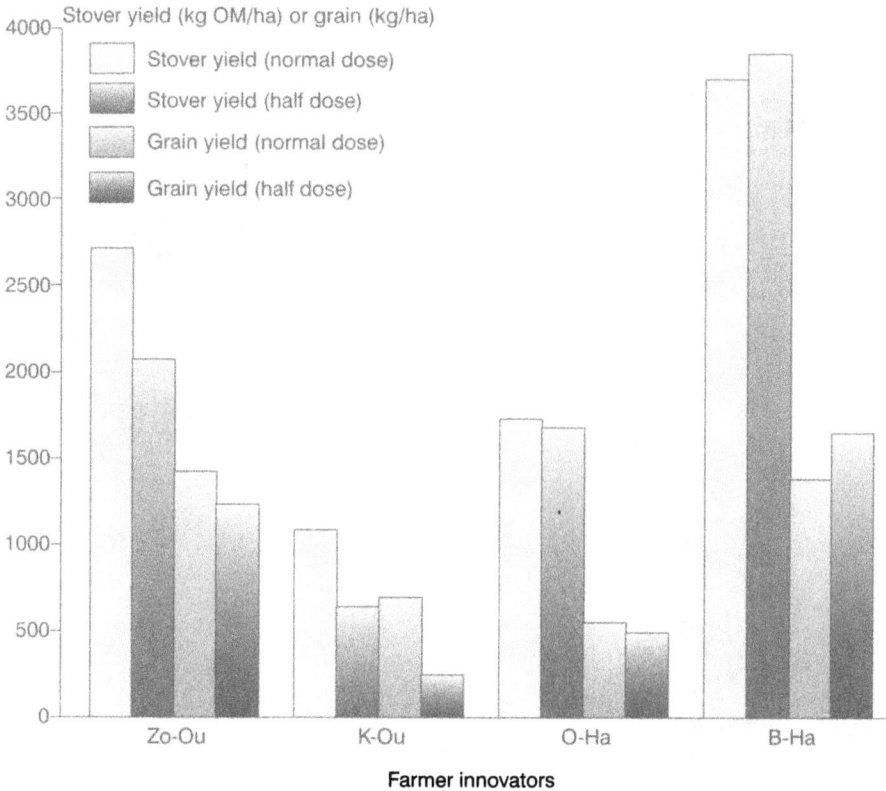

Figure 24.3 *Average yields of millet according to the amount of compost applied to fields recently rehabilitated with* zaï

least not in the first year. The reasons for this are the subject of further experimentation and analysis. On the basis of the results obtained and visualized by the farmers themselves, some of the farmers have started to question old certainties. They had always assumed that they applied the right dose of compost, but now they are no longer convinced that this is the case. The question that keeps coming back is: 'Can we obtain similar results with smaller quantities of OM?'

Table 24.3 *Quantity of compost and number of pits per hectare*

Farmer experimenter	OM/pit (dry weight in g)	No of zaï/ha	Compost application (t/ha)
Abdoulaye Zall	354	16,051	5.6
Ousséni Zoromé	556	18,138	10.1
Ousséni Kindo	834	9,463	7.9
Ali Ouédraogo	1462	8,300	12.1
Harouna Ouédraogo	758	14,650	11.1
Hamadé Bissiri[1]	710	15,883	11.1

Key
Half a circle: half a workday
Full circle: entire day
W: coin of 5 (West african) francs

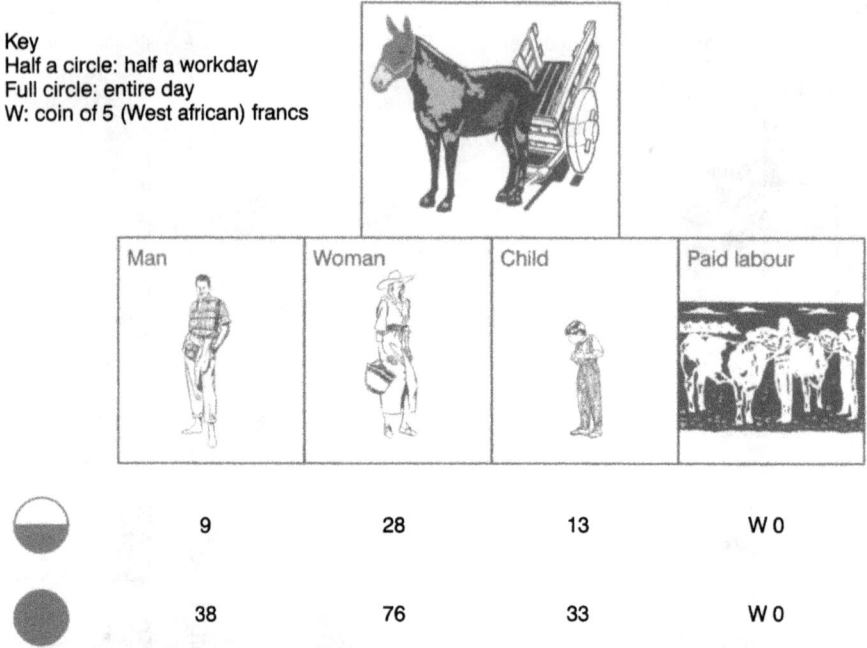

Man	Woman	Child	Paid labour
9	28	13	W 0
38	76	33	W 0

Figure 24.4 *Overview of labour inputs (family and hired) for transporting harvest from Ousséni Zoromé's fields*

With regard to the comparison between the two *zaï* systems, it was found that the yields tended to be higher in the *zaï*-ridge system than in the simple *zaï* system. However, in the 1999 season when rainfall was relatively good, the differences were not statistically significant. This is being monitored further.

SOCIOECONOMIC EVALUATION

Figure 24.4 presents some elements of the booklets that were designed for and with illiterate farmers, which allows them to monitor and to record data themselves about their agricultural operations. The example given shows the use of labour and money by one of the farmer experimenters for transporting the harvested crops from the fields.

This example shows the importance of women's and children's labour in carrying the harvest to the compound. The data recorded by each farmer allowed the scientists to estimate the time spent by each family or household on different agricultural activities, to calculate the financial costs of these activities (using the costs of local labour) and to obtain the quantitative data needed to make an input/output balance for each farm in biophysical and financial terms. The farmers were able to compare the time and other investments they had made on specific activities in specific fields and what this produced in terms of yields. They are now in a better position to plan their farming activities. Thus, the record books can be regarded as decision-support tools for farmers.

DISCUSSION AND METHODOLOGICAL ISSUES

This action research has generated data that are useful to the farmers, the extension agents and the scientists. The experiment with a normal rate of compost application and half of this rate has revealed much to the farmers about crop responses to compost and the possibility for economizing on scarce resources. It also revealed that there is a link between the density of sowing (number of pits per ha) and the quantity of organic matter used. Two main groups of *zaï* practitioners can be distinguished: one uses relatively high densities of pits (15,000–18,000 pits/ha) and the other much lower densities (9000–13,000 pits/ha). When the number of pits per hectare is high, the number of plants per pit is lower. Another finding is that the higher the quantity of manure used, the lower its quality tends to be, although this depends on the livestock holdings of the family and the conditions of composting (particularly labour inputs).

The joint experimentation generated good scientific data which, when fed back to the farmers, stimulated much interest on their part. What used to be certain for them is no longer so. The farmers themselves raised new research questions, such as comparison of the different sources of compost and investigation of compost quality, the residual effect of a half dose of OM, and the relationships between the amount of compost used, crop yield and low rainfall.

Coupled with exchange visits between the farmer experimenters, the joint experimentation opens new perspectives for agricultural extension in general and for the spreading of innovations in particular. One of the questions often mentioned in connection with participatory research concerns the scientific validity of the results and the use that scientists can make of these data for publications to advance their careers. This experience in Burkina Faso has shown that it is possible to do good scientific research and, at the same time, to improve technologies that are relevant to farmers. Much depends on the capacity of the scientists and extension agents to adapt their objectives and their research instruments to the needs of the farmers and to the local environmental conditions.

NOTES

1 It was a great shock to all the scientists, NGO staff and farmers involved in this research that Hamadé Bissiri died suddenly in November 2000

REFERENCES

Roose, E, Dugue, P and Rodriguez, L (1992) 'La GCES: une nouvelle stratégie de lutte anti-érosive appliquée à l'aménagement de terroirs en zone soudano-sahélienne du Burkina Faso', *Revue Bois et Forêts des Tropiques*, no 233, pp49–63

Roose, E, Kaboré, V and Guénat, C (1994) 'The *zaï* practice: a West African traditional rehabilitation system for semi-arid degraded lands – a case study in Burkina Faso', seminar on the Rehabilitation of Degraded Land, November 1994, Médenine, Tunisia

25

Sowing maize in pits: farmer innovation in southern Tanzania

*Zacharia Malley, Anderson Temu, Norbert Kinabo, Salome Mwigune and Anunciata Mwageni**

Wilbert Mville, a farmer in his mid-30s living in Itulike village in Njombe district of southern Tanzania, has developed a technique for growing maize in pits instead of rows. It is quickly becoming popular in and beyond his village. This chapter describes how he developed the innovation and how he is collaborating with neighbours and scientists in experimentation to explore its merits.

The undulating land in Njombe district in the Southern Highlands of Tanzania lies between 990 and 2200m above sea level. Annual rainfall (mainly from November to April) ranges from 600mm at lower altitudes to over 2000mm at higher altitudes. The growing season lasts six to eight months. The dominant soils are red kaolinitic clays (humic ferralsols) with moderate natural fertility and medium-to-high water-holding capacity. Under continuous cropping, they degrade quickly through compaction and plant rooting is shallow. The average farm size is about 1.5ha.

In this area, Wilbert Mville and his wife Emelita have developed an integrated farm of about 2ha with zero-grazed cows to produce milk as well as farmyard manure (FYM) for maize sown in pits. The slurry is carried to various parts of the farm through a network of pipes that Mville designed and built on his own and serves as a source of nitrogen for crops of maize, beans, potatoes, wheat and vegetables. From a reservoir that Mville built on higher ground, water flows by gravity through pipes to the plots.

* Zacharia Malley is agricultural research officer at the Agricultural Research Institute (ARI) Uyole in Mbeya, Tanzania; Anderson Temu is principal agricultural research officer at ARI Uyole, and Norbert Kinabo, Salome Mwigune and Anunciata Mwageni are extension officers with the Ministry of Local Government and Regional Administration in Mbeya, Tanzania

Map 25.1 *Tanzania (Njombe district action area)*

SCIENTISTS ENCOUNTER FARMER INNOVATORS

In Njombe and three other districts in the Southern Highlands, ISWC-Tanzania set out in 1997 to identify farmer innovators as a first step in establishing a process of PTD. Scientists and extensionists were trained in tools for farmer-led research which differs greatly from the scientist-led research dominating the official system in Tanzania. Two PTD training workshops in March 1998 and April 1999 in Njombe were crucial for changing the attitudes and behaviour of the participants from conventional transfer-of-technology to a more participatory approach. They learned how to plan and carry out PRAs, and learned about informal experimentation by 'research-minded farmers'. It was stressed that these should not be confused with 'progressive' or 'contact' farmers, who had the resources to adopt techniques suggested by extension agents. Farmers who are less responsive to such messages often have fewer resources, but may still be very active in trying out new techniques in their farming system (van Veldhuizen et al, 1997).

A field trip during the 1998 workshop exposed scientists to farmer innovation. Godson Lupenza, a village extension officer in Njombe who had seen Mville's maize pits, suggested that a fieldwork group visit him. The group marvelled at Mville's knowledge and integrated innovations and at his ability to explain, listen to and answer questions. The scientists were keen to analyse Mville's innovations and to start joint research with him.

When the two co-authors from the Agricultural Research Institute (ARI) Uyole visited Mville again in February 1999, he explained how he came to

Plate 25.1 *Wilbert Mville digging and manuring a maize pit (Tanzania)*

develop his idea of sowing maize in pits. Originally, he had been growing maize according to the extension recommendations at the time: one plant per hole in a row. In 1992, two of his neighbours attended a course organized by Sasakawa Global 2000 and started growing two plants per hole. This recommendation had been based on experiments at ARI Uyole in the late 1970s that had shown that, with the same population of maize plants, differences in sowing pattern did not lead to statistically significant differences in yield, if the level of nutrients per planting hill was high. Since then, agricultural programmes in the area had promoted multiple seeding and many farmers without the equipment that is needed to plant one seed per mound at the recommended closer spacing of 75cm (between rows) and 30cm (in rows) had adopted the practice of sowing two or three seeds per hole with wider spacing.

After hearing of this new practice, Mville became curious and visited his neighbours to ask for an explanation. They told him what they had learnt: that two or three seeds could be sown together if the soil had enough nutrients and that the spacing within the rows could be doubled or tripled, depending on the number of seeds per hill.

Mville reasoned that it must be possible to sow even more seeds together, if the soil was fertile, and still obtain a good yield. However, the soils on his farm were exhausted. In his head, he started to design a way of growing maize in pits but, to be able to experiment with this idea, he needed more organic matter. He had some manure from his pigs and local cattle, but this was not enough. Therefore, in 1998, he acquired a cross-bred heifer on credit from Heifer Project International (assisted by the NGO Caritas) and kept it under zero grazing. In this way, he could accumulate more FYM, feed leftovers and slurry (semi-liquid manure mixed with urine).

He concentrated this organic matter in pits and tried growing four to six plants per hole, with different spacing regimes. His first informal experiments with this, in the 1993–1994 season, were on a small scale and he continued despite the ridicule of his neighbours. In 1998 two of his sisters had an opportunity to attend a course offered by Caritas. Upon their return, they told him that it covered some of the practices with which he was experimenting. Later, Mville was invited to join a Caritas group of farmers. From that time onwards, he has been working closely together with this NGO in further developing his innovation and integrated farm. He now has five cows and two heifers, probably the maximum number that he and his wife can manage on their 2ha farm. Neighbours who practise zero grazing have fewer animals; those who have more animals practise free-range grazing.

THE MAIZE PIT INNOVATION

Mville's technique involves digging pits 60–120cm in diameter, 30–60cm deep and 75–100cm apart. He puts crop residues and manure (one 20-litre bucket) into each pit and mixes this with topsoil. Then he sows 20–25 seeds in each pit and later thins them to 15–18 plants, depending on the size of the pit. He collects a mixture of slurry from the *kraal* floor and urine collected by pipes leading from the stable and dilutes this mixture 1:1 with water. Over each of three consecutive days, he applies 2 litres per pit as a top-dressing for the maize. The following season, he makes new pits on the undisturbed soil between the pits of the previous season. In this way, he hopes eventually to saturate the field with organic manure and thus improve the soil. Using this technique, Mville harvests about 20 bags of maize per acre (5t/ha). Previously, when he used to sow in rows, he obtained less than 5 bags per acre (1.25t/ha).

Mville's wife Emelita is very knowledgeable about her husband's innovations and all the field operations as she has constantly worked together with him. She is articulate in answering visitors' questions concerning the innovations when her husband is absent. She has also introduced her own experiments, conducted out of curiosity. For example, after maize harvest, she planted leafy vegetables irrigated by the pipe system to see how the residual fertility could be used. The vegetables grew very well and provided food for the family. In a few pits, she is trying out an improved grass species (Rhodes grass, *Chloris gayana*) to be multiplied on larger contoured fields near the stable. She and her husband are also trying out Napier grass (*Pennisetum purpureum*) which they plant on the ridges running along the contours to stabilize them as well as to obtain fodder, particularly in the dry season.

FARMER-LED EXPERIMENTATION

When the scientists visited Mville during the 1998 training workshop, he was already full of ideas to explore. These included:

- comparing yields from large and small pits;
- trying bigger pits, each seeded with up to 30 plants, without thinning;
- sowing on raised seedbeds in old pits (from the previous season) to observe yield response to residual fertility;
- using compost instead of manure and crop stover as organic fertilizer;
- one top dressing of slurry compared with three top dressings; and
- relay cropping in the pits with beans, vegetables or pasture grass.

In 1998–99, Mville began a trial to compare the effect of pit size on maize yield. He and his wife jointly monitored the trials. Emelita keeps one notebook to record details about their experiments and another one for visitors. A neighbour, Rose Kitamkanga, who had also attended the Caritas course, saw what Mville was doing and decided to experiment on her own to find out whether pit planting with manure produced more maize than conventional row planting.

The scientists and VEOs joined these experimenting farmers in the middle of the growing season. While we were still trying to work out mechanisms for participatory research, farmers started their trials without us! We helped them to identify simple assessment criteria to use for interpreting the results at the end of the season. As the farmers were willing to record several parameters (eg the dates of all field operations, flowering of the crop, plant stand, plant height, number of plants affected by disease), the scientists and VEOs had to record only a few, including pit dimension, grain yield and soil analysis.

The results of these trials, plus more from other farmer innovators, were presented in two farmer experimentation workshops held in November 1999 in Iringa and Mbeya regions. Assisted by the technical staff, the farmers used flipcharts and a blackboard to present their data to other innovators. Results of the maize pit trials were discussed in a plenary session.

Farmers' comments

Mville and Emelita noted that the larger pits produced better results than the smaller ones. Rose noted that the maize yield from pits was 50 per cent higher than from rows. The other farmers made the following comments on the experiments of Mville, Emelita and Rose:

- The large pits were tested on a larger plot than the small pits were.
- The exact amount of FYM in Mville's trial was not known.
- The pits and rows were fertilized according to different schedules.
- The amount of urea applied on Rose's plot was not specified.
- Maize on Rose's plot was fertilized in the pits but not in the rows.
- The varieties of maize sown were different.

The farmers saw a clear need to standardize non-experimental factors so that comparisons could be made between treatments.

Scientists' observations

During the farmer experimentation workshops, we guided farmers to brain-storm about other rules that could improve experiments in the next season. The importance of design, replication (on site and across seasons), randomiza-tion, controls, constant factors and plot area for trials were discussed. All agreed that these principles should be put into practice when joint experiments were conducted in the 1999–2000 season.

SCIENTISTS' ANALYSIS OF FARMERS' DATA

The data from seven farmers who experimented with maize pits are shown in Table 25.1. The yields ranged from 4.5t/ha to 8.8t/ha, with a mean yield across farmers of 6.3t/ha. This is certainly higher than they could have achieved with conventional sowing in rows. Although statistical analysis is not possible with this kind of data (because the farmers did not apply experimental principles), some useful inferences, subject to confirmation in later studies, can be made.

Sources of high yield

The high yield achieved by most farmers can be attributed to high plant population in the pits: the mean plant stand was 27 per cent higher than that recommended for tall maize plants sown in rows (44,444 plants/ha). The high amount of FYM and slurry used (20–40 litres/pit) could support the high plant stand in the pits.

Table 25.1 *Results from seven farmers growing maize in pits*

Name of innovator	Yield (t/ha)	Plants/ pit	Plant population		Pit diameter (cm)	Pit depth (cm)	Pit spacing (cm)
			Actual	% of 'optimum'			
W. Mville	8.80	24	69,890	157	124	60	106
R. Kiwale	6.90	9	48,760	110	66	60	96
G. Ng'ande	6.75	17	41,250	93	99	60	148
R. Kitamkanga	6.48	12	46,400	104	59	60	115
M. Lukelo	5.85	15	54,000	122	83	60	91
C. Mville	5.13	18	43,750	98	90	60	107
A. Nguhuni	4.50	11	92,400	208	58	60	76
Mean	6.34	15	56,636	127	83	60	106

Effect of high plant population and level of fertilization

The scientists noted that high plant population without adequate fertilization had a negative impact on yield, as was the case with Nguhuni (Table 25.1), who had 108 per cent more plants than the population recommended by research and realized the lowest yield of 4.50t/ha with 20 litres of manure per pit (the amount of manure applied by the first three farmers in the table was 40 litres/ha, while the rest applied 20 litres/ha).

Effect of pit depth on maize yield

The pit depth of 60cm appears sufficient for maize, given its shallow rooting habit. Applied by all farmers, this parameter did not contribute to variations in maize yield across the farmers.

Effect of maize pit technique on soil characteristics

Scientists took soil samples in Mville's field at 10cm intervals from the surface to 50cm depth on a vertical cut of a manured pit and an adjacent non-pitted, non-manured spot. The samples were analysed for soil moisture, pH, total nitrogen, available phosphorus, organic carbon and texture at the Soil Laboratory of Sokoine University in Morogoro. The results are presented in Table 25.2.

Soil organic matter (SOM) was higher in the pitted than the non-pitted, non-manured spot and more concentrated in the upper layers. Similarly, because of the higher SOM, the moisture content in the pitted area was more than 100 per cent higher than in the non-pitted area (89 versus 34 per cent). This suggests that the manured pits preserve more moisture than does the soil in which maize is sown in rows.

Table 25.2 *Changes in soil conditions as affected by pitting*

Parameter	Soil layers (cm intervals)				
	0–10	*10–20*	*20–30*	*30–40*	*40–50*
Manured pit:					
Moisture content	89.00	3.00	27.00	28.00	34.00
pH (water)	6.58	6.52	6.44	6.65	6.63
Available P (ppm)	23.01	24.43	23.74	2.70	2.41
Total N %	0.48	0.14	0.53	0.94	0.02
OC (%)	7.13	2.14	6.93	1.58	1.40
Texture:					
Sand (%)	35.20	55.20	31.20	73.220	63.20
Clay (%)	34.80	38.80	54.80	36.80	28.80
Silt (%)	10.00	6.00	14.00	26.00	8.00
Non-pitted soil:	—				
Moisture content	34.00	29.00	25.00	23.00	24.00
pH (water)	5.94	5.76	5.77	5.37	5.21
Available P (ppm)	13.86	12.88	2.65	1.67	0.59
Total N %	0.15	0.14	0.13	0.08	0.06
OC (%)	2.18	2.38	1.90	1.46	0.99
Texture:					
Sand(%)	47.20	53.20	55.20	57.20	61.20
Clay (%)	40.80	38.80	38.80	36.80	32.80
Silt (%)	12.00	8.00	6.00	6.00	6.00

The soil pH in the pitted area appears to change from slightly acidic towards neutral. This phenomenon promotes mineralization of SOM and thus the availability of nutrients to the crop. In the three top layers of the pits, available phosphorus (P) was improved by about 300 per cent (mean 23.73ppm). The corresponding figure for the non-pitted spot was only 7.80ppm. Nitrogen (N) and phosphorus are the two major limiting nutrients in maize production in the Southern Highlands (Croon et al, 1984). The total N content in the 0–10cm layer of the pit was 0.48 per cent or 300 per cent higher than in the non-pitted spot. The sharp fall in N content in the second layer of the manured pit (from 0.48 to 0.14 per cent) and the increase in the third layer (0.53 per cent) may be explained by the leaching of N from top to bottom by rainwater, a common phenomenon with this nutrient. The various nutrient and texture changes noted in the different levels of the pits are a result of manuring and redistribution of soil particles during land preparation.

FARMERS' ASSESSMENT OF THE PIT TECHNIQUE

In order to gain farmers' more detailed assessment of the pit technique, the scientists discussed with individual experimenters on the basis of a SWOT

Table 25.3 *Farmers' SWOT analysis of sowing maize in pits*

Variable	Farmers' assessment
Strengths	Land produces more: higher maize yields
	Plant nutrients are retained in the pits
	More fertile soil: the same pits can be planted for two seasons
	Loose soil in the pits makes sowing and weeding easy
	More water is retained
	Weeding is simpler because it is done only in the pits; even children can help
	Less work because not the whole field is ploughed and the same pits can be used again the next season
	Less erosion because soil is caught in the pits
Weaknesses	Large amounts of FYM needed, but supply is limited
	Farmers must have cattle, which means high investment
	Problems in transporting FYM make it difficult to apply pit technique on plots far from homesteads
Opportunities	Compost can be substituted for FYM
	Hand-pushed or animal-drawn carts can be used to transport organic matter
	Concoctions made of plants can be used as top-dressing if slurry is not available
	Small doses of inorganic N fertilizer can be used for top-dressing
	Space between pits can be used for other low and non-competitive crops such as beans or peas to make better use of the land and to generate some cash income
	Credit to purchase animals could be sought
Threats	No threats evident thus far

analysis. The results are summarized in Table 25.3. The general consensus was that the pit technique has more strengths than weaknesses. The greatest bottleneck is acquiring FYM for the pits; however, farmers who do not own cattle can use other organic matter instead, such as compost and crop residues. The farmers did not think that the pit technique presented any threats.

The technique of growing maize in pits is spreading quickly. A quick survey made in Itulike and Wikichi villages in Njombe district in June 1999 revealed that 71 farmers had already adopted or were adapting this innovation. Three farmers in Iringa district who had seen the technique during exchange visits were trying it out for themselves. This spread is the result of many factors and actors, including ISWC-Tanzania's sharing events, visits by neighbours and training courses organized by Caritas.

Table 25.4 *Yields of maize grown in two systems of sowing in Njombe and Iringa in the 1999–2000 season*

| Name of farmer | District | System of sowing | | | |
| | | Rows | | Pits | |
		bags/acre	t/ha	bags/acre	t/ha
Mville	Njombe	5.6	1.4	17.5	4.5
Rose	Njombe	8.1	1.9	27.7	5.7
Asheri	Njombe	9.5	2.4	20.9	5.3
Bange	Njombe	25.6	6.1	26.3	6.7
Ngailo	Njombe	16.2	4.1	25.1	6.7
Kiwale	Njombe	25.4	6.5	31.4	8.0
Martina	Njombe	21.0	4.4	24.2	4.5
Winnifrida	Njombe	8.8	2.2	19.0	4.8
Glesius	Njombe	18.8	4.6	23.3	6.2
Ngilangwa	Njombe	14.0	3.6	25.1	6.4
Mfikwa	Njombe	8.0	1.8	16.5	4.0
Fute	Njombe	7.1	2.0	15.6	4.2
Mtengela	Iringa	13.7	3.4	12.1	3.0
Lufyagila	Iringa	8.4	2.1	24.2	6.1
Mean		13.6	3.4	22.1	5.5
LSD (0.01) t/ha		2.00			
CV (%)		21.4			

Note: All non-treatment variables were kept constant as agreed in the previous year: plot/pit sizes and spacing and the amount of fertilizer (FYM) applied were the same in both systems, and the same variety of maize was sown on the same day. LSD = least significant difference; CB = coefficient of variation

JOINT EXPERIMENTATION IN THE 1999–2000 SEASON

While farmers were keen on the technique, it was agreed during the farmer experimentation workshops in November 1999 that the innovation would be studied again in the 1999–2000 season and that the rules of experimentation decided upon in the workshop should be applied. Only two treatments were selected: rows versus pits. We agreed on factors to be kept constant and data to monitor. Fourteen farmers were involved in the trials which the farmers, extensionists and scientists monitored closely. Table 25.4 summarizes the findings.

The pit technique proved superior to sowing in rows, confirming farmers' observations of the previous season. The mean yield of maize from pits was 22.1 bags/acre (5.5t/ha) compared to 13.6 bags/acre (3.4t/ha) from rows. This was an increase of 61.5 per cent which is a statistically highly significant difference ($p=0.01$).

FARMERS' REFLECTIONS AND CONCLUSIONS

In a reflection workshop held in October 2000, all partners in the research discussed the results. Ten visiting farmer innovators and two VEOs from the Traditional Irrigation Improvement Project in Mwanga and Arumeru districts of northern Tanzania also attended the workshop. The participants agreed that the amount of organic fertilizer required per pit should be studied more closely under a constant plant population; the farmers suggested 15 plants per pit. They also proposed that last year's pits be sown again with maize without further fertilization in order to check residual fertility. The data will be compared with those from the 1999–2000 season.

In view of the good results obtained, the farmers remarked that they intended to expand the area with maize pits in the next season. They also want to spread the idea to other farmers. The visiting farmers from northern Tanzania also expressed interest in starting experiments, using the same procedure, to find out whether the innovation also works in their area. All participants are eagerly looking forward to the results of the next season.

REFERENCES

Croon, I, Dentisch, J and Temu, A E M (1984) *Maize production in Tanzania's Southern Highlands: current status and recommendations for the future*, CIMMYT, Mexico City

van Veldhuizen, L, Waters-Bayer, A, Ramirez, R, Johnson, D A and Thompson, J (eds) (1997) *Farmers' research in practice: lessons from the field*, Intermediate Technology Publications, London

Part 6

Raising awareness, policy lobbying and mainstreaming

Understanding and influencing policy processes for SWC

*James Keeley**

This chapter looks at why policy matters for soil and water conservation and why an understanding of issues involved in policy-making matters in particular. Using examples from different places and with reference to experiences in the country programmes of ISWC 2, it explores processes of policy change and looks at possible ways of engaging with these processes.[1]

Policy environments matter for soils management: the right policy may encourage better soil conservation, while inappropriate policies may have serious negative consequences. Few these days would argue that soils are simply a technical issue. Soils need to be understood not only in a social and economic context, but also as a policy issue. In recent years, numerous studies have deepened understandings of various aspects of this relationship between soils and policy.[2] They have shown that policy changes often correlate closely with changes in soil management practices. In some cases, it has been shown that the adoption of particular policies facilitates better soil conservation and land management. In others, it is clear that policy is a fundamental constraint to investments in land resources. Sometimes there is effectively very little policy, judged in terms of the presence and activities of state agents.

Despite growing awareness of the importance of policy, there is often a fair amount of confusion as to what engaging with policy actually entails. Development projects or participatory research activities with farmers on land management issues frequently appear to view policy as an optional extra, as something to think about in dissemination stages when findings or outcomes

---

* James Keeley is a research officer at the Institute for Development Studies (IDS) at the University of Sussex, Brighton UK and a policy adviser to ISWC 2

are communicated more widely. Talk of findings that are relevant to policy is often rather perfunctory. It may be done to please donors or it may reflect a poorly thought through sense that policy is somehow important. In other cases, there is what early analysts of development policy processes (Clay and Schaffer, 1984) describe as a 'break', as if producing a report or a briefing paper is enough; what happens beyond that is regarded as someone else's responsibility. Not unsurprisingly, this failure to think systematically about policy and about how to work for policy change generally leads to poor outcomes. Policy change needs to be put much more squarely at the centre of research and development projects. This is clearly a major challenge that requires new skills and new types of project design. Examples of new thinking are already emerging – for instance, cases in which capacity building and stake-holder reflection on aims and process are built with much greater commitment and imagination into projects from the very earliest stages.[3]

Why are current attempts to engage with policy so often unsatisfactory? Part of the problem comes down to a simple lack of understanding of what policy is and how it works. Too often the implicit thinking in response to the question 'What is policy?' would be that policy is about decisions and choices in relation to specified issues which result in clear courses of action. These are then followed to the letter, or are not, in which case the policy-maker goes back to the beginning and tries again. This chapter sets out to illustrate that policy is far less linear than is usually imagined, in either temporal terms (moving neatly from decision to implementation) or spatial terms (spreading like ripples in a pond from capital city to district office to village). Understanding the diversity of spaces and points in the policy process is the key to affecting policy change. And when they are grasped, it becomes easier to develop more potentially effective strategies for influencing the policy process.

WHICH POLICIES MATTER?

Interest in policy has been on the rise. At the same time, there has been an increasing awareness of the importance of land management and subsidiary issues such as soil and water conservation and soil fertility management for livelihoods, poverty reduction and rural development and growth. These two trends have resulted in a growing interest at national and international levels in questions of policy and institutional support. Internationally, there is a range of initiatives with soils as a more or less central focus.[4] At national level, governments in many African countries have designed soil management action plans and even soils policies. Clearly, soils are on the policy agenda in many places.

However, in some places it may be that soils initiatives are not of major significance and other policy issues have much bigger impacts on soil management. Clearly, which policies matter varies massively between countries and even between areas within countries: between high and low potential zones, and between farming systems with substantial market engagement and those without, and so on. In some cases, there may be value in engaging with inter-

Table 26.1 *Types of policy significant for land husbandry*

Policy focus	Examples
Soil-specific policies	Soil conservation legislation
	Soil rehabilitation food-for-work
Agricultural and land use policies	Research and extension policy
	Input and output pricing policy
	Land tenure reform
Macroeconomic and governance policies	Devaluation
	Decentralization
	Public sector reform

national soils initiatives; in other places, it may be that a more generic issue such as land tenure should be the focus; elsewhere, a more multifaceted approach may be appropriate.

It is not the aim here to discuss in detail how policies impact. Instead we begin with a broad typology of policies before moving to look at the policy process. Often policies that do not appear to relate to soils at all can have some of the greatest impacts on farmer practices. The devaluation of the West African franc (CFA) is one example of a policy where changes in the prices of inputs and outputs shifted incentives for farmers substantially and resulted in changes in soil fertility management practices. The diverse types of policy that may be significant for land husbandry can be divided into soil-specific policies, sectoral policies and more general or macro policies that nevertheless have important impacts.

Looking at this typology, broad macroeconomic and governance policies may be the hardest ones to influence as they are often most politicized, with debate taking place at the highest government levels. Economic policies are operationalized much more quickly than other policies, leaving little scope to influence implementation. By contrast, soil-specific policies may be important foci for policy activity, but there may also be more intermediate issues that merit more attention.

It is important to engage in processes of reflection and triangulation in deciding what policies to concentrate upon. Understanding where to focus efforts is a key stage that has to be done well before starting to think about strategies. Perspectives on which policies matter for soils will vary, depending on who is asked. Stakeholders at different levels from field to national or international levels may well hold differing views on which are the key policy issues for a particular place. Different stakeholders such as farmers, extensionists, local officials, representatives of farmers' organizations and NGOs can play an important role in enriching understandings of the effects of policy at the local level. These perspectives will vary between different areas and between different stakeholders in the same area. Farmers, for example, will have different perspectives, perhaps reflecting resource endowments, gender and positions in various social networks. It should also be noted that an understanding of the policy process and policy change for different areas may also affect the choice of policy focus. In some areas, policy may look particularly intractable whereas, in others, debate, engagement and policy innovation may look more likely possibilities.

In the ISWC 2 programme, after some reflection and triangulation, most of the country teams decided to concentrate on meso-level policies between the soil-specific policies and the broad macro policies. As thinking about policy continued, it gradually became more refined and, at the Tanzania annual review meeting in 1999, the focus was strongly on research and extension as the major policy areas. Concentrating on these areas offered the best prospects for building policy support for the PTD methodologies that lie at the heart of the ISWC 2 approach to improved land husbandry.

How Does Policy Change?

Having thought about why policy and what types of policy, the next question is 'How does policy change?'. Unfortunately, there is not one single, simple lesson or formula. Policies shift in many different ways: quickly, slowly, dramatically, very slightly, after much consultation, apparently with limited discussion, reflecting political pressure or sometimes seemingly in a technical fashion in response to voices of expertise. Despite this apparent complexity, general lessons and theories do exist which can be usefully applied. There is a big and expanding literature on policy change and the policy process that brings together a range of disciplines.[5] A gap remains, however, in terms of the distillation of messages in practical, usable ways.[6]

Does all this literature on policy change really offer much? Based on a survey of the debates in the literature, this section identifies five lessons that can be useful for those practically engaging with the policy process.

Lesson 1: The policy process is not linear

The first point that emerges from the literature is that policy change is not linear. The policy process is not a business of moving through a series of stages where, if you pull the lever in the capital city, in the office of the minister or the permanent secretary, eventually something happens in a village far away. These production-line images are highly persistent. When one looks at particular instances of policy change, what emerges is that change came from a range of places. Maybe something begins in a province and shifts substantially at district level, it then gets noticed nationally and more changes happen. If one were to map change, it would appear to dot back and forth, reflecting spaces for change appearing in different places and at different times. Change can happen from the bottom up. ISWC 2 generally started to work at provincial or district levels and deliberately bypassed central government. The idea behind this was first to compile evidence that building on farmer innovation is an alternative to, and produces better results than, the conventional transfer-of-technology approach which now characterizes agricultural research and extension in most African countries.

In Zimbabwe, the extension service AGRITEX has gradually moved from being a very technocratic, top-down, command-and-control agency to one that is moving ahead with participatory extension approaches. A key part of this policy change was that field-level workers, particularly in Masvingo province, were innovating and adapting policies locally – 'doing things their bosses wouldn't approve of', as one researcher put it to us. Eventually, higher levels picked up on this and policy change became more formally authorized.[7]

A second point is that policy change does not happen as if one were taking a blank sheet of paper and designing from scratch. Rather, much policy change is incremental, a matter of small changes to the accumulated legacy of previous innovations (Lindblom, 1959). This points to the importance of looking where one can make the incremental changes that can perhaps cumulatively result in a broader paradigm shift.

Lesson 2: The importance of policy narratives

The narrative structure of policy-making is another theme found in the literature on policy process. Problems being addressed by policy are often set up to follow a narrative storyline that moves from cause to effect to action. So it might be argued that 'This is happening because of this, therefore do this', or 'If you don't do this, then this will happen'. This narrative style has the advantage of simplifying complex problems and making them amenable for decision-making and for planning courses of action. For example, a land degradation narrative is widely recited by policy-makers or stated in policy documents, publicity materials or media items. This narrative often runs as follows: African farmers' traditional land husbandry practices are ecologically insensitive, populations are increasing, problems are thereby exacerbated and so we have a real crisis on our hands, and this crisis requires urgent action. Such narratives push issues up the agenda by creating a compelling sense of 'crisis' rather than 'business as usual' (Hirschmann, 1981).

However, the simplicity and storyline structure of such a narrative can often mean that policy interventions end up resting on assumptions that are problematic in some way, sometimes seriously so (Roe, 1991; Hoben, 1996; Leach and Mearns, 1996).[8] It may be useful to study 'successful' dominant narratives and to look at how counter-narratives might be created that reflect marginalized ideas and concerns. With respect to land husbandry, ISWC 2 and PFI are systematically building such counter-narratives with the following elements: small-scale farmers in a marginal environment are often ingenious and creative land managers who experiment and innovate on their own initiative. These innovations often lead to sustainable increases in production. The challenge is to design research and extension policies that build on and support these innovations. A wide range of examples can now be used to demonstrate these points.

Lesson 3: Policy is about practice

One response to the first question mentioned above – 'What is policy?' – might be that laws, bylaws, directives and regulations are all examples of policies. This might be extended to include programmes, formal statements and guidelines. However, it is important to see policy as much more than what is documented and codified or has some type of legal status. Lipsky (1979) argues that field agents ignore, interpret or adapt instructions. They often have to exercise considerable discretion and prioritize where 'policies' are overlapping, contradictory or unclear, or where resources are limited. In other instances, it is simply impossible for managers to monitor whether staff are actually carrying out what is officially mandated. In still other places, there may be no formal policy, perhaps because a government perceives an issue as too controversial. Land issues – whether they are related to tenure or distributional reform – are often of this nature. What this points to is that it is often valuable to hold firmly to the notion of policy as practice. This does not mean ignoring decision-making fora, as decisions and choices that matter are made there, but to pay attention to what actually happens and to which individuals and which places are key in making things happen.

Under the PFI programme, for example, farmer innovators made presentations at national seminars about National Action Plans to Combat Desertification. Their presentations were highly appreciated by all participants because they were the only ones presenting practical examples of successfully reversing land degradation.

Lesson 4: Actors and their networks matter

For a long time, policy has been seen as something that reflects an aggregation of interests within the state and society. Policy change is then seen as something that emerges as a result of shifts in the overall balance of grand political forces. This type of thinking does not really offer much to those considering how to affect policy change. More recent literature (Grindle and Thomas, 1991) has emphasized the centrality of individual actors and their agency to processes of policy change. Of course, decision-makers cannot achieve anything they want, but they do make real choices and the consequences can have significant implications. It is not only actors in powerful positions who have the potential to shape policy change. Often the actions of those with less or apparently no power can have important consequences and can be the critical variable in the policy process. It is important to note that individuals can build networks of actors who can contribute to a shift in policy discourse and policy practice. It is easy to think of many visionaries and charismatic characters who have built coalitions around an argument and gradually achieved access to those with influence, who in turn have used their status to sanction the previously unorthodox as policy.

The ISWC 2 programme in Ethiopia operates in Tigray region. The head of the Regional Bureau of Agriculture and Natural Resources is a member of the Steering Committee. He is one of the key actors in promoting the farmer innovation approach in Tigray. In February 2000, the decision was made that this would become the dominant extension approach in 16 drought-prone *weredas* (districts). Another example comes from Tunisia where the Deputy Minister of Science and Technology took an interest in the evolution of the ISWC 2 programme and urged the partners to organize a national workshop on farmer innovation.

Lesson 5: Get to grips with bureaucratic and organizational cultures

Another fruitful avenue can be to look at the way that bureaucratic and organizational dynamics shape policy initiatives and result in particular styles of managing policy issues. Hierarchical bureaucracies are organized around principles of command and control. This entails a tiered system of line managers down to field staff with precisely defined tasks and responsibilities which are reinforced by organizational policies and training practices. These vertical relationships create cultures where degrees of open communication may be limited. This produces a policy practice characterized by inflexibility, limited discretion and poor organizational learning (Chambers, 1997). The same points can also hold true of relationships between two governmental departments or ministries. Historically, they have slotted into different places within the hierarchy and this may create a dynamic that prevents particular types of policy change from happening. The lessons for those who are interested in working on policy change are to look at where there may be openness to change within and between organizations and with whom this might be associated.

The example of AGRITEX in Zimbabwe is relevant again. Over time, managers came to acknowledge that the formal line-management system was not operating effectively and they also recognized that their staff were coming up with new ideas, even if they were not relaying them, maybe from fear or maybe because no channels to do so existed. Following on from this acknowledgement, a major organizational review is under way, with new ways of dealing with policy and new policies likely to emerge.

STRATEGIES TO EFFECT POLICY CHANGE

So far we have argued that there is a consensus that policy matters for SWC, but that research and development projects need to think more carefully than has been the case so far about what it means to engage with policy issues. A skim through some key ideas from the literature has suggested some potentially useful ways of understanding policy-making. But is it possible to think of more concrete strategies based on experience and lessons from the literature? While there may be no simple recipe for effecting policy change, it is possible to generate some broad strategic principles. This is what the next section sets out to do.

Principle 1: Understand policy debates

An opening step in getting involved with the policy process is to invest time in understanding policies in detail. Familiarity may reveal possible openings, ambiguities and ways in which it may be possible to find points of complementarity with dominant policy approaches. Partners in Ethiopia emphasized that, in their context, adversarial stances were not appropriate. Their tactic was to identify the gaps in policies, as they termed it, to find ways of being seen to be adding to existing policy rather than being confrontational. The neutral term 'gaps' was much more appropriate than talking about inconsistencies or weaknesses. This may sound like common sense, but sometimes, if such a feeling of critique comes across, even though it has not been openly verbalized, drawbridges can be pulled up and prospects of policy change grind to a halt.

It is important as well to learn the language of policy dialogue. Getting to grips with the policy buzzwords can be key, or even *the* key, to participating in debates, and also may be essential for grasping the substance and nuances of discussions or policy documents and position papers. Understanding the language can make it possible to reorient a policy narrative by inserting one's own ideas within a policy frame. For example, if the policy buzzword is modernization, as in Uganda, then one avenue might be to go along with that and argue why an emphasis on farmer experimentation should be a key plank of a modernization policy. Modernization can be taken to mean empowering and building on the resources and skills of citizens, not the development of high external input agricultural systems. Another point is that mapping debates and the language used in debates may show where to concentrate efforts in pushing for policy change. This may mean concentrating on one ministry over another, one department over another or cultivating actors outside the state such as NGOs, unions or farmers' organizations.

Principle 2: Build networks

Building networks of actors is absolutely vital for policy change. While these networks should include grassroots figures, they will also need to aim for those who are influential and well connected. If these contacts come to share the vision of policy changes being sought, they can become powerful protagonists in a range of arenas. They can make the case in important meetings where it may be difficult to imagine a researcher easily having impact or even being present. So how does one expand and work on contacts? One strategy is to invite key figures to serve on project steering committees, to engage and inspire them, and to use their contacts, influence and judgement to find a place in policy debates. Other influential people may be enrolled in many different ways: through training programmes, seminars, workshops and invitations to give keynote addresses or to open workshops. Researchers might create and take advantage of opportunities for informal discussions. Making informal visits or courtesy calls when passing through a ministry may facilitate this.

In Tunisia, for example, the director of the Institut des Régions Arides (IRA) chairs the Steering Committee. Recently he admitted that, in the beginning, he did not really believe in the concept of farmer innovation, but now he is convinced and tells about the experience in Tunisia at all national and international workshops he attends.

Principle 3: Use a bottom-up *and* top-down approach

Another lesson that emerges from both practice and the literature is the value of having a strategy that involves activities at different levels. Creating dialogue at national or regional level is important but cannot stand on its own. This needs to be complemented by processes of networking at district or field level. In addition to training senior officials, training of district officials and grassroots workers can be key in terms of making space to try out new approaches. In Ethiopia, it was important to have senior figures in regional government offering support, but it was also critical to have the enthusiasm and confidence of development agents at the frontline.

Farmers are important for policy work. In Zimbabwe, a key part of the substantial change that looks to be under way in land management policy and practice has been closely tied to processes of capacity building with farmers, building their confidence to articulate their experience and demands. In one ward in Chivi district in Masvingo province, farmers claimed that their extension workers no longer came and told them what to do but rather took instructions from farmers. Site location can be critical in this regard. Working in an area where other things are going on that have caught people's attention (Masvingo in Zimbabwe is a good example) can help to build a bigger impact. Even working in the home district of a key policy-maker can be a useful tactic.

Principle 4: Develop success stories

A key piece in the jigsaw of networking and going for a top-down and bottom-up approach to policy change is to develop 'success stories'. For example, while it is possible to explain persuasively the merits of PTD, it is likely to have a much bigger effect if someone can actually see things with their own eyes. Yield increases, rehabilitated land, ingenious uses of limited resources in precarious conditions are all potentially useful success stories. In Tigray, in Ethiopia, once the green light was secured, exciting results could be developed which could then be fed back to policy-makers who were now further convinced of their own wisdom in allowing the programme to operate in their region. From this point, a discussion about policy and institutionalization questions became much more feasible. Success stories help to focus and distil the ideas and aims of a project. They allow for exposure visits of officials, and they can also be used as part of farmer exchanges to develop confidence and ideas at the grassroots.

The President and the Vice-President of Uganda as well as the Minister of Agriculture have all visited farmer innovators and were amazed and impressed. In Cameroon, examples of farmer innovation and PTD have been used to conscientize policy-makers in the National Agricultural Extension and Research Programme. This resulted in an invitation to develop a training plan for senior staff and field agents.

Principle 5: Identify spaces and develop a sense of timing

Another element that should be stressed is the importance of developing an eye for the right moment. Having a sense of timing is clearly a basic qualification for a politician. It can also be useful for researchers wanting to engage with policy debates. Timing makes it possible to make the best use of networks and success stories: it becomes clear, for example, when it would be opportune to organize a field visit or a workshop and what theme should be used as an entry point. This sense of timing can mean keeping an eye open for moments of policy reform or moments when people are open to or looking for ideas, for example, when consultation processes are underway. Hearings for the Environmental Management Bill and the Land Tenure Commission in Zimbabwe are examples of this kind of process. In addition, one may look for ways of combining with other initiatives and activities, and look for imaginative ways of positioning oneself as part of other currents.

Principle 6: Produce a range of outputs

Another concrete approach is to produce a range of different outputs that effectively document and capture project experiences and lessons. These add value to identified success stories and can help to create and disseminate alternative ideas about a policy area – the counter-narratives discussed above. Obviously, different outputs might be produced for different audiences: journals for academics, accessible and high-quality briefings for policy-makers, colourful posters for waiting rooms of government offices, newsletters, books, videos, websites for a wider public. In a world where information saturation is a real issue, investing time and effort in thinking about effective communication strategies and products is essential. Professional advice and guidance is increasingly available in these areas and can be useful.

The partners in Tanzania and in Ethiopia, for example, have produced a wide range of publications. The PFI has produced a video on Promoting Farmer Innovation in Africa, which is available in English, French and Swahili. Also several ISWC 2 partners (Cameroon, Ethiopia, Tunisia) have produced or are in the process of producing videos on farmer innovation.

Principle 7: Plan a media strategy

With outputs and stories in place, another stage may be to develop a media strategy. This would entail getting to know journalists, targeting particular papers and looking at how to get items on television and radio. Having a clear and interesting message, with good stories and engaging case studies, is key to having an impact.

In March 1999, ISWC-Tunisia initiated a weekly two-hour radio programme on agricultural innovation (see Chapter 27). For the first time, farmer innovators were systematically given access to mass media to present their innovations. When the project started in Tunisia in August 1997, many researchers and agricultural technicians still believed that only researchers could generate new technologies. The radio programme has contributed to changing their ideas and attitudes, and they now recognize increasingly that some farmers have generated new and useful technologies. ISWC-Cameroon has hired a media adviser with the task to organize maximum access of innovative farmers to the mass media. All participating countries have managed, to varying degrees, to gain access to national radio and television.

CONCLUSION

The final sections of this chapter have considered ways of engaging with processes of policy change in terms of lessons from academic literature and also in terms of practical tips culled from the experience of project partners and from discussion of activities elsewhere. It is hoped that reflection on these experiences and lessons will help in moving on from the rather perfunctory or poorly thought out approach to policy set out at the beginning of this paper.

A point of warning, however, in relation to using a project as a basis for lobbying around policy is that the aim of the activity is good policy change and sustainable change in practice, not primarily spreading the name and boosting the status of the project. This may seem obvious, but it can be missed. The ISWC, as a project, might achieve a high profile in a country, but that is not the same thing as creating a policy environment that supports farmer innovation, although it may be a good step on the way.

NOTES

1 This paper is based on research carried out as part of the ESRC-funded research project Understanding Policy Processes for Soils Management (with Ian Scoones) and as part of ISWC 2 and the Nutrient Networking in Africa (NUTNET) project, both funded by the Environment Programme of The Netherlands Ministry of Foreign Affairs

2 For summaries of a range of case studies, see Scoones and Toulmin (1999)

3 See, for instance, Babu and Khaila (1996)

4 They include the Soil, Water and Nutrient Management Initiative of the Consultative Group on International Agricultural Research, the Soil Fertility Initiative of the World Bank, FAO, IFPRI, IFDC and IFA; the CCD coordinated by the United Nations Environmental Programme, the National Environmental Action Plans promoted by the World Bank and the National Conservation Strategies promoted by IUCN.

5 Much of this work initially can seem rather inaccessible, jargon-ridden and abstract, though there are now good summaries available (eg Parsons, 1995; Hill,

1997; John, 1998) that unlock the literature. On environmental issues in particular, see Keeley and Scoones (1999)
6 To some extent, Mayers and Bass (1999) is an exception
7 For further documentation of this, see Hagmann et al (1999)
8 As with, for example, policies and projects to counter land degradation that assume that population increase necessarily results in deterioration in the status of natural resources and so requires corrective interventions

REFERENCES

Babu, S and Khaila, S (1996) 'Priority setting in food and agricultural policy research: a case study and lessons from Malawi', *Quarterly Journal of International Agriculture*, vol 35, no 2

Chambers, R (1997) *Whose reality counts: putting the first last*, Intermediate Technology Publications, London

Clay, E and Schaffer, B (eds) (1984) *Room for manoeuvre: an exploration of public policy in agriculture and rural development*, Heinemann, London

Grindle, M and Thomas, J (1991) *Public choices and policy change*, Johns Hopkins University Press, Baltimore

Hagmann, J, Chuma, E, Murwira, K and Connolly, M (1999) 'Putting process into practice: operationalising participatory extension', *Agricultural Research and Extension Network Paper 94*, ODI, London

Hill, M (1997) *The policy process in the modern state*, Prentice Hall, London

Hirschmann, A (1981) *Essays in trespassing*, Cambridge University Press, Cambridge

Hoben, A (1996) 'The cultural construction of environmental policy: paradigms and politics in Ethiopia', in: M Leach, and R Mearns (eds) *The lie of the land*, op cit

John, P (1998) *Analysing public policy*, Pinter, London

Keeley, J and Scoones, I (1999) 'Understanding environmental policy processes: a review', *IDS Working Paper 89*, Institute of Development Studies, University of Sussex, Brighton

Leach M, and Mearns R (eds) (1996) *The lie of the land: challenging received wisdom on the African environment*, James Currey, Oxford

Lindblom, C (1959) 'The science of "muddling through"', *Public Administration Review*, no 2, pp79–88

Lipsky, M (1979) *Street-level bureaucracy*, Russell Sage Foundation, New York

Mayers, J and Bass, S (1999) *Policy that works for forests and people*, IIED, London

Parsons, W (1995) *Public policy: an introduction to the theory and practice of policy analysis*, Edward Elgar, Aldershot

Roe, E (1991) 'Development narratives, or making the best of blueprint development', *World Development*, no 19, pp287–300

Scoones, I and Toulmin, C (1999) *Policies for soil fertility management in Africa*, Department for International Development, London

27

A bridge between local innovation, development and research: the regional radio of Gafsa, Tunisia[1]

*Noureddine Nasr, El Ayech Hdaidi and Ali Ben Ayed**

Several months after ISWC 2 started in central and southern Tunisia, 60 innovators in dryland farming had been discovered and the number was growing. The next questions were: How could information about these innovations be spread? How could links be forged between farmer innovators, and between them and formal research and extension? ISWC-Tunisia organized visits between innovators and visits by other farmers to innovators, and some visits were shown on national television. However, the major activity to disseminate and stimulate farmers' ideas and experiments was a weekly radio programme on agricultural innovation.

ISWC-Tunisia is coordinated by a team from the Institut des Régions Arides, a research organization working in the arid centre and south of the country. Here, there are regional radio stations operating in the cities of Gafsa and Tataouine. The team decided to work initially with the one in Gafsa, for three reasons. Firstly, the Gafsa station broadcasts to a larger part of the centre and south of Tunisia than does the Tataouine station. Secondly, two-thirds of the identified innovators live in the zones covered by this Gafsa station. Thirdly, the new programme *Agriculture and Innovation* could replace an existing programme on agricultural extension which had already been broadcast for three years.

* Noureddine Nasr is an agronomist and geographer with IRA Gabès; El Ayech Hdaidi is director of the Centre de Formation de Recyclage Agricole (Regional Agricultural Training Centre) in Gafsa, and Ali Ben Ayed is an agricultural technician with IRA Gabès, Tunisia

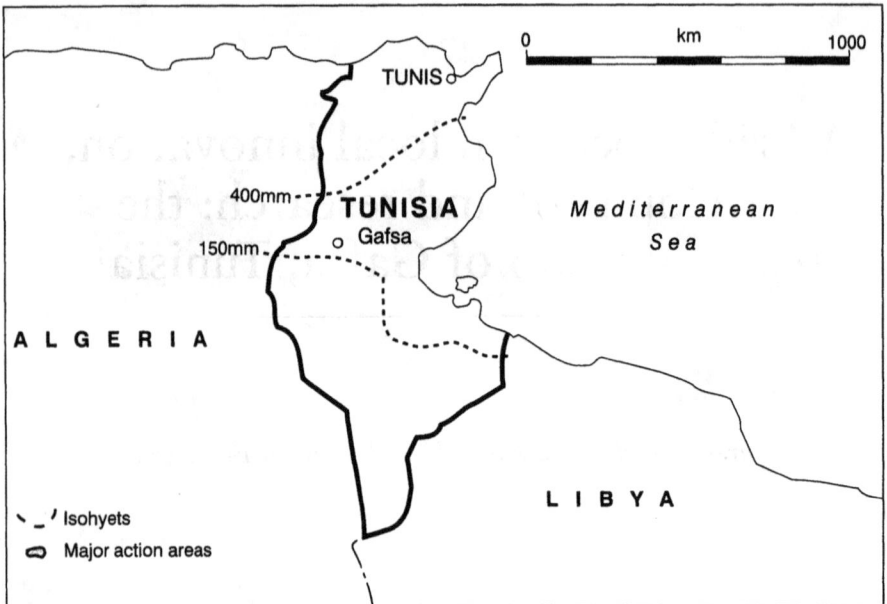

Map 27.1 *Tunisia*

The new two-hour programme went out on the same day of the week and at the same time as the old one, and the presenter of the earlier programme (El Ayech Hdaidi) took over responsibility for the new one. ISWC-Tunisia paid 5000 Tunisian dinars (circa US$3600) to cover the costs of the broadcast in its first year. During its first year, about 85 men and 15 women participated in the broadcasts and presented or discussed a wide range of innovations. The contract between the IRA and the regional radio was not renewed in March 2000. Because the broadcast is popular and the radio received many positive reactions from farmers and agricultural technicians, the regional radio decided to continue the programme, but to fund it entirely from its own resources.

BRINGING THE STAKEHOLDERS TOGETHER

When the broadcast *Agriculture and Innovation* started in March 1999, it was itself an innovation. It was the first time that a radio station in Tunisia had systematically invited farmers to present their knowledge and experience in a live broadcast. Usually it was researchers and technical advisers who passed on information and recommendations to farmers, in line with the 'transfer-of-technology' model of research and extension that dominates in Tunisia and many other countries. Agricultural extension in Tunisia meant teaching and training farmers, but not listening to and learning from them.

The radio programme not only invites farmers to present their innovations, but it also involves researchers, training specialists and development

Plate 27.1 *Mbirika Chokri presenting her innovation in the studio of the regional radio of Gafsa (Tunisia)*

agents into debates about these innovations. Sometimes, these stakeholders in development sit together in the studio, but specialists and other listeners can also call in by telephone. This means that innovators do not need to travel long distances to the radio station to share their ideas with others. Several radio programmes were presented in this way from a distance. Sometimes, innovations from different regions were presented in the same broadcast. Innovators and other listeners with telephones can take part in the debate from anywhere in the region.

To stimulate the participation of as many listeners as possible, the *Agriculture and Innovation* programme is announced in the weekly bulletin of the National Union of Agriculture and Fisheries. The IRA also makes sure that all regional Departments of Agriculture in central and southern Tunisia are informed about the topics of upcoming broadcasts and it invites staff to take part.

In its first year, about 100 farmers (85 men and 15 women) presented a wide range of innovations, including economizing on water use in rain-fed crop production, soil fertility management, fruit tree husbandry (grafting fruit trees on the root system of a shrub which indicates good soil fertility and soil humidity), small livestock rearing, animal breed improvement, and bee- and poultry-keeping.

To encourage the listeners to follow the programme closely and to get some feedback, a system of prizes was introduced three months after the first broadcast was made. Once every two weeks, a prize of 50 Tunisian dinars (circa US$35) is awarded to a listener who has responded by mail to a question posed by the presenters of the programme. The questions are usually about the innovators and innovations. Sometimes, listeners are invited to report on new innovations. This has proved to be a good way of identifying additional innovative farmers, both men and women. The prizes are provided by ISWC-Tunisia and by research and development institutions and local organizations, such as an agricultural cooperative in Gabès.

LETTERS TO THE RADIO

After each broadcast, the radio station receives 20–30 letters from listeners, mostly from rural areas and especially from women (90 per cent). In the case of the older, usually illiterate women, the letters have been written for them by their school-going children or by younger women in the village. The letters include:

- answers to the presenters' questions about the innovations discussed;
- information from listeners about new innovations, often asking if they can be described on the radio; innovations identified in this way include techniques for planting cactus and fig trees, local remedies for diseases of fowl and small livestock, and managing rain-fed vineyards to produce table wine;
- requests for more details about specific innovations, because the listeners want to try them out;
- descriptions of how listeners tried out innovations presented on the radio; these include hatching chicken eggs in piles of dry manure, grafting prunes and peaches on jujubier (*Zizyphus lotus*), planting olive trees on cactus paddles, and drip irrigation using plastic bottles;
- suggestions of new topics for the programme, such as pruning fruit trees, growing early crops in plastic greenhouses, artificial insemination, milk production, and keeping poultry and rabbits;
- congratulations and encouragement to the presenters to continue the radio programme; and
- proposals of field visits or interviews.

Some listeners have suggested starting a parallel television programme to show the best innovations.

IMPACT OF EXTENSION BY RADIO

The socioeconomist from the IRA conducted a survey to evaluate the impact of the radio programme. The letters received were analysed for content. The men and women who presented their innovations on the radio were visited to find out whether they had continued to develop their innovations and whether other farmers or extension agents had visited them. The listeners who had received prizes were visited. Farmers in villages along the Gabès–Gafsa and Gafsa–Maknassy–Mazouna roads were interviewed at random in places where farmers frequently meet, such as shops, reforestation sites and local offices of the extension service. It became evident that the radio programme had four major types of impact.

Giving incentives to continue innovating

For most of the men and women farmers who had presented their innovations on the regional radio, this experience had been an important social incentive. After the broadcast, several innovators continued to develop their innovations or started to develop new ones. For example:

- Béchir Nasri, an innovator in Médenine region (Nasr et al, 1999), invented a new system for pumping water from cisterns and a new technique for conserving wax honeycombs in beehives; he has recently developed a simple mechanical system to control the timing and amount of water used in localized irrigation, and he is working on a technique to filter sediment from run-off water in order to avoid deposition in cisterns.
- Khlifa Dadi, an innovator in Mareth region (Chahbani and Nasr, 1999), developed new techniques of localized irrigation in order to economize on the use of water; these were adaptations of an innovation he saw during a visit to another innovator featured on the radio.
- two women, Mbirika Chokri and Naziha El-Fahem (Chahbani and Nasr 1999), have increased their production efforts since they were on the radio. Naziha produces chicks and supplies them to ten other women who want to raise poultry and who are taking advantage of a women's microcredit scheme developed by a project in Mazouna as a result of her radio presentation.

Encouraging visits to innovators

Since speaking on the radio, most innovators have been visited by other farmers and agricultural technicians. During his presentation, one innovator who distils cosmetic plants made an appeal to other farmers to grow these plants on a contract basis. A few days later, he was visited by a group of farmers. This visit was organized by the team of the Presidential Pilot Project on Agricultural Extension based in Gafsa which records all broadcasts of *Agriculture and Innovation* for use in its extension workshops. A few months later, when the farmer was interviewed on radio again, he mentioned that he had already signed production contracts with 20 farmers. The director of the Gafsa Regional Department of Agriculture visited four of the innovators, including one woman. These visits were incentives to both the innovators and the extension agents, and indicate that new relationships are developing between farmers, development workers, research scientists and policy-makers.

Stimulating adoption and adaptation by listeners

Analysis of the survey results and of the letters to the radio station showed that several listeners had adopted and, in many cases, adapted the innovations presented on the radio. For example, more than 50 men and women farmers were trying out the bottle method of drip irrigation developed by Rgaya Zammouri, an elderly woman from Zammour village who had presented this on Gafsa radio and national television. Five women were hatching eggs in manure, as had been described on radio by the female innovator Mbirika Chokri.

Changing attitudes in research and extension

The broadcasts have also started to influence the attitudes of researchers and development agents. When ISWC 2 started in Tunisia in August 1997, the approach of seeking out local innovations as stimuli for rural development was strongly criticized and some research and extension staff even ridiculed it. After the first innovators had been identified and particularly since the radio programme started, it is evident that there is a growing positive interest in this new approach (see Chapter 30).

MASS MEDIA AND INNOVATION

During the first year, about 85 men and 15 women farmers took part in the radio broadcasts and presented or discussed a wide range of innovations. Because the programme was so popular among farmers and development agents and the radio station was receiving a larger number of positive letters than it had ever received for any programme in the past, Gafsa regional radio decided to continue the programme in the second year and to fund it entirely from its own resources.

Listeners requested not only that the regional radio programme be continued, but also that it be extended to other regional stations as well as to national radio. This can be done only when development agencies and, in particular, farmers' organizations accept responsibility for and 'ownership' of these radio programmes by making contact between local innovators and the radio station, encouraging farmers to listen to the programme, and so on. Also, other mass media (the press and television) should be used more systematically to convey the message that men and women farmers are taking initiatives in developing useful technologies and improving their livelihoods.

NOTES

1   This is an expanded version of an article by Nasr et al (2000) that appeared in the *ILEIA Newsletter* focused on 'Grassroots Innovation', the French version 'Promouvoir l'Innovation Paysanne' and the Spanish version 'Innovación desde las bases'.

REFERENCES

Chahbani, B and Nasr, N (1999) *Le développement participatif de technologies basé sur les innovations des hommes et des femmes en agriculture en zones arides de la Tunisie*, Rapport, IRA, Médenine/CDCS, Vrije Universiteit, Amsterdam

Nasr, N, Chahbani, B and Reij, C (1999) 'Farmer innovators in land husbandry in central and southern Tunisia', Tenth ISCO Conference, 23–29 May 1999, Indiana, USA

Nasr, N, Hdaidi, E A and Ayed, A B (2000) 'Local innovation and wider development in Tunisia: Gafsa regional radio', *ILEIA Newsletter*, vol 16, no 2, pp18–19 ('Innovation et développement agricole en Tunisie: la radio régionale de Gafsa', *Bulletin d'ILEIA*, vol 16, no 2, pp14–15; 'Innovación local y mayor desarrollo en Túnez: radio regional Gafsa', *Boletín de ILEIA*, vol 16, no 2, pp10–11)

28

Mainstreaming participatory approaches to SWC in Zimbabwe

*Kudakwashe Murwira, Jürgen Hagmann and Edward Chuma**

When the ISWC 2 programme started in Zimbabwe, there were already several islands of success in SWC that could have a positive impact on land husbandry activities beyond the borders of the communities concerned. ISWC-Zimbabwe focused on raising awareness about these successes, scaling them up and institutionalizing the approaches into major development programmes and government research and extension structures. The challenge was to spread not only the promising SWC techniques but also the participatory methods for developing, disseminating and adapting them.

BACKGROUND

The successful cases of combating degradation of natural resources in Zimbabwe had resulted from several initiatives undertaken over the years by farmer innovators, development agencies or research institutions exploring ways to improve soil fertility and water management in drought-prone areas. It started with approaches and techniques that some researchers and development agents regarded as having potential in responding to SWC problems in Southern Africa. These were tested in close collaboration with a small number

* Kudakwashe Murwira and Edward Chuma are joint coordinators of ISWC-Zimbabwe, based in Harare; the former is a community organizational development facilitator and project manager with ITDG-Zimbabwe and the latter is a SWC researcher with the Institute of Environmental Studies of the University of Zimbabwe; Jürgen Hagmann, formerly with the Conservation Tillage Project, is now based in Germany, working as a consultant in change processes in international research and development

Map 28.1 *Action areas, ISWC-Zimbabwe*

of communities. Some of the initiatives were successful, judging by the degree of farmer involvement in refining the techniques, the high level of local adoption and the effectiveness of the techniques in conserving soil and water. The land users themselves determined the success of the techniques on the basis of their own criteria.

Examples of good practice serve little benefit unless a large number of people become aware of them. It is therefore important to ensure that lessons and experiences from successful cases are documented and shared with all relevant stakeholders, particularly with those that are well placed to help spread the ideas. In Zimbabwe, such institutions include the Department of Agricultural and Technical Extension Services (AGRITEX), the Department of Natural Resources (DNR), the Department of Research and Specialist Services (DRSS), the Zimbabwe Farmers Union, NGOs and community-based organizations. If convinced, these could scale up or mainstream the good ideas and, at the same time, increase the likelihood of sustaining the process of generating technologies that are appropriate for smallholder conditions. However, institutionalizing participatory approaches is not an easy task, particularly in institutions with a track record of regarding their own staff as 'experts' who should impart knowledge by 'sending messages to farmers'.

The farmer-to-farmer approach to spreading good ideas among farmers or communities is most effective in an enabling environment in which the institutional and legal framework not only recognizes but also actively supports the approach. A good example of farmer innovation in SWC that initially failed to spread because a supportive institutional framework was lacking, was the work of Zephaniah Maseko Phiri in Zvishavane district in Midlands province of south-central Zimbabwe. Phiri developed a number of innovations in SWC, but it took more than 15 years for them to spread beyond his farm. In the 1990s, before the start of the ISWC-Zimbabwe programme, support from individuals with a good understanding of the national policy framework and from NGOs seeking alternative SWC techniques helped to spread his work by exposing other farmers to it, documenting his experiences and providing a platform for him to share these with farmers, scientists and extensionists.

Earlier, Phiri's innovations had not spread because government service providers viewed them as a 'threat' to the country's policy of natural resource management which had remained the same even after Independence. They regarded him as a mad person whose ideas should never be emulated by anyone sane. This was despite the fact that, in 1981, Phiri had proved to both the local magistrate and the Natural Resources Board that his practices were effectively reducing soil erosion and improving moisture conservation.

The government staff regarded any type of knowledge that was locally developed, ie that did not find its origin in either DRSS or AGRITEX, to be traditional and primitive and therefore not to be encouraged. The technologies disseminated to farmers by both DRSS and AGRITEX had to be 'tested and proven'. Moreover, the government staff regarded the 'master farmer' model as the best vehicle for delivering extension services. In this model, farmers are trained according to a standardized curriculum, irrespective of the trainees' access to resources or differences in agroecological conditions. The extension agent determines the criteria for 'success', which usually means following as closely as possible the guidelines and practices recommended by the extension service.

AGRITEX has been in existence for over 70 years and now has over 2000 staff posted throughout Zimbabwe to provide extension support to rural communities. The DNR, although not represented in every community, is represented in each district office. It has the task of reinforcing the sustainable management of natural resources through legislation. The researchers in the DRSS provide a range of technologies for extension workers to pass on to farmers. Despite the heavy presence of these support institutions, soil and water loss has increased on most of the cropland and pastures in communal areas. Some farmers are losing soil at a rate of up to 40t/ha annually (Chuma and Hagmann, 1995). Problems of soil erosion have worsened mainly because of the emphasis in mainstream research and extension on inflexible top-down approaches with a very strong technology focus.

It is against this background that the various development agencies mentioned at the outset had started to seek alternative approaches to and technological options for managing soil and water. These were tested and further developed at various sites with a view to scaling them up if they proved to be successful locally. In addition, a wide range of SWC techniques, including tied ridges, tied furrows, mulch tillage, infiltration pits and improved methods of organic manuring, were developed and tested jointly by farmers, NGO staff and government researchers and extension agents, with a view to promoting their adoption and adaptation by other farmers. 'Experts' from within the key agricultural services had been drawn into this process in order to increase the chances of mainstreaming the ideas subsequently.

CREATING ALTERNATIVES AT LOCAL LEVEL

Intensive work in Masvingo province

One example is the work of two institutions in Masvingo province: the GTZ-funded Conservation Tillage Project (Contil) and the Intermediate Technology Development Group (ITDG). In 1991, both of these groups started to explore alternative approaches to working with smallholder farmers. Contil was a participatory research project that initiated adaptive on-farm trials in Gutu, Zaka and Chivi districts. ITDG, whose main focus was on participatory extension to enhance household and food security, initiated work in Ward 21 of Chivi district and then expanded to Ward 4 in the same district.

In 1993 the two institutions met and realized that they had much in common. Firstly, they both understood that the problem of SWC in rural Zimbabwe was due more to social crisis than to poor access of farmers to better SWC technologies. Numerous socio-organizational and cultural problems in the rural communities undermined the process of innovation in farming. Some of the major factors affecting the level of adoption and adaptation of SWC techniques were lack of cooperation, conflict between generations, internal leadership wrangles, mistrust, jealousy, fear of trying out new ideas, and a general lack of community will and commitment. The struggle for power between modern state structures (Village Development Committees, Ward Development Committees, etc) and traditional leadership aggravated this situation.

Contil and the ITDG recognized that it was difficult for any one of them alone to institutionalize a participatory approach to research and extension within government structures. They started to network and to document lessons and experiences emerging from their work and shared these with AGRITEX. They set up programmes to support the training of extension workers in the Participatory Extension Approach (PEA) (AGRITEX, 1998). A major component of PEA is Training for Transformation, a tool for bringing about attitudinal and behavioural change in people so that they can attain new values for themselves, such as self-respect, self-confidence, self-awareness, self-esteem, mutual trust, self-reliance, inclusiveness, capacities for open criticism and equal opportunities (Hope and Timmel, 1984).

Discomfort model of training

Contil and ITDG started at community level with a process that was designed to increase local institutional and technical capacity. Farmers were exposed to Training for Transformation and, as a result, became more conscious of their needs, rights and responsibilities, as well as of the responsibilities of the public service providers. Then the extension staff, feeling uncomfortable with being challenged by farmers, asked why they were not being offered training like that given to the farmers. Indeed, they wanted to be the ones who transferred the training to the farmers. However, Contil and ITDG preferred a process that would compel AGRITEX staff to be more responsive to the needs and circumstances of smallholders. Once the extension workers realized that they needed to be exposed to the PEA, Contil and ITDG responded accordingly and facilitated the learning process at this level.

The exposure of the extension workers to new approaches of working with farmers created a crisis at the next level: they found themselves in conflict with their supervisors who, in turn, requested that they be given the same training. The demand for training in PEA continued up to the level of head office as the 'bosses' were afraid of being overshadowed by their 'juniors' and 'mere' farmers.

Parallel to this, key contact persons within AGRITEX were identified to build supportive networks. Also, the provincial heads in Masvingo were targeted. The same strategy was applied later at national level. These contact persons at different levels became instrumental in internalizing the participatory approach within AGRITEX.

Creating a strong demand pull

Through the PEA approach, farmers, especially those in Ward 21, realized the advantages of working in groups and as a community, such as being able to share knowledge and skills to help each other, to share assets, to generate ideas rapidly to solve common problems and to market their produce more efficiently. With a ratio of one extension officer to about 1000 farmers, it is almost impossible for all farmers to contact an extension worker on an individual basis. The farmers in Ward 21, with ITDG support, began to organize themselves better in order to enjoy the benefits of working together. They became more confident and were able to articulate the learning process they had gone through. They started organizing local competitions in soil and water management in order to stimulate farmers to try out new techniques.

The farming communities working with Contil and ITDG in Masvingo province also shared their experiences with visiting farmers from throughout the country as well as from South Africa, Mozambique, Zambia, Malawi and Swaziland. The visits were organized and financed by development-support institutions in these other countries. They had became aware of the initiatives in Masvingo through various intermediaries who had either read about the work in development publications or had heard about the initiatives during workshops where this information was shared by Contil and ITDG staff. The numerous foreign visitors (averaging one group of about 30 per month) to communities working with the two programmes helped to raise their profile and made it easier for programme staff to share their approach with AGRITEX staff at both district and provincial levels.

The second phase of ISWC 2, which began in 1997, fuelled this process as it strengthened the innovation and experimentation activities of farmers by linking them with scientists, and it intensified and gave more support to farmer-to-farmer exchange. In Ward 25 of Chivi district, for example, researchers from the Institute of Environmental Studies (IES) of the University of Zimbabwe developed a partnership with innovative farmers to explore alternative techniques of SWC and soil fertility management. These were techniques that combined farmers' ideas and those of scientists, sometimes based on local innovations, sometimes based on an idea from science combined with farmers' suggestions, about how it could be implemented under their conditions. The techniques, which included tied ridges, mulch ripping, strip cropping, *fanya juu* terraces, infiltration pits, stone bunds, higher quality manure and compost, were evaluated in researcher-supported on-farm trials and farmers' own experiments.

Spreading upwards and outwards

Before ISWC 2 began, communities in Ward 21 had started already to host farmers from different corners of Zimbabwe almost on a monthly basis. This had become the basis for a loose network of farmers interested in SWC. Since ISWC 2 began, this new network has spread throughout five provinces of Zimbabwe: Masvingo, Manicaland, Midlands, Mashonaland East and Matabeleland South. Farmer innovators in each province share experiences during exchange visits organized by both governmental and non-governmental institutions, with some coordination and facilitation by ISWC-Zimbabwe. In some cases, funds are made available by the programme; in others, the support institutions and even the farmers themselves have mobilized the resources to finance the travel.

The ISWC 2 programme gave an opportunity to scale up the farmer-to-farmer extension activities, with a focus on outstanding farmer innovators. These individuals were given opportunities to visit each other to learn from each other's innovations and experiments. The innovators themselves organized exchange visits within their own communities and between farmer innovators in different communities, districts and provinces. They invited government extension agents to join the visits, even though AGRITEX was not involved in their planning and funding.

The exchange visits between farmer innovators provided an opportunity for the extension staff to document information about the innovators and their new techniques or adaptations. This task, initiated by ISWC-Zimbabwe, obliged extension staff to listen carefully to the farmers and to learn from them. As a result of going through this process, the extension agents began to realize that these innovators were not mad, but rather key resource persons in their communities. The extension agents have found that documenting farmer innovation is so important that they now do this as part of their normal exten-sion work. Both field staff and managers in AGRITEX were involved in the decision to do this.

Under the ISWC-Zimbabwe programme, farmer-to-farmer visits between countries were also organized. For example, four Zimbabwean farmers went to Mozambique at the invitation of the Mozambican hosts and about 20 farmers went to Lesotho on a visit organized by the ISWC-Zimbabwe coordinators who had contacts with organizations supporting farmer innovators there. The coordinators had recognized that the climatic and socioeconomic conditions in many parts of that country are similar to those in action areas of ISWC-Zimbabwe. This meant that the lessons learnt in Lesotho could be applied readily to Zimbabwean conditions. In Lesotho, the renowned farmer innovator, James Jacob Machobane, was a special attraction, and the Zimbabwean farmers learnt about the organization of farmer innovators as exemplified by the Machobane network. They also gained a wide range of technical information, eg about the Machobane Farming System, gully reclamation, use of indigenous pesticides, the reclamation of wasteland and use of organic manures.

The continued networking of ISWC-Zimbabwe partners with key individuals at provincial and national levels contributed to incorporating these elements of more participatory research and extension into the DRSS and AGRITEX. In some cases, the spread of participatory approaches was supported by like-minded organizations such as VECO (Vreideselanden Coopibo). This Belgian organization, based in Mashonaland, East province, financed visits by groups of farmers from that province to the Zvishavane Water Projects in Midlands province, to Wards 21 and 25 in Chivi district and to Makoholi Research Station in Masvingo. Farmers and project staff assessed the impact of these exchange visits, using methods such as self-evaluation, informal observation and formal questionnaire surveys. They found that the visits increased the rates of farmer adoption of technologies such as rock catchments, water harvesting, tied ridges, infiltration pits, modified contour ridges, mulch tillage, *fanya juu* terraces, intercropping and use of termite mounds for soil amelioration. They also increased the farmers' self-organization capacity and self-confidence to experiment in order to identify their own solutions and forged wider network linkages between farmer innovators.

In Masvingo province, liked-minded individuals from a number of government services and NGOs managed to build up a strong network. These included Contil, ITDG, AGRITEX, DNR, the Farming Systems Research Unit of the DRSS, IES (University of Zimbabwe) and the Integrated Rural Development Programme. ISWC-Zimbabwe encouraged the sharing of a common understanding of the role of farmer innovators in land husbandry through workshops and seminars, and organized and facilitated training in PTD to enable practitioners to develop the capacity to identify, document and support experimentation by farmers. Jointly with partners like the Zvishavane Water Projects, the programme succeeded in making participatory approaches to research and extension more acceptable in the government agricultural services. The involvement of many of the above organizations in this informal collaboration was made possible by their flexibility in making decisions at the implementation level. This is also reflected in their use of funds which is guided to a high degree by the situation on the ground.

MAINSTREAMING THE APPROACH

Nearly all the SWC success stories were achieved through the use of participatory approaches. In an effort to mainstream such approaches in the institutions providing agricultural services, the following activities were undertaken by ISWC-Zimbabwe and its partners:

- Staff from AGRITEX and other development support institutions documented the experiences, particularly in Masvingo province. The lessons that emerged were analysed together with key staff from both AGRITEX and DRSS in order to develop a common understanding of the approach and to share it widely.
- A PEA framework and curricula for training extension staff in PEA were developed and shared with AGRITEX staff throughout the country and elsewhere in Southern Africa, such as Northern province in South Africa, Helvetas project sites in Lesotho and the Department of Agriculture in Cabo Delgado province, Mozambique.
- An organizational development process that was already underway within AGRITEX, with the support of GTZ, has helped to mainstream participatory approaches. After going through the PEA training, extension staff are exposed to successful cases in Masvingo province so that they can witness the impact and effectiveness of using participatory approaches in working with smallholders, especially in SWC.
- The identification and documentation of farmer innovators and the support given to their networking strengthened their position and they started to ask many questions to scientists and extensionists, eg 'Do you know our priorities?' 'Why do you not respond to our problems? Do you lack the necessary resources or do you lack support from your leaders?' 'How can we help you to help us?' (quoted in various workshop proceedings). It was this strong demand from farmers that pushed extension staff to demand training in participatory approaches. Under the ISWC-Zimbabwe programme, about 30 extension agents in Manicaland were trained to be able to respond better to farmers' demands. Similarly, in Shurugwi and Gwanda districts, about 60 extension agents were trained in the PEA process and are using this approach with support from ISWC-Zimbabwe.

CONSTRAINTS AND OPPORTUNITIES

The ISWC-Zimbabwe programme scored a number of successes, but also experienced some constraints, such as high turnover of staff in government departments and the 'donor syndrome' among some farming communities. Several lessons can be drawn from the experiences:

- Organizations are different; hence, what works with one may not apply to another, even if both organizations work with and for the same clients. These differences can be attributed to personalities and to organizational culture, identity and history. The process of institutionalizing the approach has been quite successful within AGRITEX: all extension agents are now required to take reorientation training in PEA. The same cannot be said of DRSS or DNR, mainly because of variations in the above-mentioned factors. Sometimes, it is the level at which these institutions interact with clients that determines their ability to transform themselves and to be more responsive. Institutionalizing participatory approaches means maintaining a learning process and continuing to integrate new practices as one learns. However, for institutions like AGRITEX, DRSS or DNR, with a tradition of hierarchy, blockages to this process can always appear. The strength of tendencies toward standardization, centralization and opposition to innovation within bureaucratic organizations should not be underestimated.
- It is important to determine the best time to move from a pilot phase to a phase of integration or mainstreaming. Moments of crisis can be the best opportunities for selling new ideas. The 1991–92 drought forced agricultural support institutions to seek alternative approaches to deal with drought-stricken communities. Government programmes like the Smallholder Dry Areas Resource Management Programme and the South East Dry Areas Programme, which sought to mitigate the effects of drought in the semi-arid areas, took the opportunity to introduce participatory approaches to SWC already in the early 1990s and thus prepared the ground for ISWC-Zimbabwe and partner programmes that focused on scaling up such activities.
- Generally, networks and alliances can be very strategic in selling new ideas within key institutions of research and extension. However, the success of these networks depends on numerous factors, such as personality, trust and shared vision, and the ability of the parties involved to compromise.
- Exposure visits stimulated by awareness created through documentation play a major role in spreading ideas, especially from farmer to farmer or community to community. Whereas many people believe that documentation can capture and share the lessons emerging out of any work, the experience of ISWC-Zimbabwe has been that the readers then want to see with their own eyes what was done. The more that the work in Wards 21 and 25 of Chivi district was documented, the more visitors these communities received from throughout Southern Africa. On the initiative of Ward 21 farmers, a follow-up study was made recently of all farmer groups hosted since 1994. This formed part of an impact assessment of several programmes (including ISWC) that was made by ITDG Southern Africa, one of the lead agencies of ISWC-Zimbabwe. Questionnaires were sent out by ITDG to all persons who had visited the sites in Ward 21 and, based on the returns, determined the rates of adoption and adaptation. The results showed that, after the farmers had returned home, nearly 95 per cent of them had tried out some of the ideas they had gained during the visit.

• Although exposure visits are an effective tool for disseminating new ideas, the great distances involved make them costly and difficult to sustain without external financial support. For the approach to be sustainable, self-help capacity needs to be developed within the rural communities. Several of the farmer groups have already started to mobilize their own resources through contributions from members. In some cases, groups have requested training from ISWC-Zimbabwe and its partners in writing project proposals and they are accessing funds directly through government programmes and from other donors.

CONCLUSION

Good progress has been made in mainstreaming participatory approaches into land management programmes in Zimbabwe, although the degree of institutionalization varies from one programme or agency to another. This success is due to a multifaceted approach. Networks on different levels were used to lobby for policy change. Key players in relevant institutions were used as entry points for lobbying. Joint documentation and analysis of success stories helped to raise awareness. Farmers joined voices to express their demands. Training for Transformation proved to be a valuable tool for creating a paradigm shift in the attitudes of all stakeholders. Exposure visits to islands of success and between islands of success stimulated the processes of experimentation and innovation by farmers. The members of the ISWC-Zimbabwe Steering Committee were drawn mainly from public institutions, policy advisers and academia and all are in a position to influence the direction of thinking in their institutions. All these different strategies contributed to the total impact of the ISWC-Zimbabwe programme.

REFERENCES

AGRITEX (1998) *Learning together through participatory extension approaches: a guide to an approach developed in Zimbabwe*, AGRITEX/GTZ/ITDG, Harare

Chuma, E and Hagmann, J (1995) 'Summary of results and experiences from on-station and on-farm testing and development of conservation tillage farming systems in semi-arid Masvingo, Zimbabwe', in S Twomlow, J Ellis-Jones and H Loos (eds) *Soil and water conservation for smallholder farmers in semi-arid Zimbabwe: proceedings of a technical workshop held 3–7 April 1995 in Masvingo*, Belmont Press, Masvingo, pp41–60

Hope, A and Timmel, S (1984) *Training for transformation: a handbook for community workers*, Mambo Press, Gweru

29

Liberating local creativity: building on the 'best farming practices' extension approach from Tigray's struggle for liberation

*Berhane Hailu and Mitiku Haile**

When the ISWC 2 programme commenced in Tigray in 1997, it found itself in the midst of streams of two different traditions of extension, one coming from the previously centralized system of technology transfer practised by the national government and one from the extension activities during the struggle led by the Tigray People's Liberation Front (TPLF) against the Communist Derg regime in Ethiopia. During the 17 years of civil war, the TPLF had built up its own system of agricultural extension in the liberated areas, focusing on disseminating the practices of good farmers who were making the most of local resources. The experience of the farmer innovation approach promoted by ISWC 2 in Tigray can be understood only in the context of this history of agricultural extension during and after the struggle for liberation.

---

* Berhane Hailu is the head of the Tigray Bureau of Agriculture and Natural Resources and member of the ISWC-Ethiopia Steering Committee; Mitiku Haile is president of Mekelle University and joint coordinator of ISWC-Ethiopia

AGRICULTURAL EXTENSION DURING THE STRUGGLE FOR LIBERATION

Berhane Hailu

The birth of the TPLF extension service

The TPLF had set out to struggle not only for liberation from the Derg regime but also for the development of the people who had suffered under the regime. There was a conscious movement to address not only political but also social, economic and cultural aspects in the people's struggle. One of the sectors of development prioritized by the TPLF was agriculture. It took some years to build up capacities to plan the development programme in detail and to make it operational in all sectors. Activities related to health and education started immediately after the first areas of Tigray had been liberated. After a few more years, the TPLF leaders felt that there was sufficient local capacity to intervene also in the agricultural sector. They selected fighters with some formal training in agriculture to form a Department of Agriculture (DoA).

At that time, in 1979, there were very few people in Tigray with advanced training in agriculture. The man appointed to coordinate the DoA held a BSc in agriculture from Alemaya University; he worked together with three men with agricultural diplomas who had been with the Ethiopian Ministry of Agriculture before the war. These four people formed the core group for agricultural development in the liberated areas. In the following three years, three more people with MSc level joined them. Most of the people in this original taskforce still work in agricultural development in Tigray today.

The core group was given the task of defining direction for the newly established DoA, in close collaboration with the TPLF leadership. They had to take into account the war situation, the conditions under which the people were living in the liberated areas, the available capacities and the extremely limited access to external resources. From the beginning, a major principle in agricultural development, as in the other sectors addressed by the TPLF, was to depend on community participation. The villagers, through their local councils (*baitos*) set up during this period, had the decision-making role, while the task of the various TPLF departments was to facilitate the implementation of the activities decided upon by the communities. Rehabilitation depended on collective self-reliance. When farmers moved back into a liberated village that had been destroyed by the enemy, neighbouring villages were encouraged to help, eg by giving seed if the seed stores had been ruined and by helping each other with labour and materials to reconstruct the farm buildings. External aid, such as handtools supplied through REST (Relief Society of Tigray) with donor funds, supplemented this ethos of mutual help.

The work of the DoA had two major objectives: 1) to motivate the farmers to continue farming and to rehabilitate the land, despite the war, and 2) to help farmers to make the most out of the available resources. As it was obvious that, under the prevailing circumstances, Tigray could not rely on introduced technologies, a major principle made at policy level was to extend the best farming practices already existing in the communities and to bring other farmers up to the level of the outstanding farmers.

Seeking the best farming practices

The first step of the small DoA team was to try to understand the existing situation of agriculture in Tigray. This involved making a critical assessment of traditional practices. At the same time, the team sought ways to identify the best farmers in the communities and to find out how they and their practices were regarded by the other farmers.

The team travelled to the *dega* (highland), *woina dega* (midland) and *kolla* (lowland) areas of Tigray. Together with the *baitos*, the team members discussed with farmer representatives chosen by the *baitos* and elders to represent both the best farmers in the area and 'normal' farmers. The discussions covered crop and animal husbandry and were recorded in writing in large notebooks.

In crop production, some of the topics discussed included:

- *Ploughing:* when and how the farmers ploughed on what type of land, and what type of ploughshare was used for different ploughings, different soils and different crops.
- *Seed selection:* how farmers selected seeds, what seed they selected for what purposes and according to what criteria, including that which was bought on the market. This revealed many interesting traditional practices of which the formally trained agriculturists had not been aware, eg that farmers placed seed in water to see how the seed reacted and that women made *injera* (the local staple dish) out of sample seed to check its cooking quality.
- *Sowing:* how and when the farmers sowed, what implements they used and how they used them, and to what depth they sowed the seeds. It was fascinating to learn how the farmers tested for soil moisture before sowing: they took a handful of soil, pressed it together and then threw it into the air to see how the soil dispersed. Also the practice of washing seed with urine to protect it against pests and noxious weeds such as striga was discovered.
- *Weeding:* which crops were weeded first, how many times they were weeded and at what stage in crop growth, what the farmers did with the weeds, such as feeding them to livestock or making traditional compost.
- *Soil fertility management:* what types of manure (eg from cattle, sheep, goats, chickens) and compost were used on what crops, how manure was stored, when it was taken to the field and ploughed in, what was the best manure to use for different purposes, how some farmers dug a channel to direct the urine of cattle in the farm compound into the maize plot immediately below it. The team also looked at crop rotation practices in relation to soil fertility management.

- *Harvesting, threshing and crop storage:* how the farmers handled the crops in the field before taking the harvest in to be stored, how they separated intact grains and those damaged by pests, how they dried the grain or forage, how they stored seed and grain for consumption, how they mixed the grains in storage, how they smoked the stores to keep out pests or used leaves and seeds of particular plant species to repel insects, how they monitored the condition of the stored grains, eg by taking samples and checking the heat.

Similarly detailed questions were asked about animal husbandry: management, feeding, haymaking, traditional veterinary medicine, etc. For each operation in crop and animal husbandry, the team recorded what the majority of farmers did and what the best farmers did. They asked the farmers to estimate the yields and storage losses, using traditional measurements. They also recorded what the 'normal' farmers thought about the practices and results achieved by the best farmers, and why they thought there was a difference.

In addition to recording the current practices, the team asked the farmers to talk about their problems and which ones needed most urgent attention. The problem most frequently mentioned by farmers was moisture stress: 'We grow the crops, but they don't reach maturity because the rains fail.' All these problems were recorded according to the traditional classification of agroecological zones in Tigray which is mainly related to altitude and includes six categories ranging from highlands above about 3500m (*dega*) to lowlands below about 1000m (*kolla*).

The direction and strategy for agricultural development were formulated on the basis of this assessment by the DoA team after consulting farmers throughout Tigray. It was decided that the initial focus should be on crop production, looking particularly at moisture conservation, SWC and soil fertility management.

The team then looked at the best practices in the various stages of crop production and storage and identified where there were the most important differences between the best farmers and others. For example, the different methods of harvesting did not have a great effect on crop yield and storability of the grain, whereas the different practices in seed selection and storage made a big difference with regard to how much grain a family ultimately had for eating or selling. The DoA decided to promote the best practices that made the greatest difference in these terms.

Training development agents and production cadres

The DoA team prepared a manual of best farming practices in the different agroecological zones. This formed the basis for the training of Development Agents to work at *wereda* (district) level. For this task, the DoA recruited fighters with a rural background who had experience in farming. In 1982, the first training was held in Atsrega, a liberated area in Western Tigray. The 24 DAs in this first batch were then sent to different postings throughout Tigray.

In addition, the DoA and the *baito* in each area selected local 'production cadres' for training. These were the best farmers who had been identified during the assessment by the DoA team. They had to be resident in the area, recognized and accepted by their communities as good farmers, and willing to disseminate their practices. Many of the best farmers were relatively old, being people with much experience. The best farmers formed the core of the production cadres and helped to train additional production cadres from villages in which the best farmers were not available for this work. A large number of men – more than 500 – were trained in each zone as production cadres in the first two years.

The DoA and the TPLF leadership paid attention not only to the content but also to the methods of training. It was decided to have as little formal classroom training and as much practical work as possible. The classroom sessions consisted often of heated discussions, based on the manual of best practices. The best farmers would explain: 'I do it like this and therefore I get these results.' Other farmers would say: 'He can get such good yields because he is blessed by God.' They had to be convinced that the practices were indeed different and that they should try out the practices of the best farmers to see if they could obtain similar results themselves. The fact that many of the best farmers were older men made it easier for them to convince the younger cadres being trained.

The training included tours during the growing season, when the best farmers could show in their fields how they selected seeds, how they weeded, etc. Some of the trainees' initial reaction was: 'What is new? This is what we have always been doing.' It took much discussion to bring about the attitudinal change that it was wise to recognize, accept and promote the best of what is local.

Phase of widespread SWC

The TPLF leaders and the cadres from the different parts of Tigray convened once a year for Farmers' Conferences to express their visions of development, to assess how far they had come in realizing them and to plan for the coming year. During these meetings, various topics were discussed, such as the *baito* system and legal matters, but also agriculture and land use. The problem of land degradation came up as a crucial issue. It became clear to the DoA and the TPLF leadership that significant interventions to address this problem could not be made by individuals on their own. Up to 1982, attention to SWC had been limited to what could be done in individual farmers' fields. The regular reviews of the extension approach eventually led to a decision by the DoA and the TPLF leadership to give top priority in the extension activities to environmental rehabilitation through widespread SWC. Not having much experience with SWC measures beyond the individual plot or farm, they decided to test large-scale SWC approaches in pilot areas, looking at both the techniques and the associated organization of the community. Two different tests were set up:

1 Large-scale water-harvesting techniques at Shewata near Tembien, an area of poor rainfall in terms of both amount and distribution.
2 Physical structures (simple stone terraces) in areas of adequate rainfall in relative terms; this was done at four sites: Edaga Arbi and Nebelet in Central Tigray, Lalalay Adiabo in Western Tigray and near Hagare Salam in Southern/Eastern Tigray (which was considered as one zone at the time).

In both tests, the approach was through community participation, involving both men and women, using Food-for-Work and handtools supplied through REST at some of the sites. In the water-harvesting area, there was a two- to threefold increase in yields in the two trial years. In the area with the simple stone terraces, the increment in yield was not as great.

After the 1984–85 drought, the TPLF leadership and the DoA evaluated the pilot activities. The water-harvesting was very labour-intensive, demanding about 1000 man days per hectare to construct the large bench terraces and to dig the main and feeder canals. A high level of engineering knowledge was required; this absorbed almost all of the DoA's advisory capacities. The TPLF leaders and the DoA decided to promote simple stone terracing because they regarded this as an activity that communities could manage on their own with relatively little external guidance. Thus, the policy decision was to focus on a SWC technology that could be spread widely more easily and leave pending, until more capacity could be created, the more impressive water-harvesting schemes with which only a very limited number of people could be reached quickly.

A massive terracing campaign was launched in 1986–87 throughout all the liberated areas where land had been distributed equitably among all farming households. An important consideration was how to organize community participation as the SWC work required combined efforts, with good local leadership and follow-up maintenance. Many farmers had to be trained from the various communities so that the communities themselves could take charge of the activities. Some high school leavers were trained by people from Oxfam UK and REST to be DAs specialized in SWC who could advise the communities. The *baitos* were responsible for coordinating the SWC activities; this often involved lengthy community discussions. The farmers were expected to terrace only within their own *kushets* (hamlets) so that they could benefit directly from their labour inputs.

This campaign resulted in an impressive pattern of stone walls along the contours of farmed and non-farmed land throughout the more accessible parts of Tigray. In some areas, these SWC measures led to improved flow of springs, reduced rates of soil erosion and increases in yield (eg up to 57 per cent higher wheat yield) recorded from terraced areas, compared with farms without terraces (Yibabe et al, 1997).

Initially, the *baitos* decided that each person who was capable of working (ie all except pregnant women, children, the elderly and the sick) should contribute three months of labour per year to the SWC work. This was generally accepted as a contribution to the struggle. After liberation, however, the *baitos* reduced the contribution of free labour to only one month per year. Still later, it was reduced to 20 days during the slack period for cropping activities; this agreement for mass mobilization to carry out SWC work is still valid today. The SWC work is governed by local bylaws agreed after discussion within the community at *tabia* level and violators (eg people who are absent from the work) are punished by the community.

The community members, with technical support from the DAs, decided where new structures would be placed on slopes or farmland and which exist-ing structures would be maintained. Many *baito* members and DAs tried to integrate local farmers' knowledge into the recommendations for aligning the SWC structures, for selecting construction materials and for managing soil fertility. Alongside this mass mobilization in SWC, the extension approach of promoting the best farming practices was continued.

Besides these attempts to conserve soil and water for agriculture, the DoA also gave attention to other natural resources, such as trees and wildlife, including birds. The *baitos* drew up and the communities approved local by-laws (*serit*) according to which certain areas were closed to allow natural rehabilitation. Each *baito* and community selected the areas and decided on the duration of closure. The DAs gave technical advice to the *baito* which was the only authority that could give permission to cut trees for construction. The grass in the closure was communal property. There were no detailed manage-ment plans; in most cases, the community members cut the grass jointly and distributed it among the households according to locally agreed principles. This approach has remained the basis for area enclosure to this day.

Attention to women in agriculture

It was during the extension phase focused on widespread SWC that the DoA started to give more attention to women in agriculture. There were many female-headed households while the men were fighting and the women were struggling to continue farming. Because the Tigrayan culture made it difficult for men to speak directly with women, female DAs had to be trained; these were female fighters from a farming background. Through a similar type of consultation as at the outset of agricultural extension under the TPLF, the DoA assessed women's problems related to agriculture and natural resource use and considered what could be done to improve their living conditions. It was decided to focus on improved traditional stoves, horticulture, creating awareness of the importance of women's participation in SWC activities and issues of women's rights.

Some women were trained to plough with oxen, although this was never a main thrust. After some time, it was realized that this idea had been overambi-tious as there were strong sociocultural pressures against it, especially from the priests in the Orthodox Church. Promoting ox ploughing clearly could not be a major solution for female-headed households and the women who ploughed had to be strong-willed to bear community criticism. Today, a small number of women can be observed in the fields doing their own ploughing, and ISWC-Ethiopia has even recorded a case in which a woman has taken the initiative to pair a donkey with an ox for ploughing (see Plate 15.1). The earlier support by the TPLF undoubtedly gave these women some of their courage to innovate.

The priests played an important role in the agricultural development activities. At Priests' Conferences organized by the TPLF, it was discussed what religious traditions must be strictly followed according to the Church and what traditions were additions from the local culture. These discussions made it possible to reduce the number of holidays when farmers traditionally were not allowed to work on the land from 126 to 62 days in a year.

Opening up to introduced technologies

A third phase in agricultural extension grew out of discussions within the DoA and the TPLF leadership as to whether technologies from other parts of Ethiopia and Africa, eg Sudan, should be introduced. This was in the latter half of the 1980s, after a period of major drought. From the outset, the TPLF had not believed that it could eliminate poverty completely by disseminating the best farming practices. Using local resources and capacities to the fullest could bring Tigray to a higher level, but still more should be possible. By the mid-1980s, many areas had been liberated, local capacities had been strengthened, there were many more professionals in agricultural extension, and the DAs had made good headway in promoting the best farming practices. While continuing this approach, the DoA thought it necessary to consider what more could be done by the best farmers and by other farmers who had reached the same level. Opportunities and demands were opening up to also integrate some introduced technologies.

The DoA studied what had been done elsewhere in Ethiopia and in neighbouring countries. Realizing that the sustained supply of fertilizer and other external inputs could not be ensured, it concentrated on what could be brought in as a prototype and then reproduced in the region, such as improved farm implements (eg mould-board ploughs) and drought-resistant sorghum seed for local multiplication. On trial plots, fighters trained as technicians tested technologies that promised to improve the existing farming system. Farmers were brought to visit these plots and to give their views on the new technologies. The DoA helped local blacksmiths to produce improved ploughshares by providing them with prototypes of equipment imported through REST.

Changes in agricultural extension after liberation

After the fall of the Derg in 1991, the TPLF brought in professionals from Addis Ababa to evaluate what had been done in agricultural extension in Tigray, to compare this with experience in the country as a whole, and to see how the local and external technologies could be integrated without losing the best of local practice and tradition. A team from the highest level of the MoA made an assessment over about three months. Based on the evaluators' advice, the DoA decided which aspects of the TPLF-initiated extension system should be maintained and what should be added. Some professionals in the DoA had visions that Tigray would be integrated quickly into a Western economy, with a system of agriculture based on 'modern' introduced technologies, now that the region no longer had to depend primarily on local resources. Others who had thought that victory would never be achieved within their lifetime still regarded modern technologies as very remote. Some people saw a need to be realistic about what could be achieved quickly. It was decided eventually to maintain the promotion of best farming practices as a principle within agricultural extension, but also to integrate some introduced technologies.

After liberation, a decentralized system of government was set up in Ethiopia. Within agricultural extension, the role of the central MoA was defined as regulatory, laying out general guidelines, while the responsibility for implementing extension activities was given to the regional Bureau of Agriculture. Each region has the freedom to modify the approaches outlined at national level to suit its particular situation and needs. The national framework initially remained in the transfer-of-technology mode, formulating packages of technologies as blanket recommendations without considering differences between agroecological zones. The packages were disseminated through various extension systems, including Training-and-Visit. Comprehensive Packages and Minimum Packages were recommended for a small number of cereal crops, including standard applications of fertilizers based on the red and black colours of Ethiopian soils, irrespective of other factors that limit agricultural productivity such as the availability of water. There were frequent reviews involving the MoA and the Ethiopian Agricultural Research Organization (formerly the Institute of Agricultural Research) to assess the functioning and efficiency of the research and extension system, and new programmes were introduced in various efforts to improve it. However, there is still a gap in technology support for the dryland areas.

A VISION OF PARTICIPATORY AGRICULTURAL DEVELOPMENT

Mitiku Haile

It was into this situation of tensions between different visions of agriculture and different approaches to extension in Tigray that the ISWC 2 programme entered in 1997. During the first phase of ISWC in the late 1980s, studies had been made about traditional techniques and systems of SWC in various regions of Ethiopia, but the political situation at that time had not permitted the participation of Tigray. For the second phase, the ISWC consortium, being aware of the history of SWC in northern Ethiopia, deliberately sought contacts in Tigray. The approach of the programme – seeking local innovators and innovations, and using these as entry points for participatory technology development and dissemination – met with resonance among some people in Tigray who had experienced agricultural extension during the struggle for liberation.

A programme Steering Committee with a common aim

ISWC-Ethiopia brought together people from organizations in Tigray who already had a vision of participatory agricultural development based on both indigenous and external knowledge, and who were already on their way towards making this vision a reality. Some scientists at Mekelle University (at that time, Mekelle University College) and Mekelle Research Centre had studied aspects of indigenous knowledge related to farming and natural resource management, and had been involved in participatory on-farm research in seed selection, pest control and land management. A UK-supported NGO, FARM-Africa, had been facilitating farmer participatory research in Central Tigray, linking experimenting farmers with individual scientists who could support the local experimentation process. A local NGO, Adigrat Diocese Development Action, had been basing its community-led development in SWC on the innovations and existing skills of local farmers (Hagos and Asfaha, 1997).

However, these were all fairly small-scale and isolated initiatives that had been operating parallel to a strong thrust by the Tigray Bureau of Agriculture and Natural Resources to transfer introduced technologies through the Sasakawa Global 2000 scheme. During the regular – previously quarterly, now semi-annual – *gumgum* (evaluation) meetings between regional government officials and the NGOs operating in Tigray to review their progress and coordinate planning, the BoANR had become aware of what was happening in these projects and was interested in seeing how the approaches could be integrated into its own extension system.

The people from the various government agencies and NGOs in Tigray who agreed to become members of the ISWC-Ethiopia Steering Committee (SC) were ready to pursue a common aim. The programme was designed to contribute to better land husbandry by recognizing and stimulating local innovativeness, and providing research and extension support to this. During the interactions of people from the government agencies and NGOs, especially in workshops and SC meetings, but also during joint documentation, such as in writing joint papers, the vision of what participatory agricultural development could be has become more clear.

Extension can become a process of PTD

The very process of extension can become a process of PTD. If DAs can recognize local innovation and informal experimentation, both they and the farmers will become more confident in the farmers' ability to experiment. They will see how farmers have always been combining their own knowledge and experience with the ideas that come from outside and, if allowed or even encouraged to do so, can select new technologies that are suitable to improve their farming systems or can adapt new technologies to make them suitable. The DAs can then support this process by suggesting promising technologies for farmers to test and adapt. They can bring together groups of farmers in a village area to evaluate the local innovations and adaptations of introduced technologies. This can stimulate ideas for improving the technologies and help to expand them within the village area. The DAs can link up farmers in one village who have already developed or adapted technologies in land husbandry with farmers in other villages or other zones who are facing similar problems and might find the technologies interesting to test on their own land. This approach builds on the 'best farming practices' approach under the TPLF.

This process of PTD in extension can be strengthened still further by bringing in scientists who are specialized in the topics being addressed by the farmers' experiments. The scientists can help to explain the farmers' results. This may demand experimentation under more controlled conditions. The scientists can add value to the indigenous knowledge by supporting farmer-led experiments that are designed to help all partners to gain a better understanding of local innovations and they can stimulate farmers to explore ideas that could further improve these locally developed technologies.

A key activity in a PTD approach to extension is monitoring and evaluation in quite a different way from what has been done in the past because the central actors in the evaluation are the farmers. When extension involves stimulating farmers to experiment with new ideas, either from innovative fellow farmers or from outside sources, the farmers' decisions about what is suitable for them need to be based on their own assessment. Farmers, of course, have always assessed technologies in an informal way, but the DAs can support a process of more systematic monitoring of farmers' experiments and evaluation of the results, and can encourage the involvement of a larger number of farmers in these assessments. These joint evaluations by experimenting and the involvement of other farmers then become a key means of disseminating information about what has worked and what has not, and the reasons why.

Strategies to pursue this vision

The SC chose to build up interest in, and support for, this PTD approach by addressing middle- to high-level staff in government institutions and administrations and in NGOs through activities such as networkshops, field visits, publications, contributions to conferences and seeking dialogue. These activities were designed to raise awareness on a broad front about the creativity of farmers and thus influence attitudes of people at all levels in extension, research, education and policy-making.

The programme decided to promote farmer innovation and PTD initially through the grassroots extension workers who, because they interact closely with farmers, are likely to be able to develop most quickly an appreciation of farmers' capabilities. A series of training sessions were offered to DAs in all four zones of Tigray. In the first round of training, the DAs were introduced to the concepts of farmer innovation and PTD, and were trained in methods of PRA to help farmers and extension staff to communicate better with each other about the current situation, problems and potentials. The training methodology was designed to stimulate discussion and mutual learning by the DAs about their own way of working and alternative approaches to extension. The DAs were encouraged to identify farmer innovators in their working areas, using a guideline in Tigrigna proposed by the ISWC-Ethiopia coordinators.

The DAs at grassroots level as well as BoANR staff and government administrations at *wereda*, zonal and regional level were also stimulated to recognize farmer innovation by means of an annual contest to honour top innovators. The discussion of criteria for selecting award winners has sharpened the awareness of the differences between local technology development and the adaptation of introduced technologies on the one hand, and unquestioning adoption of introduced technologies on the other. In addition, some BoANR staff members have been involved in accompanying or hosting field trips and travelling seminars to visit farmer innovators. The training workshops, the contest and the field visits have given the DAs numerous opportunities to learn of innovations in other areas that may be worth testing by farmers in their own working areas.

During a second round of training sessions, the DAs shared their observations directly with each other. They then reflected on the role of farmer innovators in agricultural development and the implications of this for their own extension work. They were introduced to different types of agricultural experimentation, ranging from informal experimentation by farmers without external support to highly formalized on-station research. Starting with draft guidelines for characterizing informal experiments by farmers, the DAs drew up their own guidelines to document farmer experimentation. After completion of this assignment, a third round of training sessions will give them a chance to present and discuss what they observed. On this basis, they can prepare for supporting the existing informal experimentation processes, primarily by providing promising ideas for the farmers to test ('feeding' farmer experimentation) and bringing farmers together to evaluate the results.

Staff members of BoANR at *wereda* level have been officially assigned the task of organizing workshops of farmers from several village areas to assess local innovations and introduced technologies (see Chapter 18). The intention is that, through their reflection on farmer experimentation, the DAs will expand these farmers' workshops into planning meetings for farmer-led experimentation with promising local or introduced techniques and will facilitate similar processes at the level of hamlets or farmer groups.

A further strategy to incorporate the promotion of farmer innovation and experimentation into the formal extension system is through the teaching activities of the university. Many of the diploma and degree students will later be employed by the BoANR or by NGOs and development projects in Tigray. The university also gives in-service training to people already working in agricultural development. The farmer innovation, PRA and PTD approaches have been incorporated into the curricula (see Chapter 31), and both regular and in-service students receive related field assignments.

Signs of impact within the extension service

During the course of their regular work, about 60 of the 100 DAs who were involved in ISWC-Ethiopia workshops or in-service training have documented more than 100 local innovations and brought or sent the descriptions back to the university. In response to a more recent assignment, 15 DAs have thus far documented examples of farmers' informal experimentation. Many DAs have expressed an interest in continuing this type of work. Sixteen extension supervisors at *wereda* level were so attracted by the idea that they themselves have documented local farmer innovation and experimentation.

Some activities introduced by ISWC-Ethiopia have now become regular extension activities of the BoANR, eg giving awards to top innovators and organizing farmers' workshops to assess the potentials and constraints of local innovations.

In the series of DA training sessions on farmer innovation and PTD, some participants have become quite outspoken in their critique of conventional extension activities and have requested that their concerns be conveyed to higher policy levels. Some DAs in the Southern region have openly expressed their doubts about the appropriateness of Global 2000 packages and successfully demanded a reduction in the target number of farmers to be drawn into the associated credit scheme.

There are signs of somewhat more flexibility in extension methods, moving away from the pure transfer-of-technology approach. A few cases are now being reported in which DAs are working together with farmers in experimentation, eg comparing different rates of fertilizer application to find out what is locally appropriate instead of sticking to a standard recommendation.

The seeds for the changes that were occurring at regional policy level with regard to the Global 2000 scheme were sown before the ISWC 2 programme started, as doubts about the appropriateness of the technology packages for the drylands of Tigray had already been raised. The programme, however, has heated up the debate and has offered another approach to agricultural development at a time when alternatives to Global 2000 are being sought. The Integrated Food Security Desk of Tigray has now entered into collaboration with ISWC-Ethiopia to disseminate improved low-external-input techniques developed by local innovators as options for poor farming households in the 16 most drought-prone *weredas* in the region. This is being done primarily through farmer-to-farmer extension.

In a recent strategic document presenting the Five Year Plan for agricultural development in Tigray region, concrete references are made to integrating indigenous knowledge into the process of intensifying and diversifying agricultural production. In the strategic plans for regional research, PTD is included as an approach for establishing farmer–extension–research linkages.

Challenges ahead

Better feedback is needed between farmer experimentation supported by DAs and formal agricultural research. The farmers' workshops and travelling seminars give farmers an opportunity to assess not only local innovations, but also local adaptations to introduced technologies. There is still a need, however, for better mechanisms to communicate their assessments to scientists. As only a small number (25) of agricultural scientists working in Tigray have taken part in exposure visits to farmer innovators during workshops, only 10 have made deeper studies of local innovations and an even smaller number (4) have been involved, thus far, in supporting farmer experimentation. Thus, there is still insufficient direct linkage between formal research on the one hand and local innovation and PTD by farmers and DAs at field level on the other. A larger number of scientists could receive feedback from the field at least indirectly through documentation by ISWC-Ethiopia on farmer innovation and PTD, but the programme has not yet been able to access sufficient funds to print and distribute its documents widely within Tigray, let alone within Ethiopia.

A question that the programme has almost completely ignored thus far is how to incorporate artisans and stakeholders from industry – suppliers of agricultural inputs (beehives, seed, tools, fertilizer, etc) and processors of raw materials from agriculture – into the PTD process supported by the DAs.

Future training sessions for DAs will need to address the question of how to involve farmers in assessing the impacts of local innovations on improving their livelihoods. This is already possible within the framework of the farmers' workshops and travelling seminars, but needs more targeted attention. Moreover, there is a need to strengthen the capacities of *baito* members at village level to facilitate local monitoring and evaluation of farmers' innovations, and of the process of promoting farmer innovation.

There is support already for promoting farmer innovation and experimentation at the regional policy level and enthusiasm is growing among DAs working at field level who have recognized local innovation. In many cases, however, there is still insufficient understanding of the approach at intermediate management levels in BoANR. This means that most DAs are still under considerable pressure to continue a transfer-of-technology approach, trying to convince farmers to take up credit to adopt Global 2000 technology packages, and are documenting farmer innovation and encouraging experimentation only as a sideline to their main work. Changes in the working programmes, types of targets and criteria for evaluating DAs' work still have to be made to give support to a PTD approach to extension.

CONCLUDING REMARKS

Mitiku Haile and Berhane Hailu

The TPLF was itself forced to innovate in the face of a lack of well-trained field staff and extremely limited access to external sources of information and other inputs. The approach it developed was a streamlined form of extension that stimulated farmers to learn from each other. The promotion of best farming practices generated enthusiasm to innovate among some of the people who were identified later by ISWC-Ethiopia as farmer innovators. Particularly the women referred explicitly to the encouragement they had received from the TPLF.

The farmer innovation approach of ISWC 2 struck a strong chord of recognition within the BoANR, especially among those older staff members who had experienced the struggle for liberation from the Derg regime. The similarities underlined the fact that the 'best farming practices' approach that had been taken largely out of necessity during the war had been, indeed, a progressive approach to developing smallholder agriculture. ISWC-Ethiopia thus found itself in the fortunate position of not having to fight the current; rather, it could ride on the wave of a popular movement. The pride in the knowledge, creativity and hard work of good farmers during the struggle has been carried through to the present day in Tigray, and is reinforced by the concept introduced by ISWC of recognizing and encouraging local innovation in land husbandry. The entry into PTD – bringing together the best of local knowledge and the best of modern scientific knowledge to address the problems at hand – continues in the tradition of the TPLF to draw selectively and in a carefully considered way on external resources that enhance local capacities. There can be no denying that the struggle for liberation was successful because of the enthusiasm, pride and commitment that were generated in the largely rural population of Tigray. These are the same strengths that are needed in the struggle to attain food security, and are being generated through the farmer innovation approach to PTD.

REFERENCES

Hagos W and Asfaha Z (1997) 'How to gain from erosion: catch the soil', *ILEIA Newsletter*, vol 13, no 2, pp16–18

Yibabe T, Esser, K and Mitiku H (1997) 'Impacts of conservation bunds on crop yields in Degua Tembien, northern Ethiopia', in *Proceedings of the 8th Annual Conference of the Crop Science Society of Ethiopia, 26–27 February 1997*, CSSE, Addis Ababa, pp86–102

30

Impact of the farmer innovation approach on the attitudes of stakeholders in agricultural development in Tunisia

Noureddine Nasr*

The farmer innovation approach is coordinated in Tunisia by a team from the Institut des Régions Arides, which has mobilized different partners to identify and analyse local innovations, to engage in joint experimentation with farmers, to organize exchange visits and to use mass media to promote the dissemination of farmers' innovations. When ISWC 2 started in Tunisia in August 1997, it met with some resistance from within the formal research and development agencies because the approach was entirely new to them. After three years, the resistance has disappeared and the perceptions and attitudes of all stakeholders have changed profoundly.

FROM SCEPTICISM TO CONVICTION

Attitudes of scientists

Most researchers at the IRA were initially sceptical about the farmer innovation approach. A common reaction was: 'Why go to farmers to get inspiration? Have researchers lost their own creativity?' Although the national research policy stipulates that research should contribute to solving the problems of farmers in Tunisia, many researchers appear to be more interested in producing scientific publications to advance their scientific careers. The first national

---

* Noureddine Nasr is an agronomist and geographer with IRA Gabès and PTD trainer in the ISWC-Tunisia programme

Credit: Chris Reij

Plate 30.1 *Meeting of the ISWC team at the Institut des Régions Arides (Tunisia)*

training workshop on PRA/PTD held in October 1997 in IRA Médenine opened the eyes of some researchers and development agents (see Chapter 1). Before the workshop started, even one of the ISWC-Tunisia team members from IRA said that he did not believe in farmer innovation. However, after he had participated in the fieldwork during the training, he completely reversed his view.

Among the other IRA scientists, the start of the radio programme on *Agriculture and Innovation* (see Chapter 27) was a turning point for scaling up this reversal. It contributed to making farmer innovation respectable in their eyes and led to increased interest in participatory approaches to research.

Attitudes of development agents

Also some agricultural technicians – as extension or development agents are called in Tunisia – were initially sceptical. During information sessions about the ISWC 2 programme at the headquarters of regional branches of the Ministry of Agriculture, remarks such as 'Only scientists can develop new technologies' were heard. What compounded the difficulties was that scientists and development agents were not used to working together. The latter generally felt that the researchers had little to offer to them that was of practical relevance for their daily work. The ISWC 2 programme managed to close this gap between research and development by involving all stakeholders in joint experimentation and in disseminating its results. Some examples can illustrate this.

LEARNING BY JOINTLY SOLVING FARMERS' PROBLEMS

Example of water economy in greenhouses

In Mazraa Ben Slama (Gabès region), farmers grow vegetables in greenhouses, but have no reliable source of water. The sole well in this area is managed by the National Society for the Exploitation and the Distribution of Water (SONEDE). The farmers buy water from this well and use cisterns drawn by tractor or animals to bring the water to their greenhouses. The costs of water for irrigating crops in one greenhouse during one season are in the order of 600–900 Tunisian dinars. Extrapolated to a hectare, this is equivalent to 12,000–18,000 dinars (US$8500–11,500). Thus, the cost of water for these greenhouses is excessively high when compared to irrigation schemes in Tunisia where water costs about 500 dinars per hectare.

In 1998, development agents in a pilot extension project based in Mareth had identified the farmers in Mazraa Ben Slama and, in particular, the leader of the group (Khalifa Dadi) as farmers interested in doing joint experimentation on water use in greenhouses. The development agents collaborated with the farmers and scientists in the experimentation. The major objective was to find ways to economize on the use of water and thus to reduce costs, while at least maintaining the yields. The IRA and the National Farmers' Union, which is in charge of the pilot extension programme, signed an agreement to collaborate. The Union Tunisienne de l'Agriculture et de la Pêche (Tunisian Union for Agriculture and Fisheries) made 3500 dinars available to the project team in Mareth to support farmers who wanted to experiment with the technologies that were developed in a PTD process by ISWC-Tunisia with innovators in the region of Béni Khédache: gravity-based irrigation from rain-fed cisterns and stone pockets for growing trees.

After two years of work, Khalifa Dadi stated in a national workshop on farmer innovation (5–6 September 2000 in Gabès) that he had reduced the use of irrigation water by two-thirds and that his water costs were correspondingly lower. Speaking on behalf of the entire group of farmers from Mazraa Ben Slama, he said: 'This is only one step, but we are convinced that we will do even better if we, like the other farmers, could benefit from a 50 per cent subsidy for economic use of water.' The region of Mazraa Ben Slama is not entitled to such government subsidies because it is an area of rain-fed farming. The regional president of the National Farmers' Union who attended the national workshop and arranged a bus to visit farmer innovators, was very convinced by Khalifa Dadi's presentation. Within weeks after the workshop, the MoA office in Gabès made a survey among the farmers with greenhouses in Mazraa Ben Slama and decided that they would be entitled to subsidies to stimulate the use of water-saving technologies. Several farmers who had abandoned greenhouses because of the high costs of water are now interested in starting to use them again.

Example of water harvesting, storage and distribution

In 1999 the development team of the MoA branch in Gafsa region identified an innovator by the name of Abbès Sandi who had built a small concrete checkdam in a dry streambed. The water stored behind the dam allowed him to irrigate peaches, apples, pears and other fruit trees in a marginal mountainous area with about 150mm annual (winter) rainfall. Another farmer in the Gafsa region, Amor Bou Aoun, who had already been trying out various techniques of SWC, including tree planting, to make better use of marginal land, had met Abbès when they were both making a presentation on the regional radio of Gafsa (see Chapter 27).

Based on what he heard from Abbès, Amor formulated a request to the ISWC-Tunisia team to help him build a similar dam in the foothills of the mountain range that provides run-off for his crops. During heavy rainstorms, however, this run-off sometimes causes great damage to his fields. The scientists from the IRA, the SWC specialists from the regional MoA office and the farmer innovator jointly selected the site for the dam and agreed on a division of tasks. The dam was funded mainly by the MoA. ISWC-Tunisia and MoA specialists suggested several new techniques for channelling water from the dam and distributing it over the fields, and developed a proposal for making agricultural use of 350ha of marginal land below the dam. When this was approved by the regional MoA office, it was the first time that the Tunisian government subsidized a project based on harvesting rainwater in a region with only 140mm annual rainfall. The SWC Department in Gafsa has now received several more requests from farmers in the same region and also from elsewhere to construct such dams. If the regional MoA office judges the dam and the system for storing and distributing the water over Amor's fields to be successful, then government funding for subsidizing the construction of similar dams will not be a constraint (Chahbani and Nasr, 1999).

These are two examples of good cooperation between researchers, development agents and farmers which contributed to creating mutual confidence and appreciation.

GROWING ENTHUSIASM

Attitudes of farmer innovators

As a result of their involvement in various activities supported by ISWC-Tunisia (exchange visits, joint experiments, radio broadcasts, workshop presentations, visits by policy-makers and foreign guests), the attitudes of some local innovators have changed remarkably. They no longer feel isolated and ignored, and they have discovered their own creative potential. Several innovators began to experiment more deliberately. They began to analyse what they were doing and the reasons for this so that they could better present their activities and findings to the other partners.

In September 2000, over 100 people came together for a two-day national workshop on farmer innovation in Tunisia. For several male and female innovators, this was the first time that they had ever presented their innovations to such a big audience and in such a formal setting, yet they managed to do so without any problems. Once they had had the experience of being appreciated by scientists and development agents, they felt 'liberated' and completely lost their inhibitions to speak in public. Some have developed a sense of competition with other farmer innovators and with formal researchers. 'What you can do, I can do better.' Béchir Nasri, who has become an avid inventor since the start of the programme, tends to develop something new in secret and only when it is finished and working does he show it to others. Competition between him and the scientists has become almost a game, yet there is also mutual support and appreciation and a willingness to share innovations with other farmers.

Several innovators feel the need to document their innovations. Béchir Nasri is also a good example in this respect. Although he speaks only Arabic, he managed to produce a report in both French and English about his innovations. The report is full of colour illustrations based on scanned photographs and is of better quality than what many formal researchers produce. He did this by mobilizing the help of relatives who can write French and English, and he used his own money to make copies of the report.

Attitudes of policy-makers

The support of the Director General (DG) of the IRA has been vital. When high-ranking policy-makers in research and education or even ministers visit the institute, the DG always tries to include a visit to farmer innovators in their agenda. He does the same with visitors from abroad, eg from the International Centre for Agricultural Research in Dry Areas (ICARDA), the United Nations Development Programme (UNDP), the Global Mechanism of the Convention to Combat Desertification, the Observatory for the Sahara and the Sahel, and the United Nations University. The DG closely monitors the activities of the team and, whenever he participates in national or international seminars, he always draws attention to experience in central and southern Tunisia with the farmer innovation approach. Whenever a national seminar is organized, he invites the ISWC-Tunisia team to represent the IRA and to present its activities. This is quite a change from how he perceived this approach when the programme started. When the international coordinator of ISWC 2 recently visited the institute, the DG admitted that, when the approach was first presented to him in 1997, he liked it but he did not believe it would work. Now he is a staunch supporter of the programme.

The advocacy by the DG of the IRA and by the ISWC-Tunisia team awakened the interest of the former Deputy Minister for Research and Technology who started to keep in touch with the programme and to encourage the team. He called regularly from Tunis to the institute in Médenine and inquired about progress made. In 1999 he recommended that a national workshop on farmer innovation in Tunisia be organized and insisted that the programme should be expanded to other parts of the country. This backing from national level also helped to legitimize the farmer innovation approach in the eyes of the other scientists in the IRA.

This change in attitude towards the approach was also stimulated by the fact that at least 30 articles were published in the Tunisian journals and weeklies about farmer innovation, both in Arabic and in French, as well as in international meetings (eg Nasr et al, 1999). Moreover, articles about the work appeared in the international English-language *ILEIA Newsletter* published by the Netherlands-based Information Centre for Low-External-Input and Sustainable Agriculture. This also appeared in a French version *Promouvoir l'Innovation Paysanne* (Promoting Farmer Innovation) which was widely distributed within government offices and agencies in Tunisia. In addition, the regional radios of Tataouine and Gafsa have systematically covered workshops, seminars and exchange visits of the ISWC 2 programme, and the exchanges between farmer innovators, development agents and scientists on Gafsa regional radio has been institutionalized (see Chapter 27).

FINAL REMARK

After three years of working together to solve farmers' problems and disseminate farmers' accomplishments, the barriers that existed between scientists, development agents and farmer innovators have been removed and the attitudes of all stakeholders have changed. However, the number of scientists and development agents who are actively involved in the ISWC-Tunisia programme is relatively small and therefore the basis is still fragile. The major challenge is to institutionalize the farmer innovation methodology in the national extension, research and education systems. We still have a long road to travel, but the first signs are promising.

REFERENCES

Chahbani, B and Nasr, N (1999) *Le développement participatif de technologies basé sur les innovations des hommes et des femmes en agriculture et zones arides*, IRA Médenine/CDCS, Vrije Universiteit, Amsterdam
Nasr, N, Chahbani, B and Reij, C (1999) 'Farmer innovators in land husbandry in central and southern Tunisia', Tenth ISCO Conference, 23–29 May 1999, Indiana, USA

31

Learning for sustainability: incorporating participatory approaches into education for rural development in Ethiopia and Tanzania

*O T Kibwana, Mitiku Haile and Firew Tegegne**

An approach to rural development through the promotion of local innovation will have a future only if college and university students are introduced to it and learn that this is an acceptable and effective way of working with farmers. The students will be the future researchers, extension agents, teachers and policy-makers in rural development. ISWC 2 collaborates with educationists in order to incorporate the concepts of farmer innovation and PTD into their teaching and research activities. In two of the eight countries in the programme, the lead agencies are institutions of higher learning: Mekelle University in Ethiopia and the Cooperative College Moshi in Tanzania. This chapter describes efforts in these two countries to influence teaching content and methods, and outlines the future challenges.

DEVELOPMENT-ORIENTED EDUCATION AT MEKELLE UNIVERSITY

In Tigray region in northern Ethiopia, Mekelle University was established, originally as a college, with the mandate to support dryland development.

* O T Kibwana is head of the Pilot Projects and Experimentation Department of Cooperative College Moshi, and coordinates the ISWC programme in Tanzania; Mitiku Haile is president of Mekelle University and joint coordinator of ISWC-Ethiopia, and Firew Tegegne is coordinator of the Practical Attachment Programme at Mekelle University

Since the beginning, it gave attention to farmers' knowledge. This was one reason why it became the lead agency in the second (ie farmer innovation and PTD) phase of ISWC. One of the programme coordinators is a soil scientist who had already studied indigenous soil classification systems, the other is a plant breeder with prior experience in farmer participatory research. As the former was dean of the college (now, president of the university), he could play an important role in further developing the practice-oriented system of agricultural education in Mekelle. A key feature of this system that has facilitated incorporation of an approach of learning from and with farmers is the Practical Attachment Programme (PAP). The topics and methods central to ISWC 2 are finding their way into the university courses and teaching approaches beyond the PAP. ISWC 2 has also influenced the concept and methods of agricultural research applied by MU staff and students.

The influence of ISWC-Ethiopia on the Practical Attachment Programme

Background

Before Mekelle College of Dryland Agriculture and Natural Resources (MCDANR) was set up in late 1993, almost all the Ethiopian curricula of higher learning in agriculture were reviewed. It was found that all lacked practical training. Students could graduate in agriculture without ever have set foot on a smallholder's farm. The founders of MCDANR wanted to bring about change in agriculture; this necessitated bringing about change in agricultural education so that the graduates would be able to make a real contribution in the field. For this reason, MCDANR placed great emphasis on courses related to agricultural practice. The aim was to produce graduates who understand the rural situation and can support farmers in solving practical problems. As the Ethiopian Ministry of Education (MoE) was hesitant initially to invest funds in launching this new approach, MCDANR approached the Norwegian Agency for Development (NORAD) with its proposal. NORAD understood the importance of practical education and approved a three-year trial period. Thus, with NORAD's support, the PAP could be set up at Mekelle College in 1996, starting in the Faculty of Dryland Agriculture.

The PAP cycle

The students go on practical attachment in the summer after their third year of BSc studies. By that time, they have covered most of the basic courses; in their fourth and final year, they will be taking more applied courses. The PAP gives them an opportunity to obtain first-hand experience in the field and to try to put into practice the theories they learnt in the classroom. Every year, each of the three departments in the Faculty of Agriculture – namely, Crop Science, Soil and Water Conservation, and Animal and Range Sciences – seeks organizations that will give field placements to its students. The host organizations set the topics on which they would like the students to work. The departments check the relevance of the topics for agricultural education. The students then choose and compete. Each student prepares for the fieldwork and presents a work proposal at an 'exit' workshop for discussion and approval by their supervisors, before they go to the field.

One supervisor is a person from the host organization who is judged by the faculty to be qualified to give technical guidance and to keep up the student's morale during the period in the field. The student also has a supervisor from the university who gives advice in planning the fieldwork, visits the student during the fieldwork, and guides the student in writing and presenting the report. The attachment was initially for six months, but was reduced to four and then to three months as it was difficult to fit in the full four-year programme of studies and to graduate the students on time. Originally, the supervising lecturer visited the student twice in the field but, when the period in the field was shortened, this was reduced to one visit. A system of group supervision has been introduced recently. A group made up of one lecturer from each of the three departments visits all the PAP students in a kind of travelling seminar. This became necessary when an increasing number of students were posted to other parts of Ethiopia (outside Tigray) and a single lecturer, who also gives in-service training during the summer period and might have three PAP students to supervise in three different regions, could not manage to visit them individually.

The students submit a mid-term report to their supervisors. After completing the fieldwork, they write a final report and, three weeks after returning to Mekelle, give an oral presentation about their experiences and findings at a 're-entry' workshop attended by all PAP students, the supervisors and other interested students and staff. These are usually from the departments concerned, especially the third-year students who will do the PAP the following year. The supervisors from the university help the students to edit the final reports and the faculty publishes a collection of abstracts from each batch of students. Copies of the student's report are sent to the host organization and the local office of the BoANR. The entire period of preparing for the PAP (setting the topic, arranging contacts with organizations in the field, arranging supervision), the actual period of practical attachment in the field, and the writing and presentation of the reports is referred to as one 'PAP cycle'. The supervisors from the university and the host organization evaluate the student's fieldwork (each gives 25 per cent of the mark), 20 per cent of the mark is for the oral presentation and 30 per cent for the final report.

Initial experiences with PAP

The first batch of students was sent to NGOs and governmental organizations in Tigray. The initial activities included socioeconomic surveys and studies of development project sites, eg for area enclosure or reforestation. The second batch was sent to places in Tigray as well as to other regions in Ethiopia immediately south and east of Tigray. In the third and fourth cycles, some students were attached to organizations in dryland areas in more distant parts of the country. However, most PAP students (58 per cent) are still attached to organizations in Tigray region.

The initial responses by students and host organizations suggested that, after completing the PAP, students could:

- realize the realities of farming through experiencing the lives of farm communities;
- gain confidence for employment through several months of exposure to development, research and extension activities;
- gain satisfaction from actively participating in rural development endeavours; and
- identify and/or prove themselves to potential employers.

In addition to these benefits for the students, the university benefits by gaining information for the purposes of research and teaching. The host organizations benefit immediately from the work and findings of the PAP students, and benefit later by having access to a pool of potential employees who are better trained for development work.

Over the four cycles completed thus far in the Faculty of Dryland Agriculture, the PAP has gained a good reputation and has been continuously monitored. During the exit and re-entry workshops, seminars and graduate ceremony, many interested parties in Tigray, particularly the BoANR, have proposed improvements. At the end of the trial period in 1999, two professors from other regions of Ethiopia and one professor from Norway evaluated the PAP and found it to be highly relevant for rural development. All graduates who had taken part in the PAP gained employment and are performing well in their work. Some have already secured positions as researchers. When a national workshop on Community-Based Education and Academic Practical Exposure of Students to the World of Work was held on 20–21 March 2000 at Jimma University, the example of PAP in Mekelle was found to be so valuable that the MoE decided to include such practical training in the curricula of all institutions of higher learning in Ethiopia.

Incorporating recognition of local innovation

By the time that ISWC 2 commenced in Tigray in 1997, the Faculty of Dryland Agriculture had completed one cycle of PAP. With few exceptions, the students had been working together with development projects that were still operating according to the conventional transfer-of-technology paradigm. This tended to reinforce their sense of superiority: that students and 'experts' know better than the less literate rural people and that they need to teach modern agriculture to the farmers.

During each of the PAP cycles of 1998, 1999 and 2000, three students – one from each of the departments – out of a total of 137 have done their practical attachment under ISWC-Ethiopia. They worked on the following topics in a mixture of practical participation and observation:

- the impacts of land reform on farmers' innovativeness;
- intensified agriculture and indigenous knowledge: the implications for sustainable agricultural production;
- traditional bench terrace practices on hilly farmland and community-based efforts to prevent soil and water loss;
- community-based management of common property resources (two students);
- community initiatives in agricultural practices;
- women innovators in agriculture; and
- comparison of women and men innovators (two students).

In order to be able to do their work, the PAP students need a subsistence allowance of at least 420 *Birr* (circa $US50) a month to cover food and lodging, transportation and some materials for their fieldwork. ISWC-Ethiopia provides an allowance of 500 *Birr* (circa US$60) per month, while some host organizations provide as much as 900 *Birr* (circa $US110). Thus, there is no great financial incentive for the students to focus on farmer innovation and PTD; they do so out of their own interest. Although the MoE had agreed to institutionalize the PAP as part of higher education throughout Ethiopia, no financial commitment from this source is available. Thus far, the host organizations are obliged to cover these basic expenses. In exceptional cases, if an organization is interested in hosting a student but cannot pay the subsistence allowance, NORAD covers it. Funding for students to continue studying farmers' knowledge, innovation and experimentation under the PAP will have to be sought from other sources when ISWC 2 ceases. This means that institutionalization will be assured only when other public or private organizations become willing to carry these costs. It cannot be expected that students without income could pay their own way.

The PAP students supervised by ISWC-Ethiopia work primarily with innovative farmers or communities. The close communication between ISWC and the farmers ensures that these are involved in determining the PAP topics and collaborate with the students. During their stay in the field, the students not only carry out studies, but they also experience the living and working conditions of the rural people and physically work together with the farming families.

Impact

In 2000, ISWC-Ethiopia and the PAP coordination unit assessed the influence exerted by ISWC concepts and supervision on the PAP. They found that the students who were assigned to discover indigenous knowledge and innovations were highly motivated to search in the literature in MU's well-stocked library. The students compared and married their field observations with academic scientific knowledge. Their presentations during the re-entry workshops gave a good opportunity to disseminate information about indigenous knowledge and promising local innovations to other students, lecturers and the host organizations, both government agencies and NGOs. Now that the PAP students are also posted in other parts of Ethiopia, this information is reaching different corners of the country.

The reports of the students supervised by ISWC-Ethiopia are made available to the BoANR at *woreda* (district) level. The experts and development agents (DAs) communicate the students' findings to the farmers who were involved and to other interested farmers. This feedback was introduced after the first year of PAP on the request of the DAs. The BoANR also uses reports about farmer innovation and experimentation in one village area to inform DAs and farmers in other village areas with similar economic and environmental conditions. It is hoped that this dissemination of information about farmers' ingenuity and initiatives in land husbandry will contribute to development and policy change at district and regional level in Tigray, in other regions and eventually at national level.

After returning from their practical attachment with ISWC-Ethiopia, some students commented, both in their reports and at the re-entry workshops, that the experience relieved them of an 'entrenched superiority complex'. They felt that they had learned a great deal from the rural people who have tremendous local knowledge, often far ahead of the students' knowledge. The students appreciated that many farmers are applying sophisticated and appropriate indigenous farming technologies and that they know the observable environment in great detail. The experience gained by working within the ISWC programme enriched the practical knowledge and skills of the students and changed their attitude towards rural people.

Moreover, staff members have observed that students who have made this experience during the PAP have become more active in questioning, answering and relating their studies to the local context and to specific problems at the grassroots. This makes the learning relationship between students and teachers in the university much more interactive than in the past.

Wider influence of ISWC-Ethiopia on agricultural education

The identification of indigenous knowledge and its dynamics by students and staff has influenced the process of curriculum design, the content (syllabus) and the teaching methods at MU.

Professionals conceptualize the real world in an assumed model that usually separates the 'professionals' from the practitioners. The practical world demands breadth, intelligibility and usefulness, but the academic professionals often have only peripheral knowledge regarding the practical world. Exposure to how rural people manage to be creative under great pressure to survive broadens the knowledge of the academics.

Some of the staff members in Mekelle have received their higher education overseas. They often find it difficult to translate what they have learnt elsewhere into forms that fit the local needs and to identify the local priorities for training, research and development. In addition, the curriculum has been strongly influenced by expatriates from developed countries in temperate

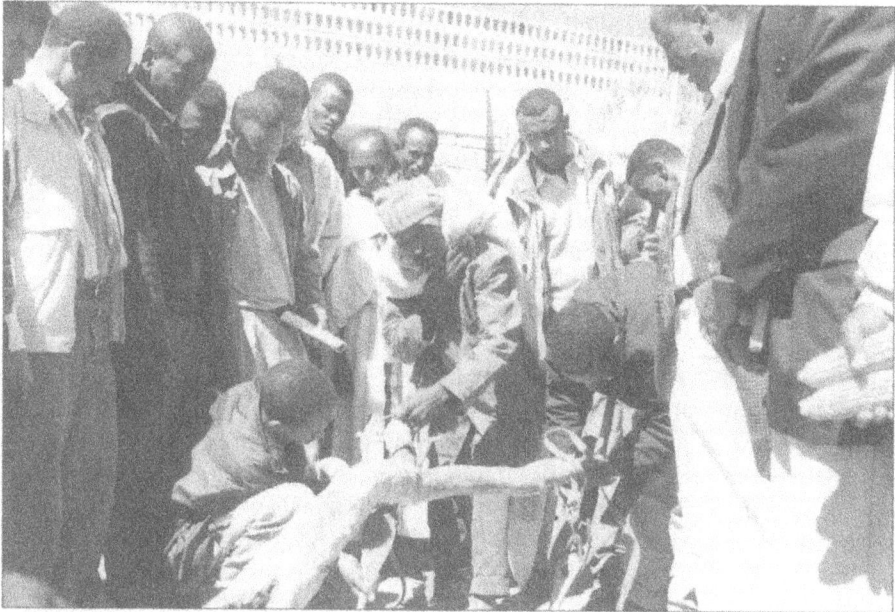

Credit: Teshome Gebrekidan

Plate 31.1 *Farmer innovator in Tigray, Ethiopia, demonstrating an improved plough to students and staff of Mekelle University*

areas. Both the Ethiopians educated overseas and the external advisers who are designing and reviewing the curriculum for teaching and who are identifying problems and setting priorities for research have tended to emphasize what they themselves have been trained in. Moreover, the methods of training have been transferred from the overseas institutions.

The introduction of concepts of farmer innovation and farmer-led experimentation into the PAP and the encouragement also of teaching staff to identify and document local innovations have led to a discernible change from a tradition of learning purely by way of lectures and reading books to a new approach of learning by listening to and working with practitioners. Academic staff members who travel to the field to supervise the PAP students, to comment on their reports and, in some cases, to do follow-up research into what the students discovered are likewise exposed to rural realities. The involvement of students and staff in the ISWC programme has made them more open to recognizing local knowledge and is bridging the divide between this and scientific knowledge. It has also helped the academics to recognize the problems, needs, opportunities and priorities of the farmers. This experience is narrowing the gap between the academic professionals and the practitioners, ie the practising professionals.

Besides the activities in identifying and documenting local innovation, the workshops organized by ISWC-Ethiopia for MU researchers have influenced the content and form of agricultural education. University staff members have participated in workshops on Concepts and Processes of Farmer Innovation (three days), Action-Oriented Research (three days) and Training in PTD (four days). Six workshops of two or three days each were held for DAs and extension supervisors on identifying and documenting farmer innovation and informal experimentation, innovation by women and methods of supporting farmer-led experimentation. All these workshops were facilitated by people from MU, the BoANR and NGOs who had been trained by the ISWC 2 programme in Zimbabwe in 1997 or were subsequently trained by ISWC-Ethiopia. A module on PTD has been incorporated into the course on Research Methods given to all students of agriculture at MU. Both national and international MSc and PhD students are making field studies on farmer innovation and experimentation. In addition, documents produced by ISWC 2 on farmer innovation and PTD are being used as learning materials in various courses, such as agricultural extension, rural sociology, plant breeding and SWC.

Perhaps just as important as the content of the workshops have been the participatory methods used during the workshops. These include brainstorming, working in small groups, sharing and jointly analysing observations and experiences, working with visualization techniques (flip-charts, cards, sheets of paper on the wall on which the participants' contributions during the workshop are written) and incorporating field activities to stimulate learning by doing and experiencing. The workshops have created situations of interactive learning in which all participants can offer what they know and think, can practise what they are learning and can reflect on this experience. Staff members who had been involved in ISWC-Ethiopia workshops are now applying these methods in some teaching situations and workshops outside the ISWC programme.

The influence of ISWC-Ethiopia on agricultural research methods

The ISWC 2 programme centres around research for development. As Mekelle University is the lead agency in Ethiopia, the country has felt an impact not only because of its teaching but also its research activities. In most colleges and universities throughout the world – and Mekelle is no exception – scientists have tended to choose research priorities on the basis of their own values, perceptions and limited experience of rural people's problems and knowledge. The results of this research, if they were conveyed at all to farmers, did not usually lead to sustainable solutions to their problems. ISWC-Ethiopia is playing an important role in bringing farmers and scientists together as partners in seeking solutions.

Twenty-two MU lecturers who also conducted some research and ten scientists from Mekelle Research Centre took part in workshops on participatory research organized by ISWC-Ethiopia. These people are incorporating the philosophy and methods of participatory research into their regular work. For some, the ISWC workshops were not their first exposure to participatory methods, but the opportunities offered by the programme allowed them to apply what they had learnt. The researchers submit proposals to make more detailed studies of innovations discovered by PAP students or DAs and to carry out joint research with farmers. ISWC-Ethiopia provides transport or travel allowances and pays the researchers a sum not exceeding half their salary. This sum, which is the equivalent to $US150 per month, is paid for the duration of the study which may take from three to six months, depending on the subject and the distance of the study area from Mekelle. Thus far, eight MU staff members have studied farmer innovation and seven proposals involving farmer-led experimentation were submitted to ISWC-Ethiopia and have recently commenced.

Among academic and research staff, there is a growing awareness of the importance of involving farmers, community leaders, DAs and scientists of different disciplines to identify problems and set priorities for research. In the past, there were strong institutional barriers in communication between these various stakeholders in agricultural research and development. Activities were often duplicated or overlapping. The participatory approach promoted by ISWC and the practical attachment of students to ISWC have helped to overcome these barriers and are building the necessary forum where the stakeholders can coordinate the planning and implementation of problem-oriented research. ISWC-Ethiopia has deliberately not focused strictly on SWC as its name would imply; it has encompassed all aspects of crop and animal husbandry and natural resource management. The forum is thus laying the foundation for a concerted research effort at the grassroots level to address rural problems in a holistic way.

As mentioned above, ISWC has influenced the content of the course on Research Methods at the university. The course instructor used to deal only with conventional methods, but his involvement with ISWC staff, in-service students, DAs and innovative farmers led him to include PTD and participatory monitoring and evaluation in the course as accepted approaches for mainstream research. He now promotes the involvement of farmers as recorders of technical data of interest to both the farmers and the scientists, and joint analysis of the data and observations. He encourages students to write proposals for this course that pertain to indigenous knowledge, innovation and experimentation. He is a prominent educator who gives the same course at other colleges and universities in Ethiopia. In this way, the ISWC programme is influencing the research methods that are being taught in other institutions of higher learning and, thus, the concepts and attitudes of people who will later join research institutes.

The university has organized interdisciplinary seminars at which the results of research into farmer innovation and PTD were presented and discussed. These seminars included staff from both MU and MRC. The international workshop on farmer innovation in anglophone Africa, which was organized by ISWC 2 in Mekelle in February 2000 and was attended by people from several countries in Africa as well as Europe and America, gave high status to the farmer innovation approach to PTD.

DEVELOPMENT-ORIENTED EDUCATION AT COOPERATIVE COLLEGE MOSHI

In the case of the ISWC 2 programme in Tanzania, the lead agency – Cooperative College Moshi (CCM) – is an institution of higher learning, but does not deal with technical aspects of agriculture. Rather, it was founded with the mandate to support the organizational and business development process of rural cooperatives in Tanzania. The focus of the college's work in the ISWC programme has been on facilitating and coordinating the work of the partner organizations in a transparent way. After all, ISWC is not just about technical innovations in land husbandry, but it is also about institutional collaboration, PTD, monitoring and evaluation, policy development and networking. The influence that the programme is exerting on the contents and methods of teaching in the college is part and parcel of an ongoing learning process about farmer innovation, participatory processes and the totality of the organizational environment that supports sustainable agriculture.

College mandate and structure

Cooperative College Moshi was established in 1963 by Act of Parliament which gives the college a semi-autonomous status as a parastatal organization. It has its own governing body (above the college management). Although placed under the Ministry of Cooperatives and Marketing, CCM has considerable room to formulate and implement its own training policy.

Essentially, CCM was established as a training, research and consultancy institution catering for the cooperative sector. In recent years, it has broadened its clientele to include other forms of people's organizations. It has become a reputable institution for cooperative education and training at community and grassroots level, as well as a leading proponent and practitioner of participatory approaches to development in general and to training in particular. It was primarily for this reason that it became the lead agency of ISWC-Tanzania.

The college is divided into four directorates: one for administration and three academic directorates, these being:

- Directorate of Studies and Programmes (DSP) which is responsible for long residential courses, ranging from two-year certificate courses (for secondary school leavers) to postgraduate courses. Awards conferred include certificates, ordinary diplomas, advanced diplomas, postgraduate diplomas and an MSc in Cooperative Studies (by distance learning in collaboration with Leicester University in the UK).
- Directorate of Field Education (DFE), the core activities of which are based at field level. Unique among the institutions of higher learning, CCM, through this directorate, has branches (or wings) in nearly all regions of the country. These branch colleges, complete with training facilities, are headed by wing tutors who are responsible for developing training materials and for training grassroots and community-based organizations.

- Directorate of Research and Consultancy Services (DRCS) which is responsible for coordinating the research and consultancy services offered (through paid contracts) by CCM. It is also offers tailor-made courses which are developed and conducted at the request of clients. These include cooperatives, other forms of people's organizations and development programmes.

Of particular interest to the ISWC programme is the Pilot Projects and Experimentation (PPE) Department which falls under the DRCS. The key function of PPE is to coordinate pilot projects aimed at testing new models of organization and new approaches in training and research activities, particularly those focusing at grassroots level. The PPE Department collects, synthesizes and disseminates the experiences generated from the pilot projects. The ultimate aim is to infuse these experiences into the mainstream of college training, research and consultancy activities. ISWC-Tanzania comes under the PPE Department and the national coordinator is the head of this department. However, the programme has taken advantage of the DFE structure to appoint wing tutors in Iringa and Mbeya (the areas in which the ISCW-Tanzania field programmes are being implemented) who work as assistants to the national coordinator.

Strategies of influence on college curricula

The curricula of CCM fall under three broad categories: those for the long residential courses (offered by the DSP), the short-term courses (offered by the DFE) and the tailor-made courses (offered by the DRCS). ISWC-Tanzania has had different degrees of influence on these three types of curriculum.

The college is categorized as an institution of higher learning because of the courses offered by the DSP. The quality standards that the college is expected to meet are set by statutory organizations at national level. The curricula of the DSP courses must be endorsed by a national committee composed of representatives from other institutions at the same level (including the universities) and government officials. The curricula must also meet standards set by various professional bodies (accountancy, materials management, etc) in order that they can recognize the degrees awarded. All of these external influences have a bearing on the content and method of delivery of the courses. These bodies are not renowned for their flexibility and innovativeness. The choice of what is included in the syllabi is determined by the requirements of the professional bodies rather than by the students' learning needs or even by the requirements for their future jobs. As the syllabi are usually tightly packed, the preferred mode of delivery is the lecture.

In the case of the curricula for courses offered under the DFE and DRCS, however, the college enjoys more latitude. These curricula do not have to be endorsed by any external organization. Curriculum development is guided by demand and the college's own philosophy.

As in all policy-influencing efforts, influencing the curricula of the college is a slow and complex process. Essentially, it involves the changing of attitudes at both institutional and individual (lecturer) level. Within a relatively short time, ISWC-Tanzania has been able to make considerable progress, largely because of two sets of factors, described below.

Conducive environment based on previous experience

The college already had considerable experience in using participatory methods. Over a period of more than ten years, CCM had executed the FAO-supervised People's Participation Programme (PPP). The basic philosophy and principles of PPP are congruent with those of ISWC 2 which encourages participatory research and development by scientists and farmers. The experiences from PPP created the desire for college and staff to be exposed to a more systematic application of participatory approaches. To that end, the college had organized relevant training for its lecturers, sometimes in collaboration with international organizations. For example, for three weeks each in 1997, 1998 and 1999, the college collaborated with the German Foundation for International Development (DSE) to train lecturers in participatory techniques and tools. During each training event, the 30 participants were exposed to theory for two weeks, punctuated by a week of practical fieldwork carried out in areas where the college already had projects on the ground. In this way, the field training did not raise undue expectations within the communities and more lasting relationships could be built up between the lecturers and the respective communities. The current national coordinator of ISWC-Tanzania was the national coordinator of the PPP.

Specific strategic measures

ISWC-Tanzania has instituted specific measures that were strategically made in order to influence the college curricula. These include:

- The principal of CCM is a member of the National Steering Committee of the programme. The NSC not only deals with policy issues related to the programme, but also discusses conceptual and methodological issues, thus allowing the principal greater insight into the farmer innovation approach. On numerous occasions and in numerous platforms involving CCM lecturers, the principal has referred to the ISWC-Tanzania experiences.
- The coordinator of ISWC-Tanzania also heads the PPE Department at the college. In this capacity, he is responsible for disseminating experiences gained from pilot projects, including ISWC-Tanzania. To that end, regular exposure workshops are organized for the lecturers.
- The ISWC-Tanzania coordination team includes the two college wing tutors at Iringa and Mbeya. These persons, like all other wing tutors, are responsible for developing and conducting training courses at the grassroots level. The experience gained by working within ISWC-Tanzania constitutes a critical input into their work.

- College staff have attended and benefited from PTD training workshops both within and outside the country. Two lecturers from CCM attended the anglophone PTD training-of-trainers workshop in Zimbabwe in 1997. Others have attended PTD workshops organized in Tanzania.
- A recent development has been the hosting and supervising of college students doing their research work. So far, four students (advanced diploma and postgraduate diploma level) have chosen action areas of ISWC-Tanzania for their research work. The programme offers no additional financial incentives. The research is part of the course requirements and the expenses are covered out of the course fees. The fees of in-service students are paid by their employers, while those of pre-service students are covered by a government bursary or other sponsors.

Areas of influence on college curricula

The influence of ISWC-Tanzania on CCM curricula is discernible in three main areas: the curriculum development process, the contents (syllabi) and the methods of teaching. The degree of influence exerted differs within the range of courses and subjects taught at the college.

Curriculum development process

The syllabus is a very important component of a curriculum. As in all training institutions, the syllabi at CCM are reviewed periodically. Traditionally, this was done by the professionals: the subject specialists. With the evolvement and nurturing of a participatory culture, the most recent (1999) syllabi revision at the college was carried out in a more participatory manner. Many more stakeholders were involved, including students, prospective employers and lecturers from other specializations. The entire process was consultative, including the organization of joint workshops.

For the short-term and tailor-made courses, the process starts with a Learning Needs Assessment. Such flexibility is built in; as the learning progresses, the syllabi can be revised.

Contents

In the past, the lecturers have tended to include in the courses all that they themselves had learned at university. The aim was to meet some externally defined professional standards. As a result of the more participative approach to curriculum development, the course contents now reflect better the learning needs of the students.

The contents for short-term and tailor-made courses focus on solving specific problems identified during the Learning Needs Assessment. Because of the nature of the participants, the contents are more practice-oriented than theoretical. Most participants want 'to be able to do' rather than just 'to know'. Issues covered include group processes, local organizational development and PRA tools as they are applicable in this connection.

Teaching methodology

All college lecturers are highly qualified professionals. They are specialists in fields such as economics, management, accountancy, cooperative development, etc. Most of them were not trained to be teachers. During the late 1970s and early 1980s, the college, with the assistance of the International Cooperative Alliance, organized teaching methodology courses for the college lecturers. This did help at the time. However, this initiative ended a long time ago, only a few lecturers attended and a good number of them have since retired.

Out of the PPP experience, the college had developed a training strategy for both development agents and community leaders. This strategy, known as the Comprehensive Participatory Training Process (CPTP), includes the following features:

- It is based on an action–reflection–action process of experiential learning. Course contents are based on identified needs as perceived by the participants. The course programme is of the 'sandwich' type: residential training is followed by a practical exercise, followed by another period of residential training, over four iterations.
- The whole process is designed to bring about a change in attitude in both the trainers and the trainees. The facilitation process therefore focuses on promoting dialogue and respect for all involved in the dialogue.
- Courses are facilitated, as much as possible, by a team, ideally composed of two persons, not only because two minds may work better than one but, more importantly, because it demonstrates and promotes a participatory working culture.

There is obviously much in common between the philosophy of CPTP and ISWC 2. A notable weakness of CPTP was, however, the lack of specific tools and techniques to make participation operational in the field. With its rich experience with PRA and PTD, ISWC-Tanzania has helped to fill this gap.

As a result of being exposed to the methods and tools of PRA and PTD and of related training events, CCM lecturers are now more adventurous with the teaching methods they use in their (non-ISWC) work. A growing number of them supplement the conventional lectures with other, more participative methods, such as using case studies and learning games in the class. Some sessions are even conducted outside the lecture rooms. This allows for the more active involvement of the students.

THE NEED TO BROADEN THE INCORPORATION

It can thus be seen that, in the countries where the lead agency was an institution of higher learning, ISWC 2 was able to influence the curricula and teaching practices. In Tigray, there is already evidence of change in the behaviour of teachers, researchers and trainees/students. They are abandoning the attitude of 'we know it all'. An appreciation of indigenous knowledge is growing and the stimulation of indigenous knowledge creation (local innovation) is becoming incorporated as an integral part of the agricultural education system. It is being stressed – particularly by the agricultural extension service (BoANR), the head of which is in the ISWC-Ethiopia Steering Committee – that appreciating indigenous knowledge is not an end in itself. The challenge to educators is to help students and trainees to find ways to add value to this knowledge by initiating participatory situation analysis, providing scientific validation and explanation of local technologies, proposing improvements that can be tried out by experimenting farmers, and facilitating the collaboration of farmers and scientists as equal partners in research.

In Chapter 29, it was shown how the past experience of Tigray created fertile ground for participatory approaches. Similarly, in Tanzania, the ISWC programme coordinated by the CCM could build on and complement existing programmes that had an in-built participatory philosophy, such as PPP. An important strategy taken by both country programmes was to make use of the existing mechanisms, such as the Practical Attachment Programme at MU and the Pilot Projects and Experimentation Department in CCM, when trying to incorporate participatory approaches into agricultural education. Similar mechanisms will need to be sought in other organizations of education and training to facilitate a broader incorporation of such approaches.

A promising development in Tanzania is the emerging network – as informal as it still is – among institutions that are concerned with agricultural education and training. These include the Sokoine University of Agriculture in Morogoro, INADES Formation (the training arm of the African Institute for Economic and Social Development) in Dar es Salaam, and several other NGOs. Like the CCM, they have developed the interest and the capacity to apply participatory tools in programmes that address the needs of rural communities. Exchanging the experiences of CCM with these other participatory approaches to learning in the classroom and in the field is already proving to be mutually enriching.

Through the influence of ISWC 2 primarily in the Faculty of Dryland Agriculture of Mekelle University in Ethiopia and in the Cooperative College Moshi in Tanzania, only the first step has been made in incorporating the promotion of farmer innovation and PTD into teaching and research in these countries. Great efforts must still be made to raise awareness in other colleges and universities about the effectiveness of this approach in stimulating rural development. There is a need for an exchange between universities about teaching methods and curricula and their relationship to rural realities. In this connection, it would be useful to develop guidelines that would help these institutions to monitor the extent to which indigenous knowledge, farmer innovation and PTD are being included in the curricula and reflected in research activities.

ACKNOWLEDGEMENTS

We thank Professor Suleman Adam Chambo, Principal of CCM, Laurens van Veldhuizen of ETC Ecoculture and Sara Tewolde Berhan of Mekelle University (currently at the University of Göttingen, Germany) for their valuable comments on an earlier version of this chapter.

32

An encouraging beginning

*Ann Waters-Bayer, Laurens van Veldhuizen and Chris Reij**

This is not the conclusion. A good start has been made in an exciting quest along a promising path. However, there are several barriers to be overcome and pitfalls to be avoided. Space has been created for innovation in approaches and methods of promoting farmer innovation by the partner organizations which are learning by examining their own and each other's experiences. The writing of this book has contributed to this process.

The farmer innovation programmes described in this book have led to the discovery of numerous and diverse local innovations and have furnished proof of the ingenuity, creativity and perseverance of small-scale African farmers in seeking to derive a living from the land. Farmers are experimenting and innovating on their own initiative, with a variety of motivations. However, it was often only after scientists and extensionists, and indeed farmers themselves had grasped the concept of local innovation that they were able to recognize that it was happening.

The outstanding innovators who are recognized as such today are farmers who have devoted many years to systematically building up sustainable farming systems. The process of improvement has been incremental, often unobserved until the farmer had become well established and was harvesting the fruits of decades of commitment to the land. The innovators built up their livelihood assets by improving and expanding their natural capital: by harvesting soil and water, by rehabilitating degraded land, by improving the fertility of the soil, by conserving and introducing new varieties and breeds of crops, trees and animals. In some cases, they also managed to expand their social capital by sharing their knowledge and products with others.

* Ann Waters-Bayer and Laurens van Veldhuizen from ETC Ecoculture in Leusden, The Netherlands, are PTD advisers to the ISWC 2 programme; Chris Reij from CDCS, Vrije Universiteit, Amsterdam, is international coordinator of the programme

Programmes focused on discovering and promoting farmer innovation can contribute to expanding social capital through creating or strengthening the links between outstanding farmer innovators so that they can be inspired by each other. They can also encourage recognition of the innovators by their communities, by other farmers, by development agents, scientists and policy-makers so that their ideas and their examples can stimulate more widespread innovation. They can increase human capital by facilitating joint experimentation by farmers and external agents in ways that enhance the skills and knowledge of all partners.

The increase in agricultural production and the greater diversity of production that can be achieved through innovation by smallholder families can lead to better human health. Many of the local innovations identified in the ISWC 2 and PFI programmes have buffered the risks of farming households in the face of climatic variability, especially in the drier parts of Tunisia, Burkina Faso and Ethiopia. To the extent that innovation has led to the diversification of the sources of livelihood and a reduction in dependency on external agricultural inputs, they have reduced vulnerability to market fluctuations. They have thus improved food security. In addition, they have given farmer innovators a sense of pride in being recognized as innovators.

PROMOTING INNOVATION, NOT JUST INNOVATIONS

Many of the locally developed techniques that have been described in this book are not new, in the sense of never having been done before anywhere in the world or even in the country concerned. For example, various systems of night paddocking, involving agreements between pastoralists and arable farmers, such as that described by ISWC-Cameroon, have been practised in many parts of West Africa for decades. What is important is the creativity and initiative displayed by people who, not being aware of these practices in other areas, visualized the possibility of improving the use of local resources and set out to realize these possibilities. It is this creativity and this type of initiative that the farmer innovation approach is celebrating and promoting. The particular technologies will change over time – in this case, for example, as pressures of people and livestock on the land change and as the relative prices of manure and crop products change – and new technologies will be needed. Approaches to agricultural development that take local innovation as their starting point will help to identify the ever new attempts to adjust and improve the local situation and will be able to point to useful ideas from other areas facing similar problems. Therefore, an important component of the farmer innovation approach is enhancing communication between the various people who are looking for and may have found solutions that are applicable by smallholder farmers.

It must be stressed that the farmer innovation approach seeks to promote a continuing *process of innovation* rather than only the collection and dissemination of specific and possibly time-bound innovations, although this is definitely part of it. Moreover, both biophysical and socio-organizational innovation in agriculture and natural resource management is being promoted. The example from North-west Cameroon is particularly interesting in this respect as it describes a process not only of improving agricultural production, but also of improving relations between two ethnic groups seeking their livelihoods from cropping and livestock husbandry, respectively, on the same land.

WORKING WITH INDIVIDUALS, GROUPS OR COMMUNITIES

The emphasis on innovation processes rather than single innovations also implies a challenge to look beyond the easily visible work of a few outstanding innovators who are known in each and every village. The research reported by Flemming Nielsen in Chapter 8 draws attention to the fact that innovativeness is widespread among both small-scale and larger-scale farmers and among both women and men. The outstanding innovators who feature in several chapters of this book can and do play an important role in promoting local agricultural development. But the challenge remains to link this potential with the innovative capacity that is available within the wider farming community.

A central question to those who apply a farmer innovation approach is whether it is more constructive to give recognition to individuals or to communities as local innovation builds on the collective local knowledge and will have a wider and more positive impact if it stimulates enthusiasm rather than jealousy among others. Particularly with respect to experiments defined by farmers and planned together with external agents (scientists and/or extensionists), decisions need to be made whether these should be conducted with only a few outspoken innovators (often men) or with their families or in collaboration with a larger number of farmers or farm families in the community. The challenge is to find ways to mobilize and encourage the creativity of individuals, tune in on interests shared by these individuals and other farmers, and encourage a wider sharing and social cohesion. In practice, this means that an eye must always be kept open for the position of individual farmers in the village, for the question of wider relevance and the applicability of innovations of individual farmers, and for ways to facilitate sharing among farmers within and outside their households. Farmer innovators have themselves shown ways to do this, for example, by sharing experiences with family members or initiating small neighbourhood groups.

By analysing their own experiences of working with individuals and groups of farmers, the programme partners have explored some basic questions concerning the farmer innovation approach. Who should identify the outstanding innovators? Should this be done by scientists, development agents, community leaders, interest groups within the farming community or the community as whole? Likewise, who should identify those farmers who will engage in joint experimentation with specialists from outside the community?

For whom are they doing the experimentation? To whom are the experimenters, both farmers and scientists, answerable? Are the experiments meant to provide 'hard' scientific data, or are they meant to give farmers a better basis for selecting the appropriate technologies for their own conditions? What experimental skills, including monitoring and evaluation, do farmers need to increase their capacity to innovate successfully? For each of these many questions, no single 'true' answer can be defined. The programme partners in both research and development are recognizing the value of discussing these questions openly and on a basis of equality with the collaborating farmers. This is, in itself, an important lesson. If farmers are expected to remain in the driver's seat – ie to continue to be the prime movers of local innovation – their voices and interests should be heard clearly in whatever answers are found to these questions.

DISSEMINATION OF BOTH TECHNOLOGY AND METHODOLOGY

If the farmer innovation approach is to be effective, the wider spread of valuable farmer innovations remains a key challenge. The cases in this book show the effectiveness of both farmer-to-farmer sharing and the mass media (radio, television and newspapers). Particularly the exchange visits were valued very much by participating farmers and directly stimulated experimentation and diffusion after their return to their home villages. In many cases, it was not necessary to set up new mechanisms for dissemination because good use could be made of existing infrastructure and events such as rural radio programmes, regular farmer field days organized by the extension service, or even an existing journal of a national farmers' network.

In dissemination, the emphasis has not been merely on the locally developed technologies, but also on creating awareness that farmers indeed experiment and innovate. This generates pride in individuals and communities about their accomplishments and celebrates the spirit of innovation. Partner organizations have been disseminating the methods of discovering farmer innovation, the methods of facilitating joint experimentation and the methods of enhancing communication between farmers and between informal (farmer) and formal (scientific) researchers. This is all part of the process of stimulating a continuing process of innovation.

Again, the task of dissemination should not, and is not, carried by government agents and programme staff alone. Farmer innovators have been sharing what they have learnt during cross-visits through their own informal communication with family members, with neighbours, at church meetings and the like. In the cases of this book, only anecdotal evidence is given of these processes. The challenge remains to become more aware of these efforts, to give greater credit to them and to pay more attention to indigenous systems and channels for sharing agricultural information and inputs. They are valuable in their own right, but might become even more powerful if they are more systematically enriched and linked with 'modern' means of communication.

The issue of intellectual property rights poses, in this respect, another major challenge. How should a scientist or development agent react when an innovator is secretive with his or her fellow farmers, but willing to share with scientists (a situation that arose, for example, in Tunisia)? Is publication of this innovation a breach of confidence? How should scientists, development agents and communication specialists acknowledge the people who originally developed the new ideas? Are these the property of individuals or of the community, as they have been built on the collective indigenous knowledge of the community? Do we even have the right to publish this book? There are many proponents of indigenous rights who would question this. We have not been able to make sure that each and every farmer and community mentioned in this book gave their permission for publication. Would such a demand mean the end of programmes to promote farmer innovation?

ASSESSMENT OF INNOVATIONS AND PROCESSES

The question of impact and its assessment also needs to be addressed in discussing the farmer innovation approach. The collection of cases of this book shows how concerted attention to M&E led to the development of a variety of approaches, methods and tools. Most combine methods to quantify the performance of technologies and methods with qualitative farmer-level assessments of the same. Many are part of the regular collaboration between farmers, scientists and development agents (eg monitoring of joint experiments), while others are set up as special studies (eg studies on gender, on the diffusion of innovations by radio or on the overall impact of five years of intervention). Together these efforts clearly contribute to quantifying the impact of certain innovations and the impact of farmer-to-farmer visits in stimulating other farmers to try out those innovations, as well as documenting the completion, if not already the impact, of training and of activities designed to influence policy.

This is not to say that developing effective M&E has been easy. A central struggle remains the challenge of designing and implementing M&E in true partnership with farmers, rather than imposing a system to extract information. Where data are needed to convince scientists and policy-makers, it cannot be expected that farmers will take on the extra work only for this reason. It proved to be possible, however, to facilitate farmers' recording of data in ways that the farmers themselves found useful and that also provided some basis for quantitative scientific analysis. It appears that the key is to achieve clarity about who wants what information and why, and then to base the division of work on the principle that farmers, extensionists and scientists collect those data that they regard as relevant for their respective purposes.

Certainly more thought and effort will have to be put into monitoring the farmer innovation process and measuring the change in the general capacity of farmers and communities to experiment and innovate, as well as in other technologies and systems related to land husbandry.

The very preparation of this book proved to be a contribution to evaluation as it mobilized numerous insights and helped all participants in the programmes to learn from our own and others' experiences. It has helped to clarify the approach and methods and to identify where we were falling into traps. For example, the exercises that each country carried out to analyse the farmer innovators and innovations they had identified and documented led to the recognition of biases in some countries to seeing more conspicuous innovations and listening to more vocal innovators, often men. It highlighted the dangers of focusing primarily on outstanding individuals rather than regarding and promoting farmer innovation as a widespread activity.

The characterization of farmer innovators revealed the importance of exposure to other areas as a source of both test-worthy ideas and test-worthy materials from other farmers, and underlined the role that exchange visits and study tours can have in stimulating innovation. It also revealed that, in order to discover innovations that are likely to be useful to poorer farm families, including female-headed households, a more deliberate focus on low-cost innovations is needed. Also the need for the classification of innovations in categories that are meaningful to farmers and development agents was recognized.

INSTITUTIONALIZING THE FARMER INNOVATION APPROACH

Strengthening the innovative capacities of farmers is a precondition for sustainable agriculture and natural resource management. At a smaller scale, this could be done by working through farmers' groups, community-based organizations and development-oriented NGOs. However, the national agricultural research and extension services can and should make an important contribution. They will be able to do so only if the roles of formal researchers and extension agents are redefined. There is a need for a much larger number of scientists who appreciate farmers' knowledge and creative capacities and are prepared to work together with farmers in their fields on questions that farmers are trying to investigate themselves. Extension agents could play major roles in identifying local innovations, organizing farmers' workshops to examine innovations and to identify those of interest to different categories of farmers, supporting farmers in organizing their own exchange and study visits, linking farmers with sources of ideas with which they can experiment and linking them with technical specialists who can help them to interpret their experimental findings. Extension agents need training in the skills required to fulfil these roles. Also local people living in the rural communities can fulfil some of these roles, with the appropriate training, as has been shown in this book. Strategies still need to be devised to organize training at these different levels.

Training has been a major component in all the country programmes. The experience accumulated over the years has made it clear that effective training in the farmer innovation approach includes at least the following characteristics:

- It consists of a sequence of training events rather than single, isolated courses.
- Field experiences are systematically fed back into classroom training.
- The training approach is strongly participatory.
- The 'trainees' are encouraged to innovate and adapt the methods being learned.
- Farmer innovators are involved in the training, either in the field or as resource persons or as co-participants in the classroom.
- A central place is given to attitudinal aspects.

To build up an in-country capacity for providing such training on a regular basis to (future) agricultural professionals, the major institutions of agricultural education and training need to be reoriented. It is encouraging to note that good progress has been made in incorporating farmer innovation approaches into institutions of agricultural education and training, particularly in those countries – Tanzania and Ethiopia – where such institutions were the lead agencies in the country programmes.

While this book shows how farmer innovation programmes have raised wide awareness about farmer innovation and have stimulated scientists and extensionists to support the processes of farmer-led experimentation, there is clearly a need for more deliberate strategies to influence policies to foster farmer innovation. Such strategies would include efforts to transform structures, procedures and behaviour within the research and extension institutions.

Perhaps the most resounding impact of the farmer innovation approach documented in this book is that it has stimulated the excitement of virtually all people who have been directly involved in the programmes. The scientists and extensionists involved at field level, in particular, became eager to continue when they started to recognize the detailed knowledge and amazing creativity of local innovators. The ISWC 2 and PFI programmes have contributed to building up capacities in the partner countries to train and coach people in the farmer innovation approach to participatory research and extension. This small but talented, experienced and enthusiastic pool of facilitators can form a basis for further development and expansion of activities within their own countries and also in new countries in Africa and elsewhere.

AN INNOVATIVE APPROACH AT PROGRAMME LEVEL

The farmer innovation programmes were set up in a very decentralized manner. The individual country programmes have enjoyed great autonomy in planning and budgeting; in most cases, the National Steering Committees consisting of the key collaborating organizations assumed these tasks. This clearly led to a feeling of programme ownership by all the key partners as they could make their own choices. However, it also implied that different choices were made and that many of the questions posed above gave cause for heated debates at programme meetings and workshops. This has been a major strength: the partners in the different countries, with their different experiences, could point out both the strengths and the weaknesses in each other's activities and methods. Numerous discussions revolved around questions of equity (rich versus poor, men versus women), questions about what constitutes an innovation – in whose eyes (scientists, extensionists, local elites, 'normal' farmers, very poor farmers?), questions of culture and ethics (How to integrate the creative energies of people who go against local cultural norms? Should the individual or the community be given recognition for an innovation?). Enriched by these discussions, each country programme gradually developed an approach that suited its particular history, circumstances and capacities.

A good example is how the various country programmes found ways to work with farmer groups. Where some countries focused on forming groups or networks of farmer innovators, other countries gave more emphasis to bringing together innovators and 'normal' farmers in an approach that was designed to encourage all farmers to innovate. ISWC-Zimbabwe, building on several years of experience in farmer-led experimentation, followed a different approach again and worked with existing and self-selected groups within the rural communities. They feel that joint experimentation is, in fact, a social process – a contribution not only to developing technologies, but also to developing community capacities to adjust to changing conditions. By allowing partners in the different countries to experiment with what they thought were locally the most appropriate approaches and by encouraging the systematic sharing and evaluation of these experiments across countries, the ISWC 2 and PFI programmes tried to practise at programme level what they preach: the creation of space for innovation and learning and a collaboration on the basis of equal partnerships.

Index

adaptation 298
adoption 203–205, 298
age
  country comparisons 79–80, 101
  Ethiopia 174
  Tunisia 126, 133–134
  Uganda 203, 204
agricultural cadres 235
agroforestry 23–27, 88–89, 141, 205
agrosylvopastoralism 137–143
Akol, Florence 117
Amhara Region, Ethiopia 28–34
anecdotal evidence 93
animal production 97, 141
annual meetings 193
arid areas 35–46, 122–131, 144–154
assessment
  community 171–177
  exchange visits 306
  farmer-led experimentation 241
  impact 200–212, 351–352
  technology 193–194
attitudes
  culture 102
  gender 118–120
  negative 160
  partnerships 49–50
  policy-makers 246, 329–330
  radio 298
  stakeholder 325–330
  see also perceptions
audiovisual aids 18, 188, 192, 215,
  290–291
autonomy 56
awards 68, 142, 158, 165, 197
awareness raising
  Ethiopia 69, 156–157
  gender 156–157
  methodology 18–19
  Tunisia 124

Banda, Cleopas 84–85

Bandjoun, Cameroon 248–255
Bao game 99
Barochy, Askwal 161–162
best farming practices approach
  310–324
Beza, Gebremichael 195
biodiversity 31, 41–42, 142
biophysical innovation 104–109, 349
bottom-up approaches 284–285, 289
Bura, Grace 89
bureaucracy 287
buried stone pocket technique 129
Burkina Faso
  extension models 213–217
  joint analysis 256–266
  Namwaya Sawadogo 137–143
  tree planting pits 35–46

cadres 235, 314
Cameroon
  Barthelémy Kameni Djambou 23–27
  follow-on innovations 104–109
  maize breeding 248–255
  participatory technology development
    221–233
categorization 96
characteristics
  common traits 77–86, 352
  Ethiopia 147–149, 151–152, 173–175
  Tunisia 126, 133–134
  Uganda 203–205
  women 133–134
checkdams 146–147, 188–189, 328
chicken manure 231–232, 251, 253
clusters 181–184
communities
  action 149
  assessment 171–177, 243
  leaders 69
  participation 311–312, 315–316
  statement to outsiders 175–176
competitions 59–60, 295–296, 321

compost *see* organic matter
conflicts 107–108, 176–177
conservation of biodiversity 142, 316
constraints 205, 303, 307–308
constructive criticism 195
contour stone bunds 29–30, 38–39, 139
Cooperative College Moshi 340–346
coordinating agencies 50–51, 57
cost-benefit analysis 209
counter-narratives 285
creativity
  celebrating 67–69, 348
  education 82
  stimulating 178–184, 349
credit 163–164, 194, 197
cross-breeding 250
culture
  challenging 66–67, 160–161, 163,
    165, 303, 354
  compliance with 176–177
  elder's achievements 80, 101–102
  policy change 287
curriculum development 341–344

Dadi, Khalifa 327
daily seminar evaluations 191
daldal 146–147
databases 65–66
debates 288, 294–295
demonstrations 216
desertification 35–36
development 3–22, 69, 70–71
development-oriented education
    331–346
devil's tie 90, 150–151
discouragement 160
dissemination
  communication strategies 290
  farmer-to-farmer 16–18, 190, 210,
    213–217
  innovation suitability 89–91
  land-use conflicts 108
  radio 293–299
  silt-harvesting systems 149–150,
    152–153
  technology and methodology 350–351
  see also awareness raising; extension;
    sharing of experience
district-level meetings 196, 236, 238,
    243–244

Djambou, Barthelémy Kameni 23–27
documentation
  anecdotal evidence 93
  extension 305, 307
  Farmers' Fora 194–196
  women 165

East Africa
  interactive groups 178–184
  study 92–103
  women 110–121
economic policies 283
education
  creativity 82, 126, 134
  level of 203, 204
  participatory approaches 331–346
egg incubation 134–135, 136
elders' achievements 80, 101–102
emic view 98–99
Ethiopia
  best farming practices 310–324
  community assessment 171–177
  facilitating communication 185–197
  farmer-led experimentation 234–247
  learning for sustainability 331–346
  partnerships 58–73
  water management 144–154
  women 28–34, 155–167
ethnicity 101–102, 103
etic view 100–102
eucalyptus 139–140, 190
evaluation
  community 171–177
  farmer-led experimentation 242–244
  joint 71
  maize yield trials 253–254
  organic matter flows 259, 260–262
  participatory 258
  socioeconomic 264
  systems 12–13, 14–15
  travelling seminars 190–192
exchange of information *see* sharing of
    experience
exchange visits
  Burkina Faso 265
  Cameroon 24
  Ethiopia 69
  farmer innovation methodology
    13–14, 350
  Tanzania 54, 275

Tunisia 127
Zimbabwe 304, 305, 308–309
expenses 182, 187
experience
    innovator characteristics 79–80
    non-governmental organizations 71
    travel 81
    value of 20
    see also sharing of experience
experimentation
    farmer-led 234–247, 271–272, 339
    participatory 70–71
    see also joint experimentation
experts 49–50, 197, 245–246
extension
    attitudes 118, 298, 326
    best farming practices approach 71,
        310–324
    Burkina Faso 213–217
    dissatisfaction with 4
    farmer-to-farmer 16–18
    institutionalization 352
    methodology 209
    non-adoption of 63, 234
    radio 294–299
    technology transfer 196, 197
    training 24–25, 113–115
    see also awareness raising; dissemina-
        tion; sharing of experience

facilitation 185–197
fallowing 159–161, 188–189
families 37, 78, 174, 204
farm size 80, 100, 103, 126, 204
farmer-led experimentation 234–247,
    271–272, 339
farmer-to-farmer communication
    evaluation 14–15
    facilitating 16–18, 24, 185–197
    visits 199, 200, 205–209, 305–306
Farmers' Fora 192–196
fertilization
    application of 239, 240, 244,
        260–262
    credit 164, 194, 197
    levels of 228–230, 273
    maize 251–254
    subsidies 83
field level partnerships 52–53
field visits 193, 202, 298, 314

fieldwork 332–333
Fikre, Ayelech 28–34
first-generation innovators 10
flexibility 20, 56, 233, 322, 341
follow-on innovations 271
food security 40, 217
Fora, Farmers' 192–196
full-time farmers 81–82

Gafsa, Tunisia 293–299
Gebrehiwot, Hailu 189–190
GebreMedhin, Zigta 146–150, 166
gender
    adopter comparisons 203–205
    analysis 112–113
    differences 77–78, 100, 103, 126
    farmer clusters 182
    imbalance 110–112, 156
    sensitization 113–121, 156–157
    tabia-level meetings 236
Gobena, Leteyesus 159–161, 188–189
grassroots level development-oriented
    education 341
group discussions 207–209
gullies 88, 89, 90, 188–189, 195

handicrafts 134, 136
higher education 331–346
home gardens 164
households
    gender related constraints 113, 114
    head of 77–78, 100, 103, 166
    innovation variations 97

identification processes
    best farming practices 312–313
    challenges in 172–173
    contests 59–60, 295–296, 321
    East Africa 95, 179–180
    gender balance 110–112
    guidelines 60, 62
    methodology 9–11, 61–65
    participatory 53, 307, 349
    women 78, 124–125, 132–133,
        155–158
illiterate farmers 264
impacts
    assessment 200–212, 351–352
    farmer innovation approach 325–330
    farmer-led experimentation 245–246

PAP 335–336
radio 294–298
improvization 233
incentives 295–296, 297
inclusive approach 72
income
   increase in 140, 142
   off-farm 81–82, 100, 126, 174
   Uganda 204
   see also wealth, poverty
indigenous knowledge 65, 67–69,
   332–339, 345
Indigenous Soil and Water Conservation
   in Africa (ISWC) 6–7
innovation, definitions 10, 94–95, 172,
   354
innovators, definitions 10, 60, 203
inspiration 85–86
institutionalization
   farmer innovation approach 4, 19–20,
      352–353
   methodology development 209
   participatory approaches 130,
      300–309
   partnerships 50–53
   student research 335
integrated farming systems 83, 84–85,
   137–143
intellectual property rights 351
interactive groups 178–184
intercropping 139
interdisciplinary seminars 339
Irobland, Ethiopia 144–154
irrigation
   joint experimentation 127, 128
   nightshade production 106–108
   women 134, 161–162, 164–165
ISWC *see* Indigenous Soil and Water
   Conservation in Africa

jessour technique 122, 128–129
joint experimentation
   Burkina Faso 256–266
   farmer innovation methodology
      15–16, 350, 354
   maize production 250, 251–254,
      276–277
   negotiating 54–56
   stimulating and supporting 221–277,
      320, 348, 349

Tunisia 127–128
water use in greenhouses 327

Kabale, Uganda 198–212
Kamgouo, Emmanuel 248–255
Kenya 115–116
key informants 61
Kindo, Ousséni 37, 40
Kiteng'u, Kakundi 116

labour cooperation 29–30
land rehabilitation
   Ethiopia 175
   motivation 138
   tree planting pits 36–46, 213–217,
      256–266
land-use conflicts 108
lead agencies 6, 331–346
liberation 165, 311–324
livestock production 40–41, 134–135,
   141–142, 260
low-risk innovation 97, 102

M&E *see* monitoring and evaluation
macro policies 283
maize 248–255, 267–277
manure 104–105, 228–232, 348
market day model 214–215
market demand 106–107
mass mobilization 315–316
master farmer model 302
media
   awareness raising 18–19
   mass 298
   policy change 290–291
   public recognition 68
   radio 293–299, 326
medicinal products 41, 43
meetings
   annual 193, 314
   extension 196
   farmer clusters 181
   participatory situation analysis
      235–238
Mekelle University 331–340
millet 213, 215, 263
minor innovations 97, 102, 162, 166,
   271
monitoring
   farmer-led experimentation 271–272

maize yield trials 252
setting-up systems 12–13
soil fertility improvement 225
monitoring and evaluation (M&E)
    12–13
    farmer-led experimentation 242–244
    PTD and extension 320
    systems 351–352
    water conservation 200–201, 210
motivation
    characteristics 83–85, 98–99,
        147–149, 151–152
    land rehabilitation 42–45, 138
    socioeconomic background 80
multidisciplinarity 56–57
Mville, Wilbert 268–276

narrative style 285
Nasri, Béchir 125, 128, 329
national level partnerships 50–52
natural regeneration 38–39
natural resource management 28–34
Ndong, Phillip 105–106
networks
    age 79
    Association for the Spreading of Zaï
        215
    farmer-to-farmer 26, 51, 178–184,
        199, 354
    policy change 286–287, 288–289
    Tanzania 345
    visits 205–209
    women 166
    Zimbabwe 305, 306, 308
night paddock manuring 104–105,
    221–233, 348
nightshade production 105–107,
    226–233
non-innovators 63–64
non-treatment factors 240–241

observations 61, 272
off-farm income 81–82, 100, 126, 174
on-farm experimentation 259, 262–264
organic matter
    application rates 262–264, 265
    flows evaluation 259, 260–262
    management 205
    pit techniques 258, 270, 274, 277
organizational aspects

culture 287
development 307, 340–346
structure 308
Ouédraogo, Ali 37, 38–39, 40, 216
outsiders
    community statement to 175–176
    farmer-led experimentation 233
outstanding individuals 352
ownership 354

PAP *see* Practical Attachment
    Programme
participatory approaches
    agricultural development 318–323
    education 331–346
    experimentation 70–71
    impact assessment meetings 202–203
    institutionalization 130
    situation analysis 235–238
    soil and water conservation 300–309
Participatory Rural Appraisal (PRA) 8–9
Participatory Technology Development
    (PTD) 5, 8–9, 221–233, 268,
        319–320
partnerships
    Ethiopia 58–73
    extension and research 49–57
pastoralist conflicts 107–108
perceptions 99, 120, 245–246
    see also attitudes
personality 79
PFI *see* Promoting Farmer Innovation in
    Rainfed Agriculture
physical structures 173, 174
pit techniques *see* sowing pits; tree
    planting pits
planning 239–242
plant production
    biomass 35–46
    breeding 248–255
    population density 273
ploughing 159–160, 188–190, 316–317
policies
    change 284–291, 18–19
    college curricula 341–342
    extension 322, 323
    links 72
    local land 191
    soil and water conservation 281–292
    travelling seminars 197

policy-makers 329–330
population 83, 182
poverty 67, 163–164
PRA *see* Participatory Rural Appraisal
Practical Attachment Programme (PAP)
    332–336
practical experience 336–338
problem trees 237
production cadres 314
programmes 6–8, 353–354
Promoting Farmer Innovation in Rainfed
    Agriculture (PFI) 7–8
PTD *see* Participatory Technology
    Development
publications 4, 330

radio 293–299, 326
rain-fed farming 124–128
rainwater harvesting 30–31
recognition
    individual or collective 349, 354
    PAP 334–335
    travelling seminars 191, 195
    women 32–33, 166, 167, 245
recording data 264
reforestation 35–36
regional workshops 54
regulations 176–177
relative scales 206–207
renewal 71
research
    attitudes 298
    joint experimentation 221–233
    land husbandry 3–22
    PAP influence 338–339
    partners 223–225, 236
    regional radio 293–299
    supportive on-station 228–230
roles
    defining 352
    joint experimentation 252
    research partners 223–223
rural reality 336–338

Sandi, Abbés 328
Sawadogo, Namwaya 137–143
Sawadogo, Yacouba 37, 38, 40, 41, 44,
    214–215
scientists
    attitudes 325–326

data analysis 272–274
farmer-led experimentation 233
indigenous techniques 128–129
involvement of 210
second-generation innovators 10
sedentary farmer conflicts 107–108
self-confidence
    exchange visits 306
    travelling seminars 191
    women 157–158, 161, 166, 167
self-identification 11
seminars
    interdisciplinary 339
    travelling 185–192, 323
sensitization
    women 157–158
    workshops 112, 113–115
seytan madewa see devil's tie
sharing of experiences
    agroforestry 25–26
    field visits 304
    gender sensitization 120
    innovation dissemination 14, 17, 91,
        149–150, 166
    traditional 175
    universities 345
    workshops 32–33, 54
    see also awareness raising; dissemina-
        tion; extension
silt-harvesting systems 146–147,
    150–151
site visits 187–188
small-scale innovations *see* minor
    innovations
social aspects
    capital 347–348
    community pressures 160–161, 163,
        165, 167
    mutual inspiration 107–108
    status 45
socio-organizational innovation 303,
    349
sociocultural aspects 66–67
socioeconomic aspects
    change 123–124
    evaluation 259, 264
    innovators and adopters 203–205
    motivation 80
soil fertility
    conservation 198–212

erosion 189–190, 302
  management 83–85, 87
  management of 30, 32
  night paddock manuring 104–105
  participatory technology development
    221–233
  pit techniques 273–274
  women 164
soil specific policies 283
soil and water conservation (SWC)
  Burkina Faso 256–266
  Ethiopia 314–316
  ISWC 6–7
  policy processes 281–292
  Zimbabwe 300–309
solution testing 238, 239
sowing pits 267–277
spatial diversity 172
stakeholders attitudes 325–330
status 49, 143
steering committees
  Ethiopia 59, 319
  programme level 353–354
  Tanzania 51–52
stone bunds 29–30, 38–39
strengths, weaknesses, opportunities and
    threats (SWOT) analysis 190–191,
    207, 274–276
study tours 13–14, 205–209
subdistrict-level meetings 235–236
subsidies 83, 327, 328
success stories 289–290
sugarcane pitting technique 116
support
  decision tools 264
  development agencies 69, 70–71
  extension service 71
  financial 150
  innovation dissemination 90
  student research 335
  survival 175
  travelling seminars 186–187
  women 157
supportive on-station research 228–230
sustainability
  exchange visits 309
  farmer clusters 183
  learning for 331–346
  soil and water conservation 256–266
SWC *see* soil and water conservation

SWOT *see* strengths, weaknesses, oppor-
    tunities and threats
Sylvester, Susanna 119

T&V *see* Training-and-Visit
tabia-level meetings 236–238, 243
Tanzania
  gender balance 116, 119
  learning for sustainability 331–346
  maize sowing pits 267–277
  partnerships 49–57
teacher–student models 216–217
teaching methodology 344
technology
  dissemination 350–351
  introduced 317–318
  transfer 3–4, 58, 193–197, 294–296,
    318
teff production 239, 244–245
telephone debates 295
tenure 190
terracing 315
Tesfaye, Yohannes 150–153
theft 102–103
Tigray, Ethiopia
  best farming practices 310–324
  facilitating communication 185–197
  farmer-led experimentation 234–247
  partnerships 58–73
  women 155–167
timing 172, 290, 308
Toh, Samuel 104–105
top-down approaches 289, 302
ToT *see* transfer-of-technology
Touroum, Burkina Faso 137–143
traditional participatory methods 210
traditional practices 199
training
  agroforestry 24–25
  best farming practices 313–314
  development agents 320–321
  Djambou, Bartelémy Kameni 26–27
  extension 303–304, 307, 352–353
  identification 132–133
  joint experimentation 56, 57
  participatory methods 342, 344
  partner organizations 60–61
  PRA and PTD 8–9
  tree planting pits 216–217
  see also workshops

Training-and-Visit (T&V) 4
transfer-of-technology (ToT) model 3–4,
58, 193–197, 294–296, 318
trash-line management, impact
assessment 198–212
travel experience 81
travelling seminars 185–192, 323
tree planting pits
dissemination 89–90, 139–140,
213–217
reforestation 35–46
sustainability 256–266
Tsagay, Abrahat 162
Tunisia
land husbandry 122–131
regional radio 293–299
stakeholder attitudes 325–330
women 132–136

Uganda 117–118, 198–212
universities
development-oriented education 72,
331–346
innovation promotion 321–322
urine measurement 228–232

validation 66
values 175–177
verification process 9–11, 180
village-level fora 193–194, 196
visual tools 233
Vitsuh, Christopher 106–107, 230–232

water conservation *see* soil and water
conservation
water harvesting
bananas 117
Ethiopia 144–154, 315
greenhouses 327
innovations 87–88
jessour 122–124

storage and distribution 328
water management 192–196, 198–212
wealth
Ethiopia 65, 67, 174
food security 217
innovator characteristics 12, 80–81,
102
Uganda 90–91, 203, 204
women 162, 163
weed growth 228
women
awards for 68
East Africa 110–121
Ethiopia 155–167, 316–317
Fikre, Ayelech 28–34
identification 64–65, 78, 124–125,
132–133
physical structures 173–174
radio 296
travel experience 81
see also gender
workshops
awareness raising 69
farmer-led experimentation 271–272
Farmers' Fora 192–196, 323
field level 52–53, 54
Fikre, Ayelech 32–33
innovation assessment 15
joint experimentation 55–56, 71,
222–225, 277, 327
PAP 334, 335, 337–338
participatory approaches 66, 124, 268
sensitization 112, 113–115, 156–157
travelling seminars 185–192, 323
see also training

zaï see tree planting pits
Zaï school model 215–216
Zimbabwe 300–309
Zoromé, Ousséni 37, 40, 42, 215–216

For Product Safety Concerns and Information please contact our EU
representative GPSR@taylorandfrancis.com
Taylor & Francis Verlag GmbH, Kaufingerstraße 24, 80331 München, Germany

www.ingramcontent.com/pod-product-compliance
Lightning Source LLC
Chambersburg PA
CBHW060756220326
41598CB00022B/2456